Hydroclimate in a Changing World: Recent Trends, Current Progress and Future Directions

Hydroclimate in a Changing World: Recent Trends, Current Progress and Future Directions

Editor

Haibo Liu

Basel • Beijing • Wuhan • Barcelona • Belgrade • Novi Sad • Cluj • Manchester

Editor
Haibo Liu
Lamont-Doherty Earth Observatory
Columbia University
Palisades
United States

Editorial Office
MDPI
St. Alban-Anlage 66
4052 Basel, Switzerland

This is a reprint of articles from the Special Issue published online in the open access journal *Atmosphere* (ISSN 2073-4433) (available at: www.mdpi.com/journal/atmosphere/special_issues/hydroclimate_changing_world).

For citation purposes, cite each article independently as indicated on the article page online and as indicated below:

Lastname, A.A.; Lastname, B.B. Article Title. *Journal Name* **Year**, *Volume Number*, Page Range.

ISBN 978-3-0365-9665-5 (Hbk)
ISBN 978-3-0365-9664-8 (PDF)
doi.org/10.3390/books978-3-0365-9664-8

Contents

Preface

The Special Issue of *Atmosphere*, Hydroclimate in a Changing World: Recent Trends, Current Progress and Future Directions, has published fifteen papers discussing the influences of global warming on hydroclimatic variables on both a regional and global scale. We thank all authors who contributed to this Special Issue, the anonymous reviewers, and the editors, especially Villanelle Wang, from *Atmosphere*, whose assistance made this Issue possible.

Global warming is imposing tremendous challenges upon human and otherwise biotic life on Earth. A warmer atmosphere holds more moisture. The consensus is that the moisture transport by the atmospheric circulation strengthens and makes already wet areas of moisture convergence wetter and already dry areas of moisture divergence drier. Therefore, the tropics and mid-to-high latitudes will get wetter and the subtropics will get drier. Without any change in the interannual variability of hydroclimate, the change in the mean hydroclimate would equally increase drought risk in some places and flood risk in others. Moreover, global warming will cause the interannual variability of the hydroclimate to intensify, which will induce more droughts and floods. Furthermore, the changing atmospheric circulation interaction with the land surface may cause storm track alterations and may play an important role in shaping moisture redistribution. The authors contributions have documented the precipitation trends in southeast of the US, the Nile River Basin Ethiopia, Iraq, the Huai River Basin of northern China, and the Qilian Mountains of western China. The precipitation predictability on both global and regional scales are also studied. The interaction among climate systems in southeast Asia is also explicitly documented.

Because of the limitation of space, this Special Issue left many important aspects of hydroclimatic changes unexamined. We hope we will have more opportunities in the future to include more studies on this topic.

Haibo Liu
Editor

atmosphere

Editorial

Special Issue Editorial: Hydroclimate in a Changing World: Recent Trends, Current Progress and Future Directions

Haibo Liu

Lamont-Doherty Earth Observatory, Columbia University, New York, NY 10960, USA; haibo@ldeo.columbia.edu

Citation: Liu, H. Special Issue Editorial: Hydroclimate in a Changing World: Recent Trends, Current Progress and Future Directions. *Atmosphere* **2023**, *14*, 1725. https://doi.org/10.3390/atmos14121725

Received: 24 October 2023
Revised: 23 November 2023
Accepted: 23 November 2023
Published: 24 November 2023

The sixth report of the Intergovernmental Panel on Climate Change (IPCC) has confirmed that human-induced climate change is already affecting many weather and climate extremes in every region across the globe [1]. Increasing evidence of observed changes in extremes such as heatwaves, heavy precipitation, droughts, and tropical cyclones have emerged, and point to human influence. Under all emissions scenarios considered, global surface temperature will continue to increase until at least the mid-century [1]. A warmer atmosphere holds more moisture thus global warming leads to strengthening moisture transport via atmospheric circulation and makes areas of moisture convergence wetter and areas of moisture divergence drier [2,3]. Global warming also causes the interannual variability in the hydroclimate to intensify, which induces more droughts and floods. These changes are imposing tremendous challenges upon human lives and other lives on Earth.

Observations have shown that precipitation patterns are changing globally. While some regions experience increased rainfall and more intense precipitation events, others are facing prolonged dry spells and increased drought conditions [4,5]. These changes have significant implications for water resources, agriculture, and ecosystems. Rising temperatures have led to reductions in snowpack and accelerated the melting of glaciers in many mountainous regions [6]. This affects the availability of water resources, especially in regions that depend on snowmelt for freshwater supply and irrigation. There is a growing body of evidence suggesting an increase in the frequency and intensity of extreme hydroclimatic events such as hurricanes, floods, and heatwaves [1]. In some places, drought conditions may extend well into the next few decades [7]. These events can have severe societal and environmental consequences, and they are leading to extremes such as crop failure, infrastructure damage, and even humanitarian crises.

It has become increasingly evident that present-day human activity is driving the global warming that induces global climate change. In turn, a changing global climate is impacting human lives. On one hand, we need to mitigate the level of warming; on the other hand, we must adapt to the continued increase in global temperature and the changing hydroclimate. We need to deepen our understanding of human-induced global warming, natural variability, and the subsequent changes in the hydroclimate, including extreme floods, droughts, and storms. This will enable a better prediction of these events so that society can make more informed choices to accommodate our changing physical environment.

This Special Issue of *Atmosphere*, "Hydroclimate in a Changing World: Recent Trends, Current Progress and Future Directions", is devoted to further understanding and better predicting hydroclimatic changes across the globe. Many thanks to the enthusiastic authors who have made contributions to this Special Issue and the many anonymous reviewers and editors from the Editorial Office who have made this successful.

Conflicts of Interest: The author declares no conflict of interest.

References

1. IPCC Sixth Assessment Report—Working Group I—The Physical Sciences Basis. Available online: https://www.ipcc.ch/report/ar6/wg1/downloads/report/IPCC_AR6_WGI_Headline_Statements.pdf (accessed on 27 September 2023).
2. Seager, R.; Vecchi, G.A. Greenhouse warming and the 21st century hydroclimate of southwestern North America. *Proc. Natl. Acad. Sci. USA* **2010**, *107*, 21277–21282. [CrossRef] [PubMed]
3. Cawdrey, K.; Carlowicz, M. Available online: https://www.nasa.gov/feature/warming-makes-droughts-extreme-wet-events-more-frequent-intense (accessed on 27 September 2023).
4. Seager, R.; Neelin, D.; Simpson, I.; Liu, H.; Henderson, N.; Shaw, T.; Kushnir, Y.; Ting, M.; Cook, B. Dynamical and Thermodynamical Causes of Large-Scale Changes in the Hydrological Cycle over North America in Response to Global Warming. *J. Climate* **2014**, *27*, 7921–7948. [CrossRef]
5. Xie, Q.; Gu, X.; Li, G.; Tang, T.; Li, Z. Variation Characteristics of Rainstorms and Floods in Southwest China and Their Relationships with Atmospheric Circulation in the Summer Half-Year. *Atmosphere* **2022**, *13*, 2103. [CrossRef]
6. Available online: https://www.climate.gov/news-features/understanding-climate/climate-change-mountain-glaciers (accessed on 10 October 2023).
7. Seager, R.; Ting, M.; Alexander, P.; Liu, H.; Nakamura, J.; Li, C. Ocean-forcing of cool season precipitation drives ongoing and future decadal drought in southwestern North America. *NPJ Clim. Atmos. Sci.* **2023**, *6*, 141. [CrossRef]

Article

ENSO Impact on Winter Precipitation in the Southeast United States through a Synoptic Climate Approach

Jian-Hua Qian [1,*], Brian Viner [1], Stephen Noble [1], David Werth [1] and Cuihua Li [2]

[1] Savannah River National Laboratory, Aiken, SC 29808, USA; brian.viner@srnl.doe.gov (B.V.); stephen.noble@srnl.doe.gov (S.N.); david.werth@srnl.doe.gov (D.W.)
[2] Lamont-Doherty Earth Observatory, Columbia University, Palisades, NY 10960, USA; cli@ldeo.columbia.edu
* Correspondence: jianhua.qian@srnl.doe.gov

Abstract: The ENSO impact on winter precipitation in the Southeast United States was analyzed from the perspective of daily weather types (WTs). We calculated the dynamic contribution associated with the change in frequency of the WTs and the thermodynamic contribution due to changes in the spatial patterns of the environmental fields of the WTs. Six WTs were obtained using a k-means clustering analysis of 850 hPa winds in reanalysis data from November to February of 1948–2022. All the WTs can only persist for a few days. The most frequent winter weather type is WT1 (shallow trough in Eastern U.S.), which can persist or likely transfer to WT4 (Mississippi River Valley ridge). WT1 becomes less frequent in El Niño years, while the frequency of WT4 does not change much. WTs 2–6 correspond to a loop of eastward propagating waves with troughs and ridges in the mid-latitude westerlies. Three WTs with a deep trough in the Southeast U.S., which are WT2 (east coast trough), WT3 (off east coast trough) and WT6 (plains trough), become more frequent in El Niño years. The more frequent deep troughs (WTs 2, 3 and 6) and less frequent shallow trough (WT1) result in above-normal precipitation in the coastal Southeast U.S. in the winter of El Niño years. WT5 (off coast Carolina High), with maximum precipitation extending from Mississippi Valley to the Great Lakes, becomes less frequent in El Niño years, which corresponds to the below-normal precipitation from the Great Lakes to Upper Mississippi and Ohio River Valley in El Niño years, and vice versa in La Niña years. The relative contribution of the thermodynamic and dynamic contribution is location dependent. On the east coast, the two contributions are similar in magnitude.

Keywords: El Niño; winter weather types

Citation: Qian, J.-H.; Viner, B.; Noble, S.; Werth, D.; Li, C. ENSO Impact on Winter Precipitation in the Southeast United States through a Synoptic Climate Approach. *Atmosphere* **2022**, *13*, 1159. https://doi.org/10.3390/atmos13081159

Academic Editor: Bryan C. Weare

Received: 1 June 2022
Accepted: 21 July 2022
Published: 22 July 2022

Publisher's Note: MDPI stays neutral with regard to jurisdictional claims in published maps and institutional affiliations.

1. Introduction

The influence of the El Niño–Southern Oscillation (ENSO) on the U.S. climate is strongest during the winter [1,2]. The problem is usually studied based on seasonal means, such as the ENSO impact on the mean precipitation in December–February [3]. As we experience in everyday life, climate and its variability are actually manifested by daily weather. Therefore, it is equally important to learn how the ENSO impacts daily weather. The problem will be examined and quantified using a weather typing (WT) analysis in the Southeast United States (SEUS) for the winter in the current study. A Similar method was applied in a study on the impact of the ENSO in other places in the world, such as in the Northeast U.S. [2,4] and in Southeast Asia [5–8], as well as on spring precipitation in the SEUS [9]. The rationale for the WT analysis and reference therein can be found in Qian el al. [9].

From the large-scale point of view, winter WTs in the SEUS are closely related to the North Atlantic Subtropical High (NASH) south of the Gulf coast (along about 30° N) and westerly winds north of 30° N, as shown in the monthly climatology in Figure 1. However, the monthly precipitation and 850 hPa circulation of the four months from November to February look almost the same, especially in the deep winter months of December to February, with the anticyclonic circulation of the NASH withdrawn further to the south.

Winter rainfall is often produced by southwesterly flows ahead of a cold front that brings moisture from the warmer Gulf of Mexico. The location of precipitation should depend on the weather regime of the day, especially the longitude of the trough. However, the ridges and troughs cannot be seen from the monthly or seasonal mean figures. As will be seen shortly, the weather typing analysis brings up the dynamic view of these propagating ridges and troughs. Cold air damming east of the Appalachian Mountains can also bring cold weather to the south in the winter [10,11]. Geographically special for the east coast, the strong temperature gradient between the warm Gulf Stream off the east coast and the cold coastal land in the cold seasons also forms a baroclinic belt, which corresponds to the quasi-stational east coast trough and storm track, as seen from the heavy precipitation along the east coast in Figure 1. All these are important factors that affect winter weather in the SEUS.

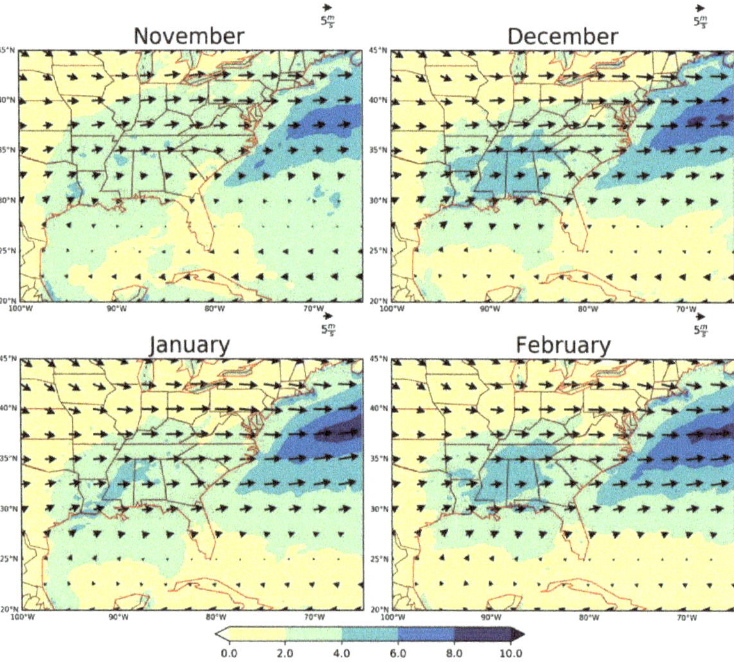

Figure 1. Monthly climatology of precipitation (shades, mm/day) and 850 hPa horizontal circulation (arrows, m/s) in winter in the Southeast U.S. The mean winds are based on the NRP reanalysis of 1948–2022. The mean precipitation is based on the CMORPH (NOAA Climate Prediction Center-CPC Morphing Technique) satellite estimated data for 1998–2020.

In the following, we will first describe the data and method briefly, then examine the spatial pattern and frequency of the winds and precipitation of the WTs. Finally, we will use the characteristics of the WTs to interpret the ENSO impact on precipitation in the SEUS.

2. Data and Weather Typing Analysis

2.1. Data

As in Qian et al. [7,9], the 850 hPa wind components of the NCEP-NCAR Reanalysis Project-I (NRP) data [12], from 1 January 1948 to 28 February 2022, are used for the WT analysis. To examine the precipitation pattern of each of the WTs, the satellite estimated 3-hourly precipitation data from CMORPH (NOAA Climate Prediction Center-CPC Morphing Technique, $0.25° \times 0.25°$, 1998–2020, covering both land and sea) [13] were used to plot the precipitation maps. The CPC-unified gauge-based analysis of daily precipitation over the contiguous United States (CONUS) at quarter-degree grids in 1948–2022 (https://psl.noaa.gov/data/gridded/data.unified.daily.conus.html, accessed on 20 July 2022, covering only land area) is used for analyzing the ENSO impact on precipitation.

2.2. WT Analysis

The domain used for the WT analysis, 100–65° W and 20–45° N, is the same as was used to analyze the WTs in the warm season in the SEUS [9]. In this paper, we analyze November to February. We also extend the years of analysis by including all currently available NRP data: 1948–2022.

As in Roller et al. [2] and Qian et al. [7,9], the classifiability index (CI) is calculated to determine the optimum number of clusters for the dataset, with large CI indicating a better separation between clusters. The optimum number of clusters (k) corresponds to the peak in CI following the initial decrease. The daily weather in the SEUS in the winter can be best represented by six clusters (k = 6) (Figure 2).

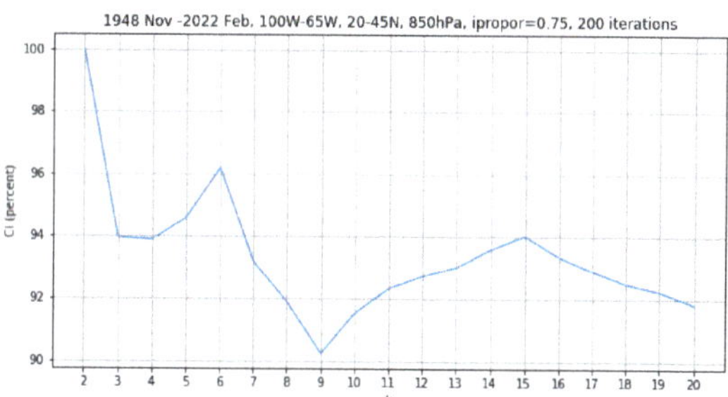

Figure 2. The classifiability index (CI) of the k-means clustering analysis using 850 hPa zonal and meridional wind component (u and v) in the Southeast U.S. and the surrounding region (100–65° W, 20–45° N) in the winter months (November–February) of 1948–2022; the optimum number for the k-means analysis is k = 6.

3. Results

3.1. Winter Weather Types in the SEUS

The spatial distribution of the 850 hPa winds for the six WTs, along with the associated mean precipitation of each, is shown in Figure 3 and a brief description is given in Table 1. WT1 represents a broad and shallow trough over central-east US. WT2 is an east-coast trough with precipitation ahead of the trough along the east coast from Florida to the mid-Atlantic coast, but with maximum precipitation over the costal ocean along the Gulf Stream. Behind the trough, northwesterly dry flow takes place over the Midwest and Great Plains. The Gulf coast is also quite dry. WT3 is an off-east-coast trough with maximum precipitation over the ocean, while the land in the Central and Eastern U.S. is dry. WT4 features a ridge over the Mississippi River Valley, with a trough and rainy area in the Atlantic Ocean quite far away from the east coast. The east coast and the Great Lakes area are dry. The southern Great Plains get some rainfall with moisture supplied by the southerly flow from the Gulf coast. In WT5, a trough is over the western Great Plains, with the southwesterly flow bringing forth moisture and precipitation over the Mississippi River Valley. An anticyclone is centered off the east coast of the Carolinas. The east coast is quite dry. In WT6, a deep trough is over the eastern Great Plains and the Mississippi and Missouri river valleys. The strong southwesterly flow east of the trough transports moisture from the Gulf of Mexico and dumps heavy precipitation in the Mississippi River Valley and SEUS.

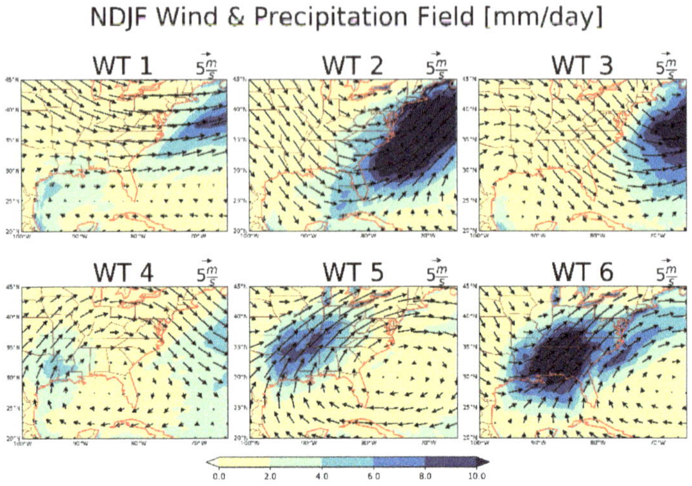

Figure 3. The average precipitation (shade, mm/day) and 850 hPa winds (arrows) of the six WTs. WTs 1–6 can be briefly described as Shallow Trough (ST), East Coast Trough (ECT), Off East Coast Trough (OECT), Mississippi River Valley Ridge (MRVR), Western Plains Trough (WPT) and Plains Trough (PT), respectively.

Table 1. Six weather types in the winter and their 850 hPa flow patterns.

Weather Type	Description
WT1	Shallow trough (ST) over central-east US
WT2	East coast trough (ECT)
WT3	Off-east-coast trough (OECT)
WT4	Mississippi River Valley Ridge (MRVR)
WT5	Western plains trough (WPT)/Off coast Carolina high (OCCH)
WT6	Plains trough (PT)

Comparing the winter WTs to those WTs in March–October, given in [9], some WTs are similar to or even identical to each other. For instance, the WT2 in Figure 3 here is quite similar to the WT1 (ECT) in [9]. WT4 here in Figure 3 is almost identical to the WT2 (MRVR) in Qian et al. [9], while WT6 here in Figure 3 is similar to the WT3 (PT) in Qian et al. [9]. It is reasonable that some WTs in November–February may share similar features of the WTs in the transitioning seasons of spring and fall, contained in the period of March–October in Qian et al. [9].

WT5 bears some similarity to the WT6 (Figure 3), but with a shallower trough in the western Great Plains. WT5 is also somewhat similar to the warm season WT6 (Carolina High, [9]), but the anticyclone center in the WT5 here is off the Carolina coast, while the warm season WT6 has the Carolina High centered on the coast.

We cannot find similar warm season WTs similar to the winter WT1 and WT3; therefore, they are distinctively winter WTs. In the winter, with the contrast of cold continent and warm ocean, the general circulation favors a quasi-stationary trough along the east coast of the continents. This is manifested by the occurrence of a shallow trough in coastal land (WT1) and a deep off-shore trough (WT3).

It is also worth noting that precipitation is heavy over the Great Lakes (near the northern border of the domain in Figure 3), as compared to the surrounding lake shore region, in WTs 5 and 6, in which southwesterly flow brings moisture toward the Great Lake region. In the winter, surface temperature over the lakes should be warmer than that over the land. Therefore, moist air and the warm lake surface both work to increase the conditional instability of the atmosphere, thus, enhancing precipitation over the lakes.

3.2. Frequency of the WTs

From 1 November to 28 February of 1948–2022 (120 days per year for 74 years, 8880 days in total), the percentages for the total number of days with WT 1–6 are 21.3%, 11.7%, 13.9%, 19.0%, 17.5% and 16.7%, respectively. The occurrence of the daily WTs through the season is shown in Figure 4. All six WTs sporadically occur in the winter season, as can be seen from the top panel in Figure 4. WT1, a distinctive winter WT, occurs quite frequently in every month of the winter, especially in December and January; therefore, it is the most frequent WT in the center of the winter. In the winter, westerly winds are the strongest in the mid-latitude due to the strongest baroclinicity (Figure 1), which corresponds to the quite zonal flow of WT1. WT2 (ECT), the least frequent winter WT, occurs slightly more in January and early February. As mentioned before, the WT2 (ECT) here is similar to the WT1 (also ECT) in [9] that frequently occurs in March–April. Therefore, ECT is common in the winter and spring in the SEUS. WT3 (OECT), the second least frequent winter WT, occurs more in January and February. WT4 (MRVR, the second most frequent WT) occurs more frequently in November (especially in the early part of the month) and least frequently in January. A precipitation map of WT4 indicates that it corresponds to fair weather in most of the SEUS, except for the southwest portion of the region. In other words, the fair weather in late fall in the SEUS is due to frequent WT4 (MRVR). WT5 (WPT) is more frequent in the first half of the winter (from mid-November to early January) with a contrast of rainy weather west of the Appalachian and fair weather along the east coast. Finally, WT6 is less frequent before late November and then somewhat evenly distributed after that. Comparing to the warm season WTs with a dominant summer WT (Florida High Zone that occurs about 60% of days in July, see [9]), there is no single dominant WT in the winter.

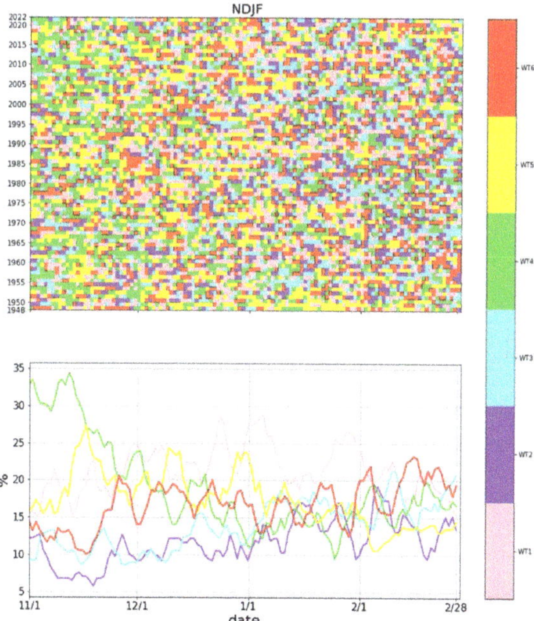

Figure 4. The daily weather types (WTs) in the winter of 1948–2022 (top). The 74-year averaged frequency (percentage) of the 6 WTs from November 1 to February 28 (120 days/year); a 5-day running average was conducted (bottom). Color: WT1 (pink), WT2 (purple), WT3 (light blue), WT4 (green), WT5 (yellow), WT6 (red).

Figure 5 shows the persistence of the WTs. For all six WTs, over 50% of their respective occurrences persist for only one day (and progress to other WTs the next day) and 20% persist for 2 days, 5–10% persist for 3 days and very few can persist for 4–12 days. These are short compared to the summer WTs that can persist for up to 20 days [9]. Winter weather is mostly controlled by eastward propagating baroclinic waves and, therefore, they have short persistence and fast transition/progression from one WT to another.

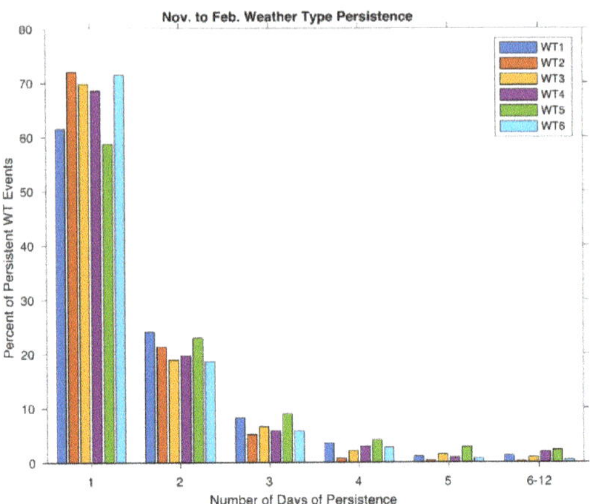

Figure 5. Persistence of the 6 WTs presented by different color bars, respectively. The x-axis is the length of consecutive days with the same WT. The last group of bars is the sums of persistent events of 6 to 12 days.

The progression of the WTs is shown in Figure 6. WT1 significantly progresses to itself (i.e., persistence, due to its nearly zonal flow) and also likely progresses to WT4 (shown by red bars). Actually, persistence is statistically significant for every WT (red bar). Besides persistence, the preferred progression loop is from WT2 (ECT) to WT3 (OECT), to WT4 (MRVR), to WT5 (WPT), to WT6 (PT) and then back to WT2. The circulation patterns of WTs 2–6 (Figure 3) indicate that the above progression loop corresponds to the eastward propagation of westerly waves.

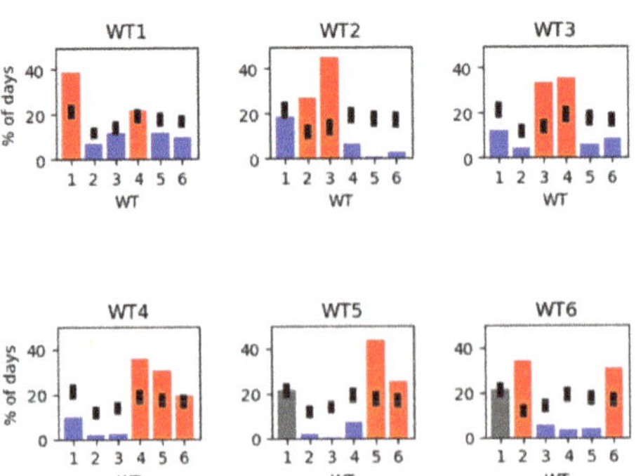

Figure 6. The progression of the WTs. The 95% confidence interval of the WT background frequency (the occurrence of WT transition due to chance) is shown with black shading. Frequencies significantly greater than the background frequency are shown with red bars; frequencies significantly less than the background bars are shown with blue bars.

We also checked the long-term trend in the 1948–2022 time series for the frequency of each of the WTs in November to February (figure not shown) and did not find a significant trend based on the Mann–Kendall test at the 95% significance level.

3.3. ENSO Impacts on Rainfall in the SEUS in the Winter

After analyzing the frequency and spatial pattern of the WTs, we can then use them to interpret the ENSO impact on the winter precipitation in the SEUS. As can be seen from Figure 1, the monthly climatology is similar among each month in deep winter months of December to February (DJF); therefore, we will focus on DJF.

As conducted by Ohba and Sugimoto in [14] for the ENSO impact on precipitation in Japan, we also quantified the dynamic and thermodynamic contributions of ENSO to the winter precipitation in the SEUS. El Niño and La Niña have the opposite impact, but may not be totally asymmetric, so we will check their impact separately. The anomalous precipitation associated with El Niño (denoted by En) can the written as:

$$\delta P = P^{En} - P^{Clim} = \sum_{i=1}^{K} p_i^{En} f_i^{En} - \sum_{i=1}^{K} p_i^{Clim} f_i^{Clim} = \delta P_{thermo} + \delta P_{dyn} \tag{1}$$

$K = 6$ is the number of the WTs; P^{En} and P^{Clim} are the precipitation in El Niño years and the climatology (all year average), respectively; p_i^{En} and p_i^{Clim} are the precipitation of WT i (i = 1 to 6) in El Niño years and climatology, respectively; f_i^{En} and f_i^{Clim} are the frequency of WT i in El Niño years and climatology, respectively; and δP_{thermo} and δP_{dyn} are the thermodynamic and dynamic contribution to the anomalous precipitation, respectively, as follows:

$$\delta P_{thermo} = \sum_{i=1}^{K} (p_i^{En} - p_i^{Clim}) \frac{f_i^{En} + f_i^{Clim}}{2} \tag{2}$$

$$\delta P_{dyn} = \sum_{i=1}^{K} \frac{p_i^{En} + p_i^{Clim}}{2} (f_i^{En} - f_i^{Clim}) \tag{3}$$

The δP_{thermo} in (2) is resulted from the changes in the basic (environmental) field, such as temperature and moisture. The δP_{dyn} in (3) is caused by the changes in frequency of the WTs. The thermodynamic and dynamic contribution to precipitation anomalies in La Niña (denoted by Ln) can be calculated similarly, by replacing En by Ln in (1) to (3).

Figure 7 shows the frequencies of the winter WTs in El Niño, La Niña and all years in DJF (f_i^{Clim}, f_i^{En} and f_i^{Ln}, $i = 1, 6$). In El Niño years, the frequencies of WTs 1 and 5 decrease, those of WTs 2, 3 and 6 increase and that of WT4 slightly increases.

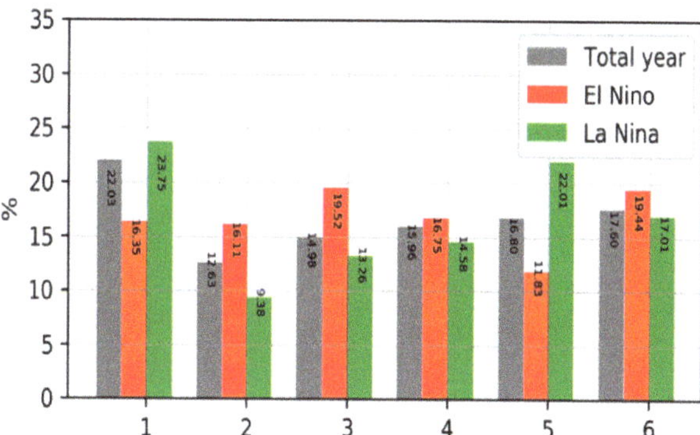

Figure 7. Frequencies of the WTs in all years (grey), El Niño years (red) and La Niña years (green) in December–February of the ENSO years.

Every term in Equations (2) and (3) is calculated and shown in Figure 8. The thermodynamic contribution (Equation (2)) and dynamic contribution (Equation (3)) to the anomalous precipitation in El Niño years are shown in the middle and right column of Figure 8, respectively, and the sum of the thermodynamic and dynamic contributions by each WT is shown in the left column. The total contribution of the six WTs is shown in the bottom row.

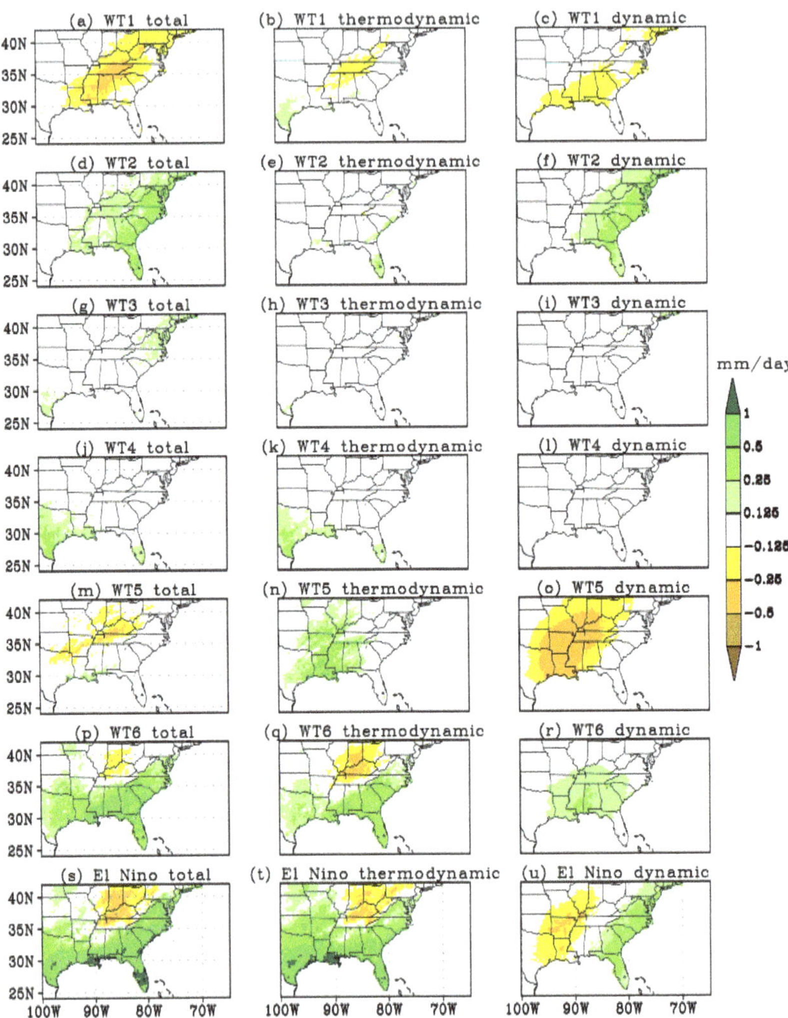

Figure 8. Precipitation anomalies in El Niño years in December–February (panel (**s**)). Panels (**a–r**) are the contribution of the WTs 1–6 to the anomalous precipitation. Left, middle and right panels are the total, thermodynamic and dynamic contribution from each WT, respectively. Panels (**t,u**) are the sum of thermodynamic and dynamic contribution for all WTs, respectively.

The dynamic contribution (Equation (3)) is approximately the precipitation pattern shown in Figure 3 multiplied by the frequency difference between El Niño years and all years (red and gray bars) in Figure 7. The WTs with a shallow trough (WT1) or a trough of more western location (WT5, WPT) are less frequent, producing negative dynamic contributions from WT1 in the coastal plains and from WT5 in the Mississippi River Valley, respectively (Figure 8c,o). The WTs with deep troughs and more eastern location (WTs 2, 3, and 6) are more frequent in El Niño years. Therefore, positive dynamic contribution is in

Eastern US and Southern US in WT2 and WT6, respectively (Figure 8f,r). The precipitation over land is very light in WT3 (Figure 3), so the dynamic contribution to precipitation from WT3 is also very small (Figure 8i). The frequency difference of WT4 is small (Figure 7); thus, the dynamic contribution to anomalous precipitation is small (Figure 8l). Summing up dynamic contribution from all six WTs, the total dynamic contribution to the anomalous precipitation (Figure 8u) features a dipolar pattern of a negative anomaly in the Mississippi River valley versus a positive precipitation anomaly in the east coast states.

The thermodynamic contribution is quite different among the WTs. For each WT, the thermodynamic contribution is also quite different from the corresponding dynamic one. The thermodynamic contribution of WT1 is the negative precipitation anomaly west of the Appalachian Mountain. Therefore, the total contribution of WT1 (Figure 8a) is the southwest–northeast-oriented area of negative anomalous precipitation from the Gulf coast to the Northeast U.S. The thermodynamic contribution of WT2, 3 and 4 is small, except for some localized areas in Florida and Texas. The thermodynamic contribution of WT5 is positive anomalous precipitation in the Gulf coast and Mississippi River Valley region (implying wetter air in El Niño years, Figure 8n), with the opposite sign to the dynamic contribution (Figure 8o). The latter is stronger than the former so that the total contribution of WT5 is the negative anomalous precipitation, extending from northeast Texas to southwest Pennsylvania (Figure 8m). The thermodynamic contribution of WT6 is strong (Figure 8q), with positive anomaly in the Gulf and east coast and negative anomaly in the Ohio River Valley. The combination of thermodynamic and dynamic contributions makes the total contribution of WT6 (Figure 8p) with positive precipitation in all south and east coastal states and negative precipitation anomaly within the Ohio River Valley. The pattern of the total contribution of WT6 (Figure 8p) is quite similar to the total El Niño impact shown in Figure 8s, indicating that most of the precipitation anomaly is contributed by WT6. WT2 increased the positive precipitation anomaly in the east coast, and WT1 and WT5 enhanced the negative precipitation anomaly in the Ohio River Valley.

The relative role of the thermodynamic and dynamic contribution to the precipitation anomaly, in terms of percentage, is location dependent, as seen in Figure 8s–u. For some places, such as Louisiana, the two contributions are opposite in sign. For the east coast, the two contributions both increase precipitation with slightly stronger contribution from the thermodynamic term. In the Gulf coast, the thermodynamic contribution dominates over the dynamic one, especially in the area close to the coast. In the Ohio River valley, both contributions are negative, therefore, reinforcing each other, but the thermodynamic contribution looks stronger than the dynamic one.

The La Niña impact on the precipitation in the SEUS is shown in Figure 9. The patterns are quite similar to the El Niño impact shown in Figure 8, but with opposite signs. However, there are some asymmetries between the El Niño and La Niña impacts. For example, the dynamic contribution of WT1 is smaller (Figure 9c), the thermodynamic contributions of WT5 seems weaker and less homogeneous (Figure 9n) and both the thermodynamic and dynamic contribution of WT6 are weaker in La Niña than those in El Niño. The area of positive precipitation anomaly in the Ohio River Valley is larger in La Niña (Figure 9s) than that in El Niño (Figure 8s).

The current result is consistent with that of Nieto Ferreira et al. [15] who found stronger mid-latitude cyclones and more intense precipitation over a large area in the SEUS during El Niño than La Niña and normal years. In contrast, negative rainfall anomalies are found in the east coast and Gulf coast areas in the SEUS in La Niña years.

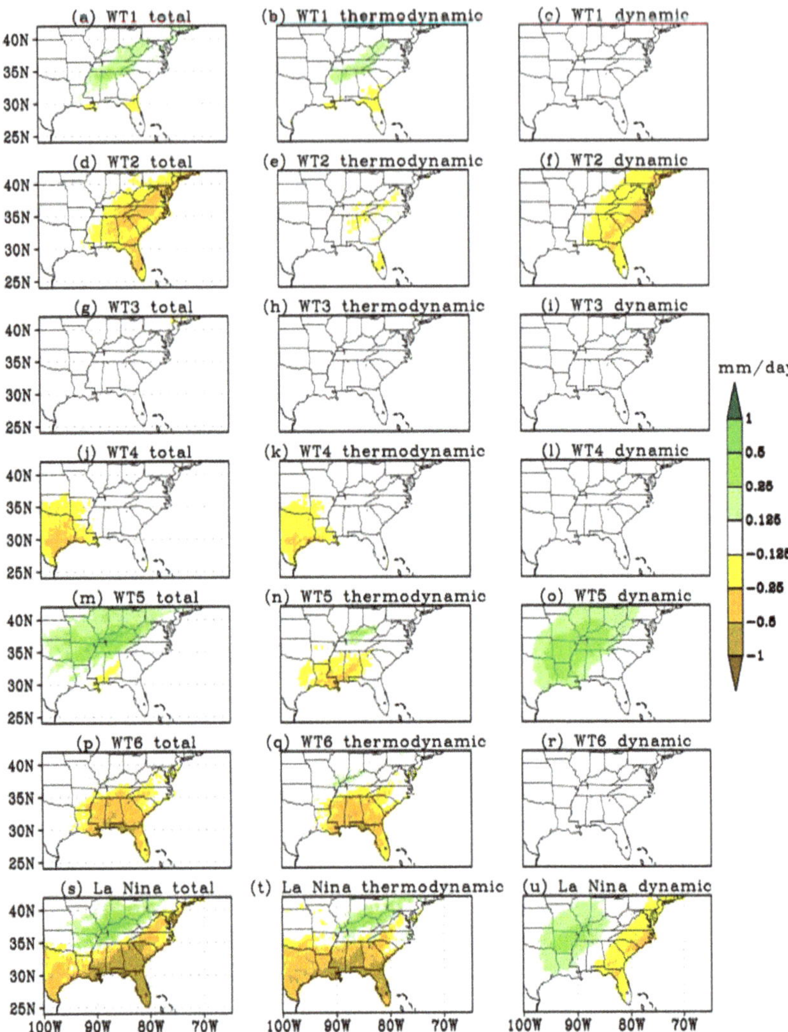

Figure 9. Precipitation anomalies in La Niña years in December–February (panel (**s**)). Panels (**a–r**) are the contribution of the WTs 1–6 to the anomalous precipitation. Left, middle and right panels are the total, thermodynamic and dynamic contribution from each WT, respectively. Panels (**t,u**) are the sum of thermodynamic and dynamic contribution for all WTs, respectively.

4. Conclusions and Discussion

In order to link weather and climate variability associated with ENSO, we analyzed daily WTs in the SEUS. Six winter weather types were obtained using k-means clustering analysis of the NRP 850 hPa wind components in November to February of 1948–2022. These six WTs have shorter persistence as compared to the summer WTs in [9]. This is because winter weather is mostly controlled by fast propagating westerly waves. A progression loop comprising WTs 2–6 corresponds to the transient westerly waves.

In our previous study, there were six WTs in the long period of eight months from March to October [9]. Here, there are also six WTs in the short period of four months from November to February. The stronger baroclinicity in the winter generates more fast-moving WTs. In contrast, the quasi-stationary NASH produces long-lasting summer WTs.

The frequency and spatial pattern of the WTs can be effectively used to understand the ENSO impact on winter precipitation in the SEUS. We quantified the ENSO impact in terms of dynamic and thermodynamic contribution to anomalous precipitation, respectively. The dynamic contribution is due to the change in the frequencies of the WTs. There are more frequent WTs of deep east coast trough types and plains trough type (WT2, 3 and 6) and less shallow troughs (WT1) and western Plains troughs (WT5) in El Niño years. The more frequent east coast trough and plains trough produce above normal precipitation in southeast coastal plains and Gulf coastal plains. The less frequent WT5 tends to reduce rainfall from the Great Lakes southwestward, producing a negative precipitation anomaly in the Central U.S., south of the Great Lakes. The thermodynamic contribution is from the change in spatial pattern of environmental variables, such as temperature and humidity. In the SEUS, the thermodynamic contribution is as strong as, and even slightly stronger than, the dynamic contribution. Particularly, the thermodynamic term of WT6 makes the largest contribution to the anomalous precipitation associated with El Niño. In general, the thermodynamic and dynamic contribution to the anomalous precipitation in La Niña is quite similar to those in El Niño, but with the opposite sign. However, the influence from WT6 in the La Niña is weaker than that in the El Niño.

Now that we have studied both warm season WTs in March to October [9] and winter WTs in November to February in the current study, respectively, we obtained a quite complete view of the WTs in the whole year. Of course, there are some overlapping WTs between warm season and winter WTs, which may be related to the WTs in the transitioning seasons. We studied the ENSO impact on precipitation from the perspective of the WTs in spring [9] and winter, respectively. It may also be interesting to study the ENSO impacts in the summer and fall using the WTs. Impacts of climate variability on other time scales, such as the intra-seasonal Madden Julian Oscillation and the long-term trend of climate change, also warrant further investigation.

Author Contributions: Conceptualization, J.-H.Q.; Data curation, C.L.; Formal analysis, J.-H.Q. and C.L.; Methodology, J.-H.Q.; Writing—original draft, J.-H.Q.; Writing—review & editing, B.V., S.N. and D.W. All authors have read and agreed to the published version of the manuscript.

Funding: This work was produced by Battelle Savannah River Alliance, LLC under Contract No. 89303321CEM000080 with the U.S. Department of Energy. Publisher acknowledges the U.S. Government license to provide public access under the DOE Public Access Plan (http://energy.gov/downloads/doe-public-access-plan, accessed on 20 July 2022).

Data Availability Statement: The data presented in this study are available on request from the corresponding author. The data are not publicly available due to database access restrictions.

Acknowledgments: We appreciate the constructive comments by the anonymous reviewers, which greatly improved the paper.

Conflicts of Interest: The authors declare no conflict of interest.

References

1. Ropelewski, C.F.; Halpert, M.S. North American precipitation and temperature patterns associated with the El Niño–Southern Oscillation. *Mon. Weather Rev.* **1986**, *114*, 2352–2362. [CrossRef]
2. Roller, C.D.; Qian, J.-H.; Agel, L.; Barlow, M.; Moron, V. Winter weather regimes in the Northeast United States. *J. Clim.* **2016**, *29*, 2963–2980. [CrossRef]
3. Nigam, S.; Sengupta, A. The full extent of El Niño's precipitation influence on the United States and the Americas: The suboptimality of the Niño 3.4 SST index. *Geophys. Res. Lett.* **2021**, *48*, e2020GL091447. [CrossRef]
4. Coe, D.; Barlow, M.; Agel, L.; Colby, F.; Skinner, C.; Qian, J.-H. Clustering analysis of autumn weather regimes in the Northeast U.S. *J. Clim.* **2021**, *34*, 7587–7605. [CrossRef]
5. Qian, J.-H.; Robertson, A.W.; Moron, V. Interaction among ENSO, the monsoon and diurnal cycle in rainfall variability over Java, Indonesia. *J. Atmos. Sci.* **2010**, *67*, 3509–3524. [CrossRef]
6. Qian, J.-H.; Robertson, A.W.; Moron, V. Diurnal cycle in different weather regimes and rainfall variability over Borneo associated with ENSO. *J. Clim.* **2013**, *26*, 1772–1790. [CrossRef]

7.   Qian, J.-H. Mechanisms for the dipolar patterns of rainfall variability over large islands in the Maritime Continent associated with the Madden-Julian Oscillation. *J. Atmos. Sci.* **2020**, *77*, 2257–2278. [CrossRef]

8.   Ohba, M.; Nohara, D.; Yoshida, Y.; Kadokura, S.; Toyoda, Y. Anomalous weather patterns in relation to heavy precipitation events in Japan during the Baiu season. *J. Hydrometeorol.* **2015**, *16*, 688–701. [CrossRef]

9.   Qian, J.-H.; Viner, B.; Noble, S.; Werth, D. Precipitation characteristics of warm season weather types in the southeastern United States of America. *Atmosphere* **2021**, *12*, 1001. [CrossRef]

10.   Bell, G.D.; Bozart, L.F. Appalachian cold-air damming. *Mon. Weather Rev.* **1988**, *116*, 137–161. [CrossRef]

11.   Rackley, J.A.; Knox, J.A. A climatology of southern Appalachian cold-air damming. *Weather. Forecast.* **2016**, *31*, 419–432. [CrossRef]

12.   Kalnay, E.; Kanamitsu, M.; Kistler, R.; Collins, W.; Deaven, D.; Gandin, L.; Iredell, M.; Saha, S.; White, G.; Woollen, J.; et al. The NCEP/NCAR 40-year reanalysis project. *Bull. Amer. Meteor. Soc.* **1996**, *77*, 437–471. [CrossRef]

13.   Xie, P.; Joyce, R.; Wu, S.; Yoo, S.-H.; Yarosh, Y.; Sun, F.; Lin, R. Reprocessed, bias-corrected CMORPH global high-resolution precipitation estimates from 1998. *J. Hydrometeorol.* **2017**, *18*, 1617–1641. [CrossRef]

14.   Ohba, M.; Sugimoto, S. Dynamic and thermodynamic contributions of ENSO to winter precipitation in Japan: Frequency and precipitation of synoptic weather patterns. *Clim. Dyn.* **2021**, 1–16. [CrossRef]

15.   Nieto Ferreira, R.; Hall, L.; Richenbach, T.M. A climatology of the structure, evolution, and propagation of midlatitude cyclones in the Southeast United States. *J. Clim.* **2013**, *26*, 8406–8421. [CrossRef]

atmosphere

MDPI

Article

Assessing the Impacts of Land Use/Land Cover Changes on Water Resources of the Nile River Basin, Ethiopia

Mohammed Gedefaw [1,2,*], Yan Denghua [3,*] and Abel Girma [1]

[1] Department of Natural Resources Management, University of Gondar, Gondar P.O. Box 196, Ethiopia
[2] College of Environmental Science and Engineering, Donghua University, Shanghai 201620, China
[3] State Key Laboratory of Simulation and Regulation of Water Cycle in River Basin, China Institute of Water Resource and Hydropower Research, Beijing 100038, China
* Correspondence: mohammedgedefaw@gmail.com (M.G.); yandh@iwhr.com (Y.D.)

Abstract: Land use/land cover change and climate change have diverse impacts on the water resources of river basins. This study investigated the trends of climate change and land use/land cover change in the Nile River Basin. The climate trends were analyzed using the Mann–Kendall test, Sen's slope estimator test and an innovative trend analysis method. Land use/land cover (LULC) change was examined using Landsat Thematic Mapper (TM) and Landsat Enhanced Thematic Mapper (ETM+) with a resolution of 30 m during 2012–2022. The findings revealed that forestland and shrub land area decreased by 5.18 and 2.39%, respectively. On the other hand, area of grassland, cropland, settlements and water bodies increased by 1.56, 6.18, 0.05 and 0.11%, respectively. A significant increasing trend in precipitation was observed at the Gondar (Z = 1.69) and Motta (Z = 0.93) stations. However, the trend was decreasing at the Adet (Z = −0.32), Dangla (Z = −0.37) and Bahir Dar stations. The trend in temperature increased at all stations. The significant changes in land use/land cover may be caused by human-induced activities in the basin.

Keywords: climate trend; land use land cover; Nile River Basin; water resources

Citation: Gedefaw, M.; Denghua, Y.; Girma, A. Assessing the Impacts of Land Use/Land Cover Changes on Water Resources of the Nile River Basin, Ethiopia. *Atmosphere* **2023**, *14*, 749. https://doi.org/10.3390/atmos14040749

Academic Editors: Haibo Liu and Nicole Mölders

Received: 17 February 2023
Revised: 1 April 2023
Accepted: 12 April 2023
Published: 21 April 2023

1. Introduction

Land use change and climate variability are two important factors that affect water resources and freshwater ecosystems [1]. Currently, land use/land cover (LULC) change is one of the major global environmental challenges to humanity. Land use/land cover change (LULCC) can be driven by multiple forces: demographic trends, climate variability, national policies, and macroeconomic activities which in turn have a significant impact on hydrologic systems both at basin and regional scales [2,3]. It significantly affects hydrological response [4,5], ecosystem services [6] and climate processes. LULCC has become a global concern [7] because of its diverse impacts on water resources [8–10]. The expansion of agricultural land use causes significant changes in runoff and sediment load [11,12]. Land use change can also lead to a significant change in groundwater recharge and base flow [13], flood frequency [14], peak runoff [15] and total suspended sediment concentration and can change the hydrological system of the region [16,17]. It has been one of the main contributors to climate change [18]. On the other hand, climate change has also affected the land use system through changes in agricultural productivity and forest ecosystem [19] and leads to alteration of hydrological conditions of the watersheds [20–22].

Climate change, deforestation, forested wetland shrinkage and desertification have also resulted in the spatiotemporal deterioration of vegetated ecosystems [23]. Many studies in Africa have revealed a decline in availability of water and agricultural productivity within catchments as a result of changes in land use and land cover [24,25]. To understand the impacts of land use/land cover change on water resources, it is essential to know the historical effects in land use/land cover on the hydrological system [26]. Human activities alter the hydrological systems of river basins mainly due to the expansion of cultivated lands and LULCC [2,23,27]. Such changes were not well understood by society. For instance, conversion of forestland to cultivated land between 1985 and 2011 in the Angereb watershed has increased the mean wet flow by 39% and reduced the dry mean flow by 46% [2]. The spatial temporal variability of climate change, land use/land cover change and management practice in the watershed are extremely challenging for sustainable water resource management in the river catchment. Surface runoff is lower and groundwater flow becomes higher in forestlands due to the infiltration of rainfall into deep aquifers. However, surface runoff becomes higher in bare lands and groundwater flow is lower [28,29]. Different studies have so far shown that LULC changes had adverse effects on the water resources of river basins. For example, LULC changes induced by human activities and rainfall variability have adversely affected the condition of water resources in the Great Ruaha Sub-catchment of the Rufiji Basin [24], and decreased base flows due to land modifications in the Upper Great Ruaha river basin [30]. Qiu et al. (2011) [31] simulated the effects of the Conversion of Cropland to Forest and Grassland Program on the catchment water budget in the Jinhe River using the SWAT model. The results showed that LULC changes had adverse impacts on the water resources of the Jinhe river basin. The increase in agricultural land activities is associated with transformation of the land use and an increase in water abstraction for irrigation purposes as a result of an increase in surface runoff following rainfall events. Land use change decreased the blue water and green water flow of the Weihe River Basin [32]. The spatial distribution showed an uneven change. The LULC changes in the Blue Nile River Basin, which occurred during the period of 1985 to 2015, increased the annual flow (2.2%), wet seasonal flow (4.6%), surface runoff (9.3%) and water yield (2.4%). Conversely, the observed changes had reduced dry season flow (2.8%), lateral flow (5.7%), groundwater flow (7.8%) and ET (0.3%) [33]. Human activities and climate changes were the main driving factor for the reduction in discharge [34] and mean annual stream flow in the middle reaches of the Yellow River basin [35].

Since land use change has a significant and profound effect on water quality and quantity, there is an urgent need to understand the interaction between land use change, hydrology and water resources management [36,37]. This calls for the need to understand the extent to which alterations of the land use and land cover have impacted on water availability in river sub-catchments.

The increasing demand for water abstractions by industries, households and irrigation projects within the river basin has aggravated these problems. Thus, investigating the correlation between land use/land cover change and hydrological systems plays a critical role for the management of water resources in the river basin. These enable policy makers, local government bodies and decision makers to formulate and implement effective and appropriate response strategies to reduce the adverse impacts of land use/land cover change on water resources. Hence, the aim of this study is to investigate the observed impacts of LULCC on water resources in the Nile River Basin, Ethiopia. Furthermore, the study investigated the trends in climate from 1980 to 2016.

2. Materials and Methods

2.1. Description of the Study Area

The Nile River Basin is the Nile's largest tributary; the largest in terms of discharge volume, and the second largest in terms of area in Ethiopia. Geographically, the Nile River Basin is found in the northwestern part of Ethiopia which lies between 70°40' N and 120°51' N latitude, and 340°25' E and 390°49' E longitude. It covers an area of 199,592.17 km^2 [36]. The basin drains much of the central, north–central and northwestern Ethiopian Highlands. The basin is subdivided into 16 sub-basins based on major rivers in the basin and its tributaries [38]. The topography of the Nile River Basin is extremely complex, with elevations ranging from 435 m in the lowlands near the Sudan border to 4229 m in the basin's upper section. The upper plateau near Lake Tana and the lower elevations close to the Sudan border both have flat areas. Furthermore, the highland areas in the eastern, northern, southeastern, and northeastern sub-basins are distinguished by high altitude, steep slope, hilly, gully, rugged terrain, troughs and mountainous landforms. As far as LULC is concerned, agriculture, shrub land, and forest are the major LULC categories, followed by water bodies, grassland, settlements, bare land and wetland [39]. Agriculture is the dominant LULC class in the eastern, northern, northeastern, and southeastern sub-basins. The forest LULC class is predominant in the western, southern, northwestern, and southern central sub-basins. The climate of the basin varies from cool to hot due to topographic variations, with large variations within a limited elevation range [40]. The sub-basins are characterized by moderate to high annual mean temperature (25–30.3 °C) and high annual rainfall (1083.4–2051.4 mm). The mean annual rainfall of this river basin decreased from southwest (>2000 mm) to northeast (≥789.34 mm) with a mean annual temperature of 18.5 °C [41]. Lake Tana is the major water source of this river basin (Figure 1).

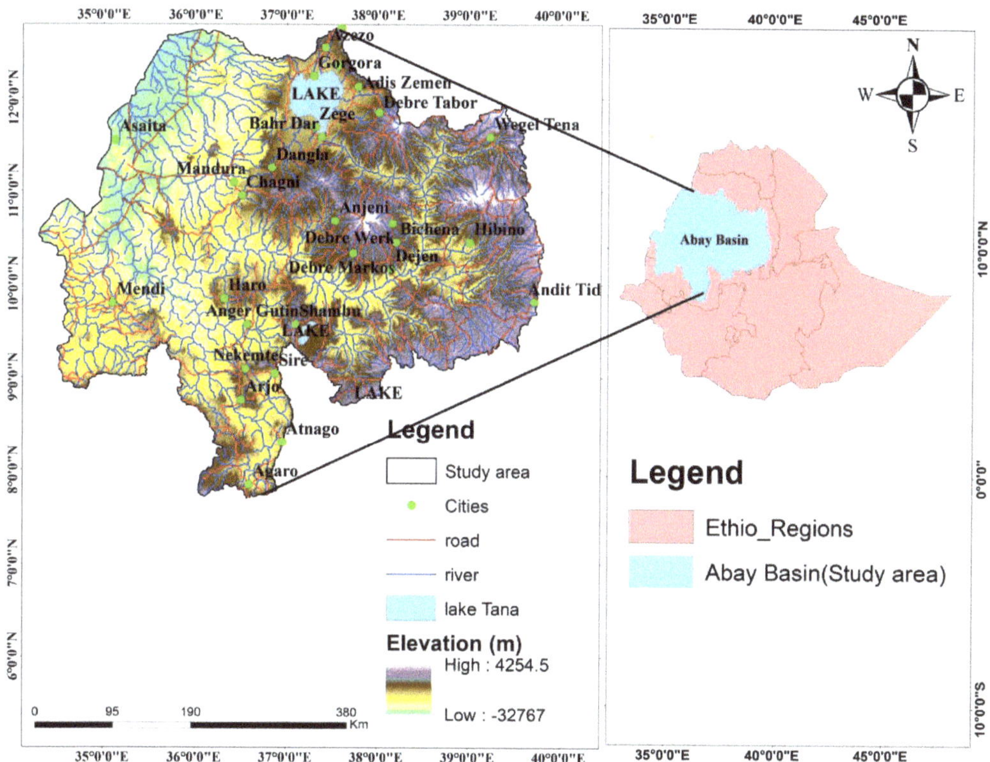

Figure 1. Location map of Abay/Nile River Basin.

2.2. Data Sources

2.2.1. Land Cover Data Sources

The land cover satellite data of the study area were collected from the Ministry of Energy and Water Resources of Ethiopia. The topographic map, soil map, ecological landscape potential map, forest map, and vegetation map were chosen for accuracy testing and validation. Landsat Thematic Mapper (TM) and Landsat Enhanced Thematic Mapper (ETM+) with a resolution of 30 m from 2012 to 2022 were used for land cover classification. The satellite data for land covers in the study area were obtained from the Global Land covers dataset (GlobeLand30) product. Land cover change was investigated between the years 2012 and 2022. The expanded classification system was adopted to capture the local characteristics of the study area (Table 1).

Table 1. Land cove types in Nile/Abay River Basin.

Land Cover Types	Code	Description
Forest land	Fl	Land covered with trees, with vegetation cover over 30%, including deciduous and coniferous forests, and sparse woodland with cover 10–30%, etc.
Grass land	Gl	Land covered by natural grass with cover over 10%, etc.
Shrub land	Sl	Land covered by small shrubs, plants less than 30 cm height.
Settlement	S	Land modified by human activities, including all kinds of habitation, residential, commercial, industrial, transportation facilities and interior urban green zones, etc.
Water body	Wb	Water bodies in the land area, including rivers, lakes, reservoirs, fish ponds, lands covered by temporary snow, glaciers and icecaps, etc.
Crop land	Cl	Land used for agriculture, horticulture and gardens, including paddy fields, irrigated and dry farmland, vegetation and fruit gardens, etc.

2.2.2. Meteorological Data

Daily precipitation and temperature data from stations during the period of 1980–2016 were obtained from the National Meteorological Services Agency of Ethiopia (NMSA) and National Oceanic and Atmospheric Administration's (NOAA) National Climatic Data Center. https://www.ncdc.noaa.gov/cdo-web/ accessed on 20 September 2023.

2.3. Methods

2.3.1. Analysis of Land Use/Land Cover Change

The overlaying operation was used to perform spatial analysis, which revealed how land use and land cover changed over time and established a connection between the two. A land cover transformation map was obtained and utilized for transformation matrix analysis by intersecting the two land cover/land use maps (2012 and 2022). The magnitudes of land cover shifts were calculated as [42].

$$A = TA(t_2) - TA(t_1), \tag{1}$$

$$CE = [CA/TA(t_1)] * 100, \tag{2}$$

where: TA = Total Area, CA = Changed Area, CE = Change Extent, and t_1 and t_2 are the beginning and ending time at which the land cover studies were conducted, respectively.

The Kappa coefficient and error matrix are standard measures of the reliability and accuracy of the maps produced. The Kappa statistics were determined in this study using methods described in detail in previous studies [43,44]. The Kappa coefficient was calculated using [45].

$$K = P(A) - P(E)/1 - P(E) \tag{3}$$

$$P(A) = \frac{(A+D)}{N}, \tag{4}$$

$$(E) = \left(\frac{A1}{N}\right) * \left(\frac{B1}{N}\right) + \left(\frac{A2}{N}\right) * \left(\frac{B2}{N}\right), \tag{5}$$

where: K is the Kappa coefficient, $P(A)$ is the number of times the k raters agree, and $P(E)$ is the number of times the k raters are expected to agree only by chance [46]. A and D are unchanged categories, $A1$ and $B1$ are the subject's categories, and N is the change in results.

2.3.2. Mann–Kendall (MK) Test Method

Trend analysis is used to identify whether the observed values of a time series data are increasing, decreasing, or show no trend. The non-parametric Mann–Kendall (MK) test has been applied in most studies to detect trends in hydro–meteorological time series since it does not require normally distributed data. This paper used the innovative trend analysis method (ITAM) to detect the trends in rainfall and temperature time series data. To evaluate the reliability of ITAM, the results were compared with the MK and Sen's slope estimator tests. The trends in rainfall time series data were assessed at 10%, 5% and 1% level of significance using the MK, ITAM and Sen's slope estimator methods. Significance was considered at the 10% threshold. The Mann–Kendall test is a non-parametric statistical test employed to detect monotonic trends in series of environmental, climate, and hydrological data. The MK test has been used to detect the presence of monotonic (increasing or decreasing) trends in the study area and whether the trend is statistically significant or not. Since there are chances that outliers are present in the dataset, the non-parametric MK test is useful because its statistic is based on the plus or minus signs, rather than on the values of the random variables, and therefore, the trends determined are less affected by the outliers [47]. Each data value is compared with all subsequent data values. If the data value from a later time period is greater than a data value from earlier time period, the statistic S is increased by 1. However, if the data value from a later time is less than a data value from earlier time period, the statistic S is decreased by 1. The net sum result would give the value of S.

The test statistics "S" is calculated by:

$$S = \sum_{i=1}^{n-1} \sum_{j=i+1}^{n} sgn\left(x_j - x_i\right) \tag{6}$$

$$sgn\left(x_j - x_i\right) = \begin{cases} +1 \text{ if } \left(x_j - x_i\right) > 0 \\ 0 \text{ if } \left(x_j - x_i\right) = 0 \\ -1 \text{ if } \left(x_j - x_i\right) < 0 \end{cases} \tag{7}$$

where x_j and x_i represent the data points at periods j and i, respectively. While the amount of data series is larger than or equivalent to ten ($n \geq 10$), the MK test is then categorized by a standard distribution with the mean $E(S) = 0$ and variance $Var(S)$, given as [48]:

$$E(S) = 0 \tag{8}$$

$$Var(S) = \frac{n(n-1)(2n+5) - \sum_{k=1}^{m} t_k(t_k - 1)(2t_k + 5)}{18} \tag{9}$$

The test's Z statistic is obtained using approximation, as follows:

$$Z = \begin{cases} \frac{s-1}{\delta} \text{ if } S > 0 \\ 0, \text{ if } S = 0 \\ \frac{s+1}{\delta} \text{ if } S < 0 \end{cases} \tag{10}$$

where Z follows a normal distribution, a positive Z and a negative Z depict an upward and downwards trend for the period, respectively.

2.3.3. Sen's Slope Estimator Test

The magnitude of the trend is predicted by slope estimator methods. In general, the slope Qi between any two values of a time series x can be estimated using the following equations.

The slope Q_i between two data points is given by the equation [48]:

$$Q_i = \frac{x_j - x_k}{j - k}, \; for \; i = 1, 2, \dots N \tag{11}$$

where x_j and x_k are data points at time j and $(j > k)$, respectively.

2.3.4. Innovative Trend Analysis Method (ITAM)

The Innovative Trend Analysis Method has been used in many studies to analyze hydro–meteorological time series data, and its reliability has been compared with the results of the MK method. In ITAM, the hydro–meteorological time series data are divided into two halves and arranged in ascending order independently. Then, the two halves are placed on a coordinate system, with (X_i: i = 1, 2, 3, ... n/2) on the X-axis and (X_j: j = n/2 + 1, n/2 + 2, ... n) on the Y-axis. If the time series data on a scattered plot are collected on the 1:1 (45°) straight line, it means there is no trend. It shows an increasing or decreasing trend if the time series data points accumulate above or below the 1:1 straight line, respectively. The magnitude of the trends in time series data can be estimated by calculating the average difference between the values of X_i and X_j. The change in trend is determined by the first half of the time series data points. Therefore, the trend indicator is estimated by dividing the average difference from the straight line to the average of the first half of the time series data points. The ITAM trend indicator is multiplied by 10 to represent the same scale as the MK method and Sen's slope estimator test at a 10% level of significance.

The trend indicator is calculated as [48]:

$$\phi = \frac{1}{n} \sum_{i=1}^{n} \frac{10(x_j - x_i)}{\mu} \tag{12}$$

where, ϕ = trend indicator, n = number of observations in the subseries, x_i = data series in the first half subseries class, x_j = data series in the second half subseries and μ = mean of data series in the first half subseries.

3. Results and Discussion

3.1. Analysis of Land Cover Change in the Nile River Basin

The 2012 and 2022 land use/land cover map of the Nile River Basin was generated from Landsat TM and ETM+ imagery classification (Figure 2). The spatial distribution of major LULC classes for 2012 and 2022 are presented in Table 2. The results show that shrubland and forestland were significantly reduced by 2.39% and 5.49% (Table 2), respectively. In a visual examination of land use maps, it was evident that from 2012 to 2022, the area under forest and shrub coverage diminished significantly in the study region. Agriculture was the predominant land use type in the Nile River Basin, which covered 41.42% in 2012 and 47.60% in 2022 (Table 2). A significant expansion of cultivated land of about 6.18% between 2012 and 2022 was observed, whereas the forest coverage diminished between 2012 and 2022, which accounted for about 5.49%. This finding is supported by the results of [49]. The reduction in water bodies and forestland area may be associated with human activities surrounding the river basin [50].

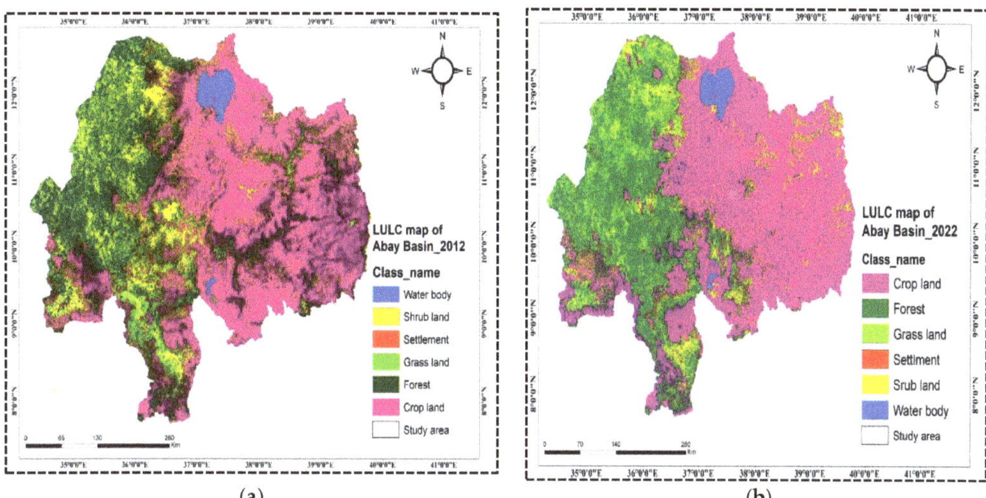

(**a**)   (**b**)

Figure 2. LULC in (**a**) 2012 and (**b**) 2022.

Table 2. Land use/land cover change detection in the Nile/Abay river basin during 2012 and 2022.

No.	Land Cover Types	Initial Area 2012 Percent (%)	Final Area 2022 Percent (%)	Changing Status Percent (%)
1	Forest land (Fl)	37.05	31.56	−5.49
2	Grass Land (Gl)	2.18	3.74	1.56
3	Shrub land (Sl)	17.53	15.14	−2.39
4	Settlement (S)	0.07	0.12	0.05
5	Water Body (Wb)	1.74	1.85	0.11
6	Crop land (Cl)	41.42	47.60	6.18

3.2. Analysis of the Impacts of Land Cover Change on Water Resources

Due to anthropogenic activities and climate change, the water resources of the Nile River Basin were significantly affected [51]. Furthermore, the dynamic changes in land use/land cover in the Nile River Basin impacted the water resources. The change detection results of the different LULC types from 2012 to 2022 are presented in Figure 3. The high demand for cultivated land for crop production in the Nile River Basin devastated the water resource potentials of the river basin. Land use/land cover changes have a great influence on the rainfall runoff process. The cultivation of forests and the demand for more agricultural land forced by urban development into settlements and infrastructure forms a sealed surface, which adversely changes the partitioning of precipitation towards increasing surface runoff and reduced ground water recharge [52–56]. The increase in surface runoff due to built-up areas which have a high proportion of impervious surfaces decreases water percolation and groundwater contribution to stream flow. This is supported by the results obtained by [57], who studied the impacts of land use and land cover changes on flow regimes of the Usangu wetland and the Great Ruaha River. His findings stated that the change in the land use and land cover within the catchment caused an increase in runoff, a decrease in base flow, an increase in sediment deposition on the bank of the river and a decrease in the width of the river channel.

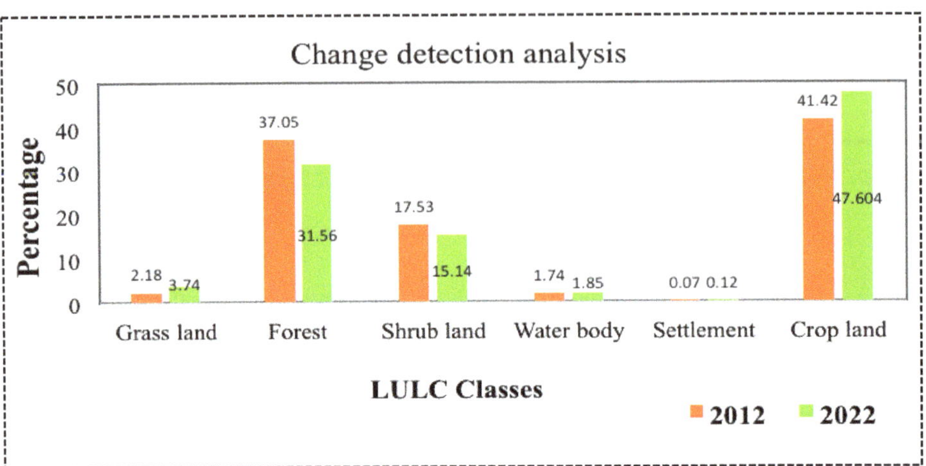

Figure 3. Change in LULC types between 2012 and 2022.

3.3. Analysis of Climate Trends

3.3.1. Analysis of Trends of Precipitation

The findings showed that the annual precipitation shows a significant increasing trend in Gondar (Z = 1.69), a sharp decreasing trend in Adet (Z = −0.32), a slightly decreasing trend in Dangla (Z = −0.37), a significant increasing trend in Bahir Dar and a significant increasing trend in Motta (Z = 0.93) (Figure 4). Significance levels at $\alpha = 0.01$, $\alpha = 0.05$, $\alpha = 0.1$ were taken to detect the trends at all stations. The ITAM test showed an increasing trend in Gondar and Motta and a decreasing trends in Adet, Bahir Dar and Dangla. The increase and decrease in innovative trend analysis (ITAM) test values represent strong and weak magnitudes, respectively. The variability in trends of precipitation across stations might be due to human activities and climate change impacts [58–60]. The trend in precipitation seen for each station could imply that the changes are more pronounced for certain locations and less so for others. The trend results are depicted in Table 3.

Table 3. Trends for stations in the Nile River Basin.

S/No	Name of Stations	Z (MK)	φ	β	Change (%)
1	Gondar	1.69 **	0.54	1.84 **	0.93
2	Adet	−0.32	−0.79	3.50	2.20
3	Bahir Dar	−0.07 *	−23.51	1.80 *	1.36
4	Dangla	−0.36	−0.39	1.26	1.27
5	Motta	0.93 ***	1.48	0.63 ***	0.79

* Trends at 0.1 significance level; ** trends at 0.05 significance level and *** trends at 0.01 significance level.

3.3.2. Analysis of Trends in Temperature

The results revealed that a statistically increasing trend was observed at Bahir Dar (Z = 2.63), Gonder (Z = 6.96) and Motta (Z = 4.58) stations. Even though there were variations in trends of temperature in Adet and Dangila stations, the trends were not statistically significant. From 1980 to 2016, the temperature increased by 0.5 °C, which shows a change in the climate system in the Nile River Basin. In general, the trend in temperature was increasing in all stations.

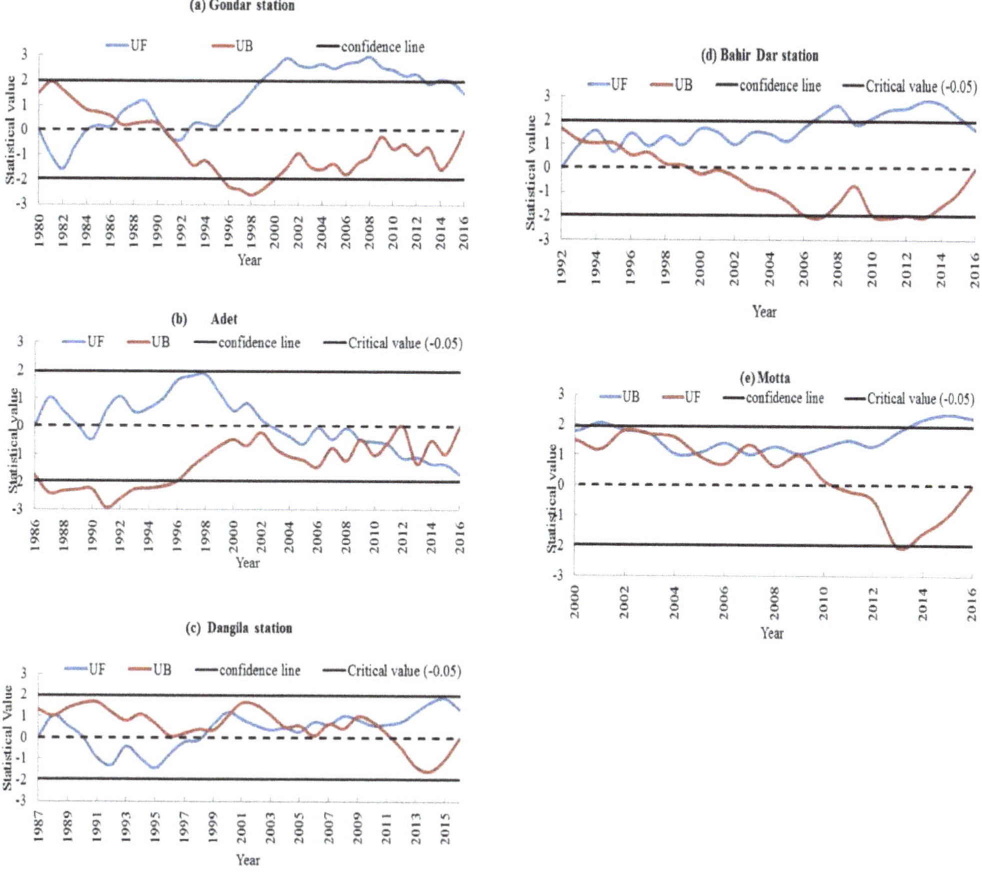

Figure 4. Trends of precipitation in the Nile River Basin.

4. Discussion

A land use/land cover change detection analysis was conducted to assess the spatial and temporal changes in land use/land cover (LULC) in Nile River Basin. The findings of this study show that, as a results of anthropogenic activities and climate change, the water resources of the Nile River Basin was significantly affected [61]. For example, the water bodies were significantly reduced by 1.72% from 2003 to 2013. Of the total water body coverage of the Nile River Basin, 5.49, 1.56, 2.39, 0.05 and 6.18% was converted into forestland, grassland, shrub land, settlement and croplands, respectively. These changes might be due to human activities and climate change in this region [62]. The intense human activities have increased artificial surfaces and cultivated land [63]. Therefore, climate change and human activity may produce a strong effect on vegetation and land cover in this region. The impact of the climatic variability on agricultural production is further aggravated by widespread soil degradation that has led to a reduction in the capacity of soils to hold moisture.

The impacts of land use and land cover change investigated by different researchers suggested that by increasing rainy season flow and decreasing dry period flow, the near future climate scenario would exacerbate extreme flow. Conversion of cultivated land on steep slopes into woodland, however, may reduce these intense flows. Under overlapping projections of potential land use/land cover and climate change, stream flow responses will be intensified at the outlet of the Tana watershed. The study region is characterized by maximum rainfall from June to August and little rainfall from March to May. There is inter-annual variability of rainfall between the stations. Some researchers tried to investigate the impacts of LULC changes on water resources. For example, Dibaba et al., (2020) [64] assessed the separate and combined impacts of climate and LULC change in the Finchaa catchment, and suggested that the changes in LULC led to an increase in surface runoff and water yield and a decrease in groundwater, while the predicted climate change indicates a decrease in the yield of surface runoff, groundwater and water yield. The combined study of the effects of LULC and climate change is a scenario considered equivalent to that of climate change. On the other hand, Berihun et al., (2019) [65] reported that, from 1982 to 2017, the observed LULC changes resulted in runoff increases ranging from 4% in Kecha to 28.7% in Kasiry, though climate variability had no significant impact on estimated runoff in terms of annual rainfall, while both LULC transition and climate variability had a significant effect on estimated evapotranspiration. Furthermore, other studies investigating the effects of historical climate and LULC change on the hydrology of the Nile River Basin showed that the impact of climate change (16.86%) on catchment stream flow was greater than that of land use/land cover change (7.25%), while the combined change effect accounted for a 22.13% increase in flow. In general, this result shows that high flow is more responsive to climate change, while land use/land cover change has shown a more substantial impact on low flow. Legesse et al., (2003) [65] indicated that a 10% decrease in rainfall resulted in a 30% decrease in the simulated hydrological response of the catchment, whereas a 1.5 °C rise in air temperature would result in a reduction of about 15% in the simulated discharge, while converting the current dominantly cultivated land in the studied river basin to woodland would reduce the discharge at the outlet by about 8% in south–central Ethiopia. When we see how the connection between human activities and climate change affects water resources, both have adverse impacts. The reduction in discharge causes challenges for global river ecosystems as a result of human activities and climate change (frequent extreme weather problems) [66,67]. Direct human activities also have an effect on variations in runoff [68]. They reduced the discharge of the Yellow River basin by 73.4% and 82.5% in 1980–2000 and 2001–2014, respectively [69]. The hydrological cycle of watersheds in both spatial and temporal changes is a complex process that is widely influenced by climate change and human activities. The IPCC (2013) [15] report indicated that climate change has led to changes in global precipitation patterns since the 20th century, which has changed the global hydrological process and directly affected the spatial and temporal distribution of global water sources; thus, it can cause changes in discharge [70]. Human activities, such as changes in land use/cover, dam construction and urbanization, have an obvious impact on all aspects of the water cycle [46], which can greatly change the spatiotemporal distribution of water resources. For example, Yuan et al., (2016) [68], reported that as a result of climate change, about 60.07–67.27% of the changes were observed during the change period I (1981–2002) and change period II (2003–2010), accounting for about 58.89–78.33% of changes due to human activities controlling stream flow changes. Furthermore, research should be carried out to determine the cumulative impact of land use/land cover and climate change on the Nile River Basin stream flow. This and other factors draw the researcher's attention to work on this subject. The hydrological study on the land use and land cover changes within the Nile River catchment showed that the flow characteristics have changed, with an increase in surface flow and reduction in base flow.

As far as the trends in climate are concerned, a significant increasing trend was observed at Gondar (Z = 1.69) and Motta (Z = 0.93) stations. However, a sharp decreasing trend was observed at Adet (Z = −0.32) and Dangla (Z = −0.37) stations. The trends in temperature were increasing at all stations. This result was in line with the global average temperature, which has increased by 0.85 °C from 1880 to 2012, and this may even accelerate in the near future [71]. The within-year and between-years variability in rainfall over the Nile Basin is high, making over-reliance on rainfed supply systems risky [48]. The high potential evaporation values in the Nile region ranging from some 3000 mm/year in northern Sudan to 1400 mm/year in the Ethiopian Highlands, and around 1100 mm/year in the hills in Rwanda and Burundi, make the basin particularly vulnerable to drought events. Drought risks are further amplified by the high variability of the rainfall between seasons and years. This is manifested by uncertainty in the onset of rains, occasional cessation of rainfall during the growing season, and consecutive years of below average rainfall. It has marked adverse impacts on the productivity of rainfed agriculture, and represents a serious constraint to rural development. However, the average temperature has fluctuated significantly in the past decade [48]. This will lead to significant environmental impacts. Gondar, Adet, and Bahir Dar stations exhibited a coefficient of variation of CV > 0.1, except for Dangla and Motta stations. This result is generally consistent with other studies, which reported increased precipitation and non-uniform precipitation changes. The Mann–Kendall test, Innovative Trend Analysis Method and Sen's slope estimator test showed the decreasing and increasing trend of rainfall across the stations. However, there was no statistically significant trend at the 95% level at any station. This result is also supported by [47], which showed that the seasonal trend in most parts of the country was decreased by 30% to 40%. This variation may influence regional climate systems as well as the hydrological cycle.

5. Conclusions

In this study, the impacts of land use/land cover change on water resources in the Nile River Basin were investigated. The trends in climate change were detected using the MK, Sen's slope estimator test and ITAM. The results showed that forestland and shrublands were significantly reduced by 6.18% and 2.39%, respectively. As far as the trends in precipitation are concerned, a significant increasing trend was observed at Gondar ($Z = 1.69$) and Motta ($Z = 0.93$) stations. However, a sharp decreasing trend was observed at Adet ($Z = -0.32$) and Dangla ($Z = -0.37$) stations. The trends in temperature were increasing at all stations. The change in trends in precipitation at each station could imply the impacts of climate change on water resources. Thus, these could indicate that climate change can affect the water resources of the river basin. Therefore, it can be concluded that there is evidence of some changes in the trend of rainfall, which has impacted the water resources of the Nile River Basin during the study period. Although further study is needed, climate change and LULC changes could impact the availability of water resources in river basins.

Author Contributions: M.G. made substantial contributions to the design, investigation, analysis, and drafting of the original manuscript. Y.D. supervised the work and A.G. participated in designing this manuscript. All authors have read and agreed to the published version of the manuscript.

Funding: This research was supported by the National Science Fund Project (Grant No. 52130907; 52109043) and the Five Major Excellent Talent Programs of IWHR (WR0199A012021).

Institutional Review Board Statement: Not applicable.

Informed Consent Statement: Not applicable.

Data Availability Statement: Not applicable.

Acknowledgments: The authors would thank the National Meteorological Service Agency of Ethiopia for providing the raw rainfall data.

Conflicts of Interest: The authors declare no conflict of interest.

References

1. Li, Y.; Fan, J.; Hu, Z.; Shao, Q.; Zhang, L.; Yu, H. Influence of land use patterns on evapotranspiration and its components in a temperate grassland ecosystem. *Adv. Meteorol.* **2015**, *2015*, 452603. [CrossRef]
2. Getachew, H.; Melesse, A. The impact of land use change on the hydrology of the Angereb watershed, Ethiopia. *Int. J. Water Sci.* **2012**, *1*, 1–7.
3. Onyutha, C.; Tabari, H.; Taye, M.T.; Nyandwaro, G.N.; Willems, P. Analyses of rainfall trends in the Nile River Basin. *J. Hydro-Environ. Res.* **2015**, *13*, 36–51. [CrossRef]
4. Wagner, P.D.; Bhallamudi, S.M.; Narasimhan, B.; Kumar, S.; Fohrer, N.; Fiener, P. Comparing the effects of dynamic versus static representations of land use change in hydrologic impact assessments. *Environ. Model. Softw.* **2017**, *122*, 103987. [CrossRef]
5. Su, Z.H.; Lin, C.; Ma, R.H.; Luo, J.H.; Liang, Q.O. Effect of land use change on lake water quality in different buffer zones. *Appl. Ecol. Environ. Res.* **2015**, *13*, 639–653.
6. Lawler, J.; Lewis, D.J.; Nelson, E.; Plantinga, A.J.; Polasky, S.; Withey, J.C.; Helmers, D.P.; Martinuzzi, S.; Pennington, D.; Radeloff, V.C. Projected land-use change impacts on ecosystem services in the United States. *Proc. Natl. Acad. Sci. USA* **2014**, *111*, 7492–7497. [CrossRef] [PubMed]
7. Woldeamlak, B. Land cover dynamics since the 1950s in Chemoga watershed, Blue Nile basin, Ethiopia. *Mt. Res. Dev.* **2002**, *22*, 263–269.
8. Gete, Z.; Hurni, H. Implications of land use and land cover dynamics for mountain resource degradation in the northwestern Ethiopian Highlands. *Mt. Res. Dev.* **2001**, *21*, 184–191.
9. Lambin, E.; Geist, H.; Lepers, E. Dynamics of land use and land cover change in tropical regions. *Annu. Rev. Environ. Resour.* **2003**, *28*, 206–232. [CrossRef]
10. Hurni, H.; Kebede, T.; Gete, Z. The implications of changes in population, land use, and land management for surface runoff in the Upper Nile Basin area of Ethiopia. *Mt. Res. Dev.* **2005**, *25*, 147–154. [CrossRef]
11. Gebremicael, T.G.; Mohamed, Y.A.; Betrie, G.D.; van der Zaag, P.; Teferi, E. Trend analysis of runoff sediment fluxes in the Upper Blue Nile basin: A combined analysis of statistical tests, physically based models and landuse maps. *J. Hydrol.* **2013**, *482*, 57–68. [CrossRef]
12. Memarian, H.; Balasundram, S.K.; Abbaspour, K.C.; Talib, J.B.; Sung, C.T.B.; Sood, A.M. SWAT-based hydrological modeling of tropical land-use scenarios. *Hydrol. Sci. J.* **2014**, *59*, 1808–1829. [CrossRef]
13. Budiyanto, S.; Tarigan, S.D.; Sinukaban, N.; Murtilaksono, K. The impact of land use on hydrological characteristics in Kaligarang watershed. *Int. J. Sci. Eng.* **2015**, *8*, 125–130.
14. Alexakis, D.D.; Grillakis, M.G.; Koutroulis, A.G.; Agapiou, A.; Themistocleous, K.; Tsanis, I.K.; Michaelides, S.; Pashiardis, S.; Demetriou, C.; Aristeidou, K.; et al. GIS and remote sensing techniques for the assessment of land use change impact on flood hydrology: The case study of Yialias basin in Cyprus. *Nat. Hazards Earth Syst. Sci.* **2014**, *14*, 413–426. [CrossRef]
15. IPCC. The Physical Science Basis—Summary for Policymakers. Contribution of WG1 to the Fourth Assessment Report of the Intergovernmental Panel on Climate Change. 2013. Available online: http://www.ipcc.ch/ipccreports/ar4-wg1.htm (accessed on 6 April 2023).
16. Mango, L.M.; Melesse, A.M.; McClain, M.E.; Gann, D.; Setegn, S.G. Land use and climate change impacts on the hydrology of the upper Mara River Basin, Kenya: Results of a modeling study to support better resource management. *Hydrol. Earth Syst. Sci.* **2011**, *15*, 2245–2258. [CrossRef]
17. Yeshaneh, E.; Wagner, W.; Exner-Kittridge, M.; Legesse, D.; Blöschl, G. Identifying Land Use/Cover Dynamics in the Koga Catchment, Ethiopia, from Multi-Scale Data, and Implications for Environmental Change. *ISPRS Int. J. Geo-Inf.* **2013**, *2*, 302–323. [CrossRef]
18. Balthazar, V.; Vanacker, V.; Molina, A.; Lambin, E.F. Impacts of Forest Cover Change on Ecosystem Services in High Andean Mountains. *Ecol. Indi-Cators* **2015**, *48*, 63–75. [CrossRef]
19. DeFries, R.; Eshleman, K.N. Land-Use Change and Hydrologic Processes: A Major Focus for the Future. *Hydrol. Process.* **2004**, *18*, 2183–2186. [CrossRef]
20. Girmay, G.; Singh, B.; Nyssen, J.; Borrosen, T. Runoff and sediment-associated nutrient losses under different land uses in Tigray, Northern Ethiopia. *J. Hydrol.* **2009**, *376*, 70–80. [CrossRef]
21. Setegn, S.; Srinivasan, R.; Dargahi, B.; Melesse, A. Spatial delineation of soil erosion vulnerability in the Lake Tana Basin, Ethiopia. *Hydrol. Process.* **2009**, *23*, 3738–3750. [CrossRef]
22. Ahn, G.; Gordon, S.I.; Merry, C.J. Impacts of remotely sensed land use data on watershed hydrologic change assessment. *Int. J. Geospat. Env. Res.* **2014**, *1*, 9.
23. Kashaigili, J.J. Impacts of Land-Use and Land-Cover Changes on Flow Regimes of the Usangu Wetland and the Great Ruaha River, Tanzania. *Phys. Chem. Earth* **2008**, *33*, 640–647. [CrossRef]
24. Jat, M.L.; Gathala, M.K.; Ladha, J.K.; Saharawat, Y.S.; Jat, A.S.; Kumar, V.; Sharma, S.K.; Kumar, V.; Raj, G. Evaluation of Precision Land Leveling and Double Zero-Till Systems in the Rice-Wheat Rotation: Water Use, Productivity, Profitability and Soil Physical Properties. *Soil Tillage Res.* **2009**, *105*, 112–121. [CrossRef]
25. Tekleab, S.; Mohamed, Y.; Uhlenbrook, S. Hydro-climatic trends in the Abay/Upper Blue Nile basin, Ethiopia. *Phys. Chem. Earth Parts A/B/C* **2013**, *61–62*, 32–42. [CrossRef]

26. Kidane, W.; Bogale, G. Effect of land use land cover dynamics on hydrological response of watershed: Case study of Tekeze Dam watershed, northern Ethiopia. *Int. Soil Water Conserv. Res.* **2017**, *5*, 1–16.

27. Gyamfi, C.; Ndambuki, J.; Salim, R. Hydrological responses to land use/cover changes in the Olifants basin, South Africa. *Water* **2016**, *8*, 588. [CrossRef]

28. Tsidu, G.M. High-resolution monthly rainfall database for Ethiopia: Homogenization, reconstruction, and gridding. *J. Clim.* **2012**, *25*, 8422–8443. [CrossRef]

29. Kashaigili, J.J.; Majaliwa, A.M. Integrated Assessment of Land Use Land Cover Changes on Hydrological Regime of the Malagarasi River Catchment in Tanzania. *J. Phys. Chem. Earth* **2013**, *35*, 730–741. [CrossRef]

30. Qiu, G.Y.; Yin, J.; Tian, F.; Geng, S. Effects of the "Conversion of Cropland to Forest Grassland Program" on the water budget of the Jinghe River Catchment in China. *J. Environ. Qual.* **2011**, *40*, 1745–1755. [CrossRef]

31. Zhao, A.; Zhu, X.; Liu, X.; Pan, Y.; Zuo, D. Impacts of land use change and climate variability on green and blue water resources in the Weihe River Basin of northwest China. *Catena* **2016**, *137*, 318–327. [CrossRef]

32. Gashaw, T.; Tulu, T.; Argaw, M.; Worqlul, A.W. Modeling the hydrological impacts of land use/land cover changes in the Andassa watershed, Blue Nile Basin, Ethiopia. *Sci. Total Environ.* **2018**, *619–620*, 1394–1408. [CrossRef] [PubMed]

33. Li, B.; Li, C.; Liu, J.; Zhang, Q.; Duan, L. Decreased discharge in the Yellow River Basin, China: Climate change or human-induced? *Water* **2017**, *9*, 116. [CrossRef]

34. Zhao, G.; Tian, P.; Mu, X.M.; Jiao, J.; Wang, F.; Gao, P. Quantifying the impact of climate variability and human activities on discharge in the middle reaches of the Yellow River basin, China. *J. Hydrol.* **2014**, *519*, 387–398. [CrossRef]

35. Yilma, A.D.; Awulachew, S.B. Characterization and Atlas of the Blue Nile Basin and its Sub basins. *Int. Water Manag. Inst.* **2009**.

36. Abate, T.; Angassa, A. Conversion of savanna rangelands to bush dominated landscape in Borana, Southern Ethiopia. *Ecol Processes* **2016**, *5*, 6. [CrossRef]

37. Haregeweyn, N.; Tsunekawa, A.; Poesen, J.; Tsubo, M.; Meshesha, D.T.; Fenta, A.A.; Nyssen, J.; Adgo, E. Comprehensive assessment of soil erosion risk for better land use planning in river basins: Case study of the Upper Blue Nile River. *Sci. Total Environ.* **2017**, *574*, 95–108. [CrossRef]

38. Cao, Q.; Yu, D.; Georgescu, M.; Han, Z.; Wu, J. Impacts of land use and land cover change on regional climate: A case study in the agro-pastoral transitional zone of China. *Environ. Res. Lett.* **2015**, *10*, 124025. [CrossRef]

39. Hussien, K.; Kebede, A.; Mekuriaw, A.; Asfaw, S.; Sitotaw, B.; Erena, H. Modelling spatiotemporal trends of land use land cover dynamics in the Abbay River Basin, Ethiopia. *Model. Earth Syst. Environ.* **2022**, *9*, 347–376. [CrossRef]

40. Gebreyes, M. Assessing the effectiveness of planned adaptation interventions in reducing local level vulnerability. Adapting to climate change in the water sector. *Work. Pap.* **2010**, *18*. [CrossRef]

41. Gedefaw, M.; Wang, H.; Yan, D.; Song, X.; Yan, D.; Dong, G.; Wang, J.; Girma, A.; Ali, B.A.; Batsuren, D.; et al. Trend Analysis of Climatic and Hydrological Variables in the Awash River Basin, Ethiopia. *Water* **2018**, *10*, 1554. [CrossRef]

42. Gwet, K. Kappa Statistic is not Satisfactory for Assessing the Extent of Agreement Between Raters. *Stat. Methods Inter-Rater Reliab. Assess.* **2002**, *1*, 1–6.

43. Manandhar, R.; Odeh, I.O.A.; Ancev, T. Improving the Accuracy of Land Use and Land Cover Classification of Landsat Data Using Post-Classification Enhancement. *Remote Sens.* **2009**, *1*, 330–344. [CrossRef]

44. Li, Z.; Liu, W.Z.; Zhang, X.C.; Zheng, F.L. Impacts of land use change and climate variability on hydrology in an agricultural catchment on the Loess Plateau of China. *J. Hydrol.* **2009**, *377*, 35–42. [CrossRef]

45. Gedefaw, M. Assessment of changes in climate extremes of temperature over Ethiopia. *Cogent Eng.* **2023**, *10*, 2178117. [CrossRef]

46. Gedefaw, M.; Yan, D.; Wang, H.; Qin, T.; Girma, A.; Abiyu, A.; Batsuren, D. Innovative Trend Analysis of Annual and Seasonal Rainfall Variability in Amhara Regional State, Ethiopia. *Atmosphere* **2018**, *9*, 326. [CrossRef]

47. Gebrehiwot, S.G.; Bewket, W.; Gärdenäs, A.I.; Bishop, K. Forest cover change over four decades in the Blue Nile Basin, Ethiopia: Comparison of three watersheds. *Reg. Environ. Chang.* **2014**, *14*, 253–266. [CrossRef]

48. Butt, A.; Shabbir, R.; Ahmad, S.S.; Aziz, N. Land use change mapping and analysis using Remote Sensing and GIS: A case study of Simly watershed, Islamabad, Pakistan. *Egypt. J. Remote Sens. Space Sci.* **2015**, *18*, 251–259. [CrossRef]

49. Arsano, Y.; Tamrat, I. Ethiopia and the Eastern Nile Basin. *Aquat. Sci.* **2005**, *67*, 15–27. [CrossRef]

50. Asfaw, A.; Simane, B.; Hassen, A.; Bantider, A. Variability and time series trend analysis of rainfall and temperature in northcentral Ethiopia: A case study in Woleka sub-basin. *Weather Clim. Extrem.* **2017**, *19*, 29–41. [CrossRef]

51. Conway, D. The climate and hydrology of the Upper Blue Nile River. *Geogr. J.* **2000**, *166*, 49–62. [CrossRef]

52. Li, R.Q.; Dong, M.; Cui, J.Y.; Zhang, L.L.; Cui, Q.G.; He, W.M. Quantification of the Impact of Land-Use Changes on Ecosystem Services: A Case Study in Pingbian County, China. *Environ. Monit. Assess.* **2007**, *128*, 503–510. [CrossRef]

53. Hwang, S.A.; Hwang, S.J.; Park, S.R.; Lee, S.W. Examining the relationships between watershed urban land use and stream water quality using linear and generalized additive models. *Water* **2016**, *8*, 155. [CrossRef]

54. Wang, X.Y.; Zhao, C.Y.; Jia, Q.Y. Impacts of climate change on forest ecosystems in Northeast China. *Adv. Clim. Chang. Res.* **2013**, *4*, 230–241.

55. Woldeamlak, B.; Sterk, G. Dynamics in land cover and its effect on streamflow in the Chemoga watershed, Blue Nile basin, Ethiopia. *Hydrol. Process.* **2005**, *19*, 445–458.

56. Yang, P.; Xia, J.; Zhang, Y.; Hong, S. Temporal and spatial variations of precipitation in Northwest China during 1960–2013. *Atmos. Res.* **2017**, *183*, 283–295. [CrossRef]

57. Yenehun, A.; Walraevens, K.; Batelaan, O. Spatial and Temporal Variability of Groundwater Recharge in Geba Basin, Northern Ethiopia. *J. Afr. Earth Sci.* **2017**, *134*, 198–212. [CrossRef]
58. Zhou, G.; Wei, X.; Chen, X.; Zhou, P.; Liu, X.; Xiao, Y.; Sun, G.; Scott, D.F.; Zhou Sh Han, L.; Su, Y. Global pattern for the e ect of climate and land cover on water yield. *Nat. Commun.* **2015**, *6*, 5918. [CrossRef]
59. Sahana, M.; Ahmed, R.; Sajjad, H. Analyzing land surface temperature distribution in response to land use/land cover change using split window algorithm and spectral radiance model in Sundarban Biosphere Reserve, India. *Model. Earth Syst. Environ.* **2016**, *2*, 81. [CrossRef]
60. Palmate, S.S.; Pandey, A.; Kumar, D.; Pandey, R.P.; Mishra, S.K. Climate change impact on forest cover and vegetation in Betwa Basin, India. *Appl. Water Sci.* **2017**, *7*, 103–114. [CrossRef]
61. Dibaba, W.T.; Demissie, T.A.; Miegel, K. Watershed Hydrological Response to Combined Land Use/Land Cover and Climate Change in Highland Ethiopia: Finchaa Catchment. *Water* **2020**, *12*, 1801. [CrossRef]
62. Berihun, M.L.; Tsunekawa, A.; Haregeweyn, N.; Meshesha, D.T.; Adgo, E.; Tsubo, M.; Masunaga, T.; Fenta, A.A.; Sultan, D.; Yibeltal, M. Exploring land use/land cover changes, drivers and their implications in contrasting agro-ecological environments of Ethiopia. *Land Use Policy* **2019**, *87*, 104052. [CrossRef]
63. Legesse, D.; Coulomb, C.V.; Gasse, F. Hydrological response of a catchment to climate and land use changes in Tropical Africa: Case study South Central Ethiopia. *J. Hydrol.* **2003**, *275*, 67–85. [CrossRef]
64. Yang, Z.; Zhang, Q.; Hao, X. Evapotranspiration trend and its relationship with precipitation over the loess plateau during the last three decades. *Adv. Meteorol.* **2016**, *2016*, 6809749. [CrossRef]
65. Wang, D.; Hagen, S.C.; Alizad, K. Climate change impact and uncertainty analysis of extreme rainfall events in the Apalachicola River basin, Florida. *J. Hydrol.* **2013**, *480*, 125–135. [CrossRef]
66. Yuan, Y.; Zhang, C.; Zeng, G.; Liang, J.; Guo, S.; Huang, L.; Wu, H.; Hua, S. Quantitative assessment of the contribution of climate variability and human activity to streamflow alteration in Dongting Lake, China. *Hydrol. Process.* **2016**, *30*, 1929–1939. [CrossRef]
67. Li, Y.; Chang, J.; Wang, Y.; Jin, W.; Guo, A. Spatiotemporal impacts of climate, land cover change and direct human activities on runoff variations in the Wei River Basin, China. *Water* **2016**, *8*, 220. [CrossRef]
68. Huntington, T.G. Evidence for intensification of the global water cycle: Review synthesis. *J. Hydrol.* **2006**, *319*, 83–95. [CrossRef]
69. Sterling, S.M.; Ducharne, A.; Polcher, J. The impact of global land-cover change on the terrestrial water cycle. *Nat. Clim. Chang.* **2013**, *3*, 385–390. [CrossRef]
70. Hou, J.; Ye, A.; You, J.; Ma, F.; Duan, Q. An estimate of human and natural contributions to changes in water resources in the upper reaches of the Minjiang River. *Sci. Total Environ.* **2018**, *635*, 901–912. [CrossRef]
71. Yan, D.H.; Wang, H.; Li, H.H.; Wang, G.; Qin, T.L.; Wang, D.Y.; Wang, L.H. Quantitative analysis on the environmental impact of largescale water transfer project on water resource area in a changing environment. *Hydrol. Earth Syst. Sci.* **2012**, *16*, 2685–2702. [CrossRef]

Article

Variation Characteristics of Rainstorms and Floods in Southwest China and Their Relationships with Atmospheric Circulation in the Summer Half-Year

Qingxia Xie [1,2], Xiaoping Gu [3,*], Gang Li [1], Tianran Tang [1] and Zhiyu Li [1]

[1] Guizhou Meteorological Observatory, Guiyang 550002, China
[2] Guizhou Institute of Mountainous Environment and Climate, Guiyang 550002, China
[3] Guizhou Key Laboratory of Mountainous Climate and Resources, Guiyang 550002, China
* Correspondence: 18985000711@sohu.com; Tel.: +86-15286002576

Abstract: Local climates are responding to global warming differently, and the changes in rainstorms in mountainous areas of Southwest China are of particular interest. This study, using monthly NCEP/NCAR reanalysis and daily precipitation observation of 90 meteorological stations from 1961 to 2021, analyzed the temporal and spatial variation characteristics of rainstorms and floods in Southwest China and their relationship with atmospheric circulations. The results led us to the following five conclusions: (1) Rainstorms and floods in southwest China mainly occur from June to August, during which time July has the most weather events, followed by August. (2) The southwest of Guizhou province, the southern edge of Yunnan province, and regions from the east of the Sichuan Basin to the north of Guizhou have experienced more rainstorms and floods, while the northwest regions of Southwest China have had fewer. (3) Over the last 61 years, rainstorms and floods have exhibited an overall rising trend, especially in the last 10 years. The year 2012 was an abrupt inflection point in rainstorms and floods in Southwest China, from low to high frequency, while the correlation coefficient between rainstorms and floods and the global surface temperature is above the 95% significance level. (4) Rainstorms and floods exhibit changes at periods of 8 years, 16 years, and 31 years. (5) Rainstorms and floods show a good correlation with multiple variables, such as South Asian high-pressure systems west of 90°E, the upper trough front, the northwest side of the western Pacific subtropical high, and the convergence of warm and wet air in the middle and lower layers with cold air on the ground.

Keywords: Southwest China; rainstorm and flood; spatial and temporal change; atmospheric circulation

Citation: Xie, Q.; Gu, X.; Li, G.; Tang, T.; Li, Z. Variation Characteristics of Rainstorms and Floods in Southwest China and Their Relationships with Atmospheric Circulation in the Summer Half-Year. *Atmosphere* **2022**, *13*, 2103. https://doi.org/10.3390/atmos13122103

Academic Editor: Haibo Liu

Received: 8 October 2022
Accepted: 24 November 2022
Published: 15 December 2022

Publisher's Note: MDPI stays neutral with regard to jurisdictional claims in published maps and institutional affiliations.

1. Introduction

Rainstorms and floods are among the most common and serious natural disasters [1]. Every year, there are rainstorms and floods of varying natures in the southwest of China, and excessive precipitation is the major cause. In the rainy season, excessive precipitation increases the soil moisture content and water level in the lakes and rivers, inducing water-logging, mountain torrents, and debris flows, and even the bursting of riverbanks, which causes damage to housing, traffic, and agriculture, as well as human life loss [2–7].

The research methods for floods can generally be divided into three categories. The first is the historical flood case study [8–13], which analyzes historical floods in a specific region. The second is the index system evaluation method [14–17], which evaluates regional rainstorms and flood disasters through a fuzzy, comprehensive evaluation method. The factors affecting regional floods include not only natural attributes such as topography, landscape, and climate but also social attributes such as social economy, population composition, and flood control construction. The third category is the precipitation category method [18–20], which takes the anomaly percentage of ten-day precipitation and monthly precipitation as the main index to measure the degree of flooding and waterlogging. Some researchers also classify categories according to the daily precipitation observations of local meteorological services. In addition, domestic studies have mainly focused on disaster assessment [12–14], emergency management [4], and individual case analysis [21–24]. The scope is usually the entirety of China or a single province; the Southwest region is rarely treated as an independent region for analysis. Previous studies have used a single threshold of 50 mm/day to quantify rainstorm flooding, which does not consider the accumulated effects of less intense rainstorms and consequently underestimates damage. Furthermore, these studies have been limited by the data period available, often having been conducted many years ago. This paper offers updated research results on the rainstorms and floods in Southwest China.

This paper focuses on the southwest region of China, including Yunnan, Guizhou, and Sichuan provinces, as well as Chongqing Municipality, which is located on the eastern slope of the Tibetan plateau. Affected by weather and climate systems, such as the plateau circulation, the southwest monsoon, and inappropriate human activities, this region is prone to serious floods and drought [21–24]. This study used the daily rainfall category method to define the rainstorms and floods in the entire southwest region. A total of 90 weather stations with continuous data records since their establishment in 1961 were selected for use in this study. We analyzed the temporal and spatial variation characteristics of rainstorms and floods in Southwest China and their relationships with atmospheric circulation. This plays a complementary role in related research, which is helpful for disaster forecasting and prevention. The rest of the paper is organized as follows: Section 2 describes the method and data; Sections 3 and 4 present the results; and Section 5 offers conclusions.

2. Materials and Methods

The monthly average geopotential height, specific humidity, and wind fields were sourced from NCEP/NCAR reanalysis [25] at a resolution of 2.5° × 2.5°. The daily precipitation was sourced from 90 surface meteorological observation stations in Southwest China (Figure 1a).

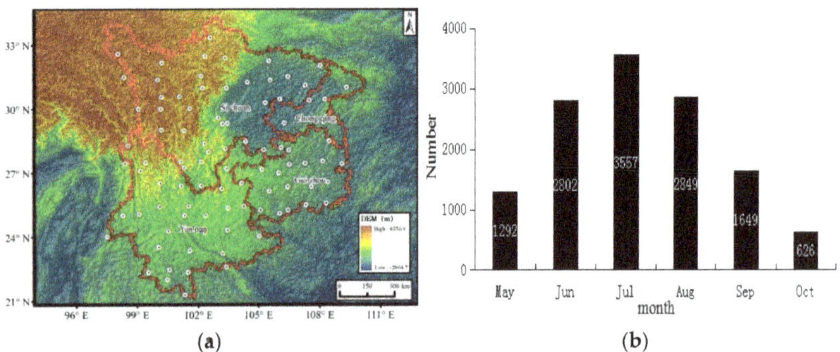

Figure 1. Distribution of the 90 surface meteorological observation stations (**a**) and the monthly cumulative frequency of rainstorms and floods during the 1961–2021 period (**b**).

Feng and Luo [3] found that 95% of rainstorm and flooding disasters were flood disasters caused by rainstorms in Southwest China. As short-term rainstorms are the main cause of rainstorm and flood disasters, the daily precipitation in Southwest China (Sichuan, Guizhou, Yunnan, and Chongqing) was chosen to define and discuss the rainstorms and floods.

According to previous studies on rainstorms and floods in Southwest China (Ye [18] and Yu [26]), a rainstorm or flood disaster occurs when the daily rainfall is greater than or equal to 50 mm or the cumulative rainfall is equal to or greater than 80 mm for three consecutive days in Sichuan and Chongqing. Guizhou and Yunnan are located on the Yunnan–Guizhou plateau. For the entire southwest region, a rainstorm or flood is defined as daily rainfall greater than or equal to 50 mm or cumulative rainfall greater than or equal to 100 mm for three consecutive days. Using the above criteria, we plotted the cumulative rainstorm and flood values from May to October of 1961–2021 for Southwest China. It is shown that most rainstorms and floods occur in the summer (June to August) (Figure 1b), and July has the most rainstorms and floods.

First, we used the Mann–Kendall method [27] and moving *t*-test [27] to discuss the time mutation points of rainstorms and floods, and we discussed its periodic variation using wavelet analysis [27]. Then, we used EOF (analysis of eigenvectors and time co-efficients) [27] to discuss the spatial distribution. Lastly, we calculated the correlation coefficients between rainstorms and floods and atmospheric circulation to discuss the dynamic mechanisms of rainstorms and floods.

2.1. Mann–Kendall Method

The Mann–Kendall method [27] is a non-parametric statistical test method that can detect not only the trend change of a sequence but also the turning points in the sequence. For a time series $x_1, x_2, \ldots, x_n$, S_k represents the cumulative count of the sample x_i greater than x_j ($1 \leq j \leq i$). Under the assumption of independence of random time series, we define UF_k as follows:

$$UF_k = [S_k - E(S_k)]/\sqrt{var(S_k)} \quad k = 1, 2, \ldots, n \tag{1}$$

Here, $UF_1 = 0$, and $E(S_k)$, $var(S_k)$ are the mean and variance of the cumulative counts; they are calculated as follows:

$$E(S_k) = n(n-1)/4 \tag{2}$$

$$var(S_k) = n(n-1)(2n+5)/72 \tag{3}$$

Given the significance level α, $U\alpha$ is a normal distribution. If $|UF_i| > U\alpha$, it indicates that there is an obvious trend change in the sequence. In the reverse order of time series $x, x_n, x_{n-1}, \ldots, x_1$, we repeat the above process while ensuring that $UB_k = -UF_k$ ($k = n$, $n - 1, \ldots, 1$), $UB_1 = 0$. We draw a graph of UB_k and UF_k. If the two curves intersect, and the intersection is between the critical boundary, then the time corresponding to the intersection is the time at which the mutation begins.

2.2. Moving T-Test

The moving *t*-test [27] aims to test whether the mean values of two segments in a climate sequence are significantly different. If the mean difference between the two sequences exceeds a certain significance level, it can be considered that the mean has undergone qualitative change and a trend change has occurred.

For a time series x with n samples, a certain point is artificially set as the reference point, and the samples of the two sub-sequences before and after the reference point are n_1 and n_2, the mean value of the two sub-sequences is x_1 and x_2, and the variance is s_1 and s_2, respectively. We define the statistics as follows:

$$T = (x_1 - x_2)/(S\sqrt{((1/n_1) + (1/n_2))}) \tag{4}$$

$$S = \sqrt{((n_1 s_1{}^2 + n_2 s_2{}^2)/(n_1 + n_2 - 2))} \tag{5}$$

The expression follows the *t* distribution of freedom $v = n_1 + n_1 - 2$.

3. Spatial–Temporal Variation

3.1. Temporal Variation

Time series analysis (Figure 2a) shows that in the past 61 years, there has been a significant upward trend in the number of rainstorms and floods, with a trend coefficient of approximately 0.5. The year with the highest frequency (rainstorms and floods per year) is 2016, followed by 2014 and 2020. The year with the lowest frequency is 1972. The polynomially fitted curves indicate that rainstorms and floods slightly increased from the 1960s to the 1980s, decreased in the 1990s, and then rapidly increased in the early 2000s.

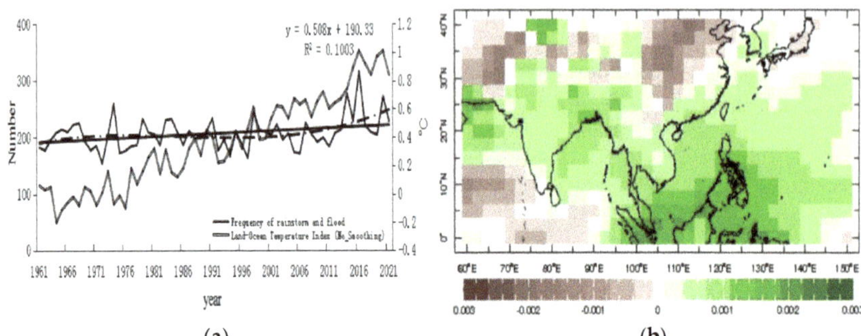

Figure 2. Trend analysis (thin solid line), polynomial fitting line (thick dashed line), linear trend line (thick solid line) and the Global Land–Ocean Temperature Index (hollow line) (**a**) and the linear trend of 850 hPa specific humidity (**b**) in the summer half-year during 1961–2021.

A linear correlation is found between the global Land−Ocean Temperature Index (No_Smoothing, https://climate.nasa.gov/vital-signs/global-temperature/, accessed on 10 March 2022) and the number of rainstorms and floods in Southwest China, which is shown in Figure 2a. The Global Land–Ocean Temperature Index shows the change in global surface temperature compared to the long-term average from 1951 to 1980. The year 2020 ties with 2016 as the hottest year on record since recordkeeping began in 1880 (source: NASA/GISS). NASA's analyses generally match independent analyses prepared by the Climatic Research Unit and the National Oceanic and Atmospheric Administration (NOAA). During the same time period, there were a large number of rainstorms and floods in Southwest China. In addition, the linear correlation coefficient of these two data sets is 0.38, passing the 99% significance level. These results show a tight link between global warming and the increase in rainstorms and floods in Southwest China.

Using NCEP/NCAR reanalysis, we calculated the linear trend of specific humidity at 850 hPa for the summer half-year (May–October) from 1961 to 2021 (Figure 2b). The local water vapor in the eastern part of Southwest China has shown a slight downward trend, but the water vapor values in the upstream, the Bay of Bengal, the South China Sea, and the Western Pacific have increased significantly. Increased water vapor is continuously transported to Southwest China through water vapor channels, which is conducive to the formation of rainstorms. This is clearly a case of atmospheric warming due to ocean areas holding more water vapor, which is a result of global warming.

Table 1 shows the interdecadal anomaly of rainstorms and floods for the summer half-year of 1961 to 2021. The interdecadal anomaly is positive in the 1980s and 2010s and negative in the 1960s, 1970s, and 1990s to the early 2000s. The highest anomaly value is 20.9, obtained in the 2010s, indicating that the storm and flood frequency has increased significantly in the last 10 years. The lowest anomaly value is −11.8, obtained in the 1970s. This is consistent with the time series polynomial fitting analysis (Figure 2a).

Table 1. Interdecadal anomaly of the rainstorms and floods in summer half-year for 1961–2021.

Years	Anomaly	Percentage of Anomaly
1960–1969	−1.7	−0.8%
1970–1979	−11.8	−5.8%
1980–1989	1.7	0.8%
1990–1999	−4.8	−2.3%
2000–2009	−4.5	−2.1%
2010–2019	20.9	10.2%

Figure 3a shows the results of the Mann−Kendall [27] analyses. There are three intersections among the positive and negative series curves in the years 1965 and 2012, indicating that these three years might be the climate change point. To verify the correctness, the moving *t*-test [27] (Figure 3b) was used. The moving *t*-test exceeds the significance interval of 95% reliability only from 2010 to 2014. As a result, the year 2012 was the climate mutation point for rainstorms and floods, changing from low to high frequency in Southwest China.

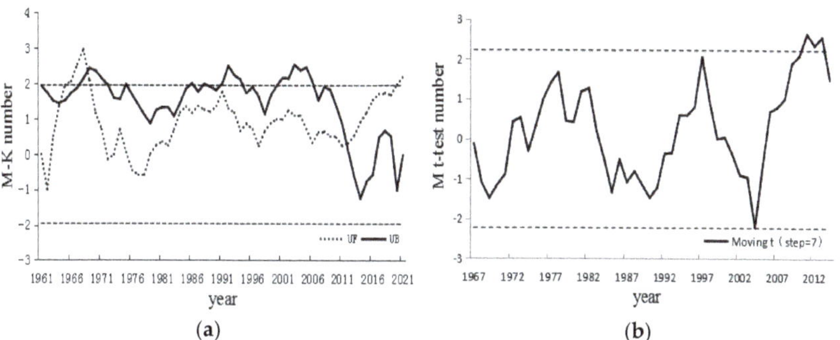

Figure 3. Mann–Kendall method (**a**) and moving *t*-test (**b**) analyses of rainstorms and floods in the summer half-year of 1961–2021 (horizontal lines are 95% reliability critical value).

To check if there were any periodic characteristics in the occurrence of rainstorms and floods in Southwest China, we applied wavelet analysis [27] to the observations. Figure 4a shows that there was an interannual cycle of 5–10 years and interdecadal cycles of 15–20 years and 30–35 years in rainstorms and floods. For example, the interannual 5–10-year cycle shows there were downward trends before 1980, a gradually increasing trend from 1980 to early 1985, and a downward trend again from 1985 until around 1990. From the wavelet variance diagram (Figure 4b), we can see that there exist three dominant time scales with period ranges of around 8 years, 16 years, and 31 years in quasi-periodic oscillations.

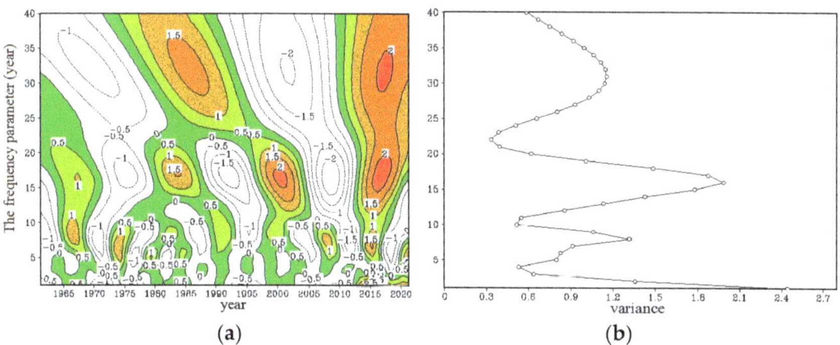

Figure 4. The wavelet real part (**a**) and wavelet variance (**b**) of the rainstorms and floods in the summer half-year of 1961–2021.

3.2. Spatial Variation

In the past 61 years, the high-incidence areas of rainstorms and floods have been mainly in the east of the Sichuan Basin, to the north of Guizhou, southwest of Guizhou, and the southern edge of Yunnan, while the low-incidence areas were in the northwest region of Southwest China (Figure 5a).

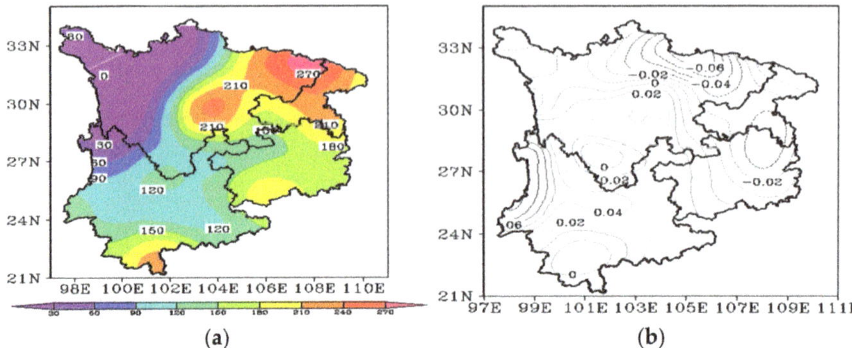

Figure 5. Spatial distribution of rainstorms and floods (**a**) and EOF (**b**) of 90 stations in the summer half year of 1961–2021.

In the EOF decomposition of the 61-year monthly cumulative rainstorm and flood time series of 90 stations, the variance contribution of the first four modes is 80.9%, with the first mode representing 68.7%. Hence, the first mode is sufficient to describe the main distribution characteristics, and its correlation analysis was performed (Figure 5b). The spatial distribution map shows a few high-value centers in the eastern and southern parts of the Sichuan Basin and Northern Yunnan, indicating that the inter-annual variation in these areas is larger than in other areas. The time series (figure omitted) shows an obvious upward trend, except for the 1990s, which is basically consistent with the time analysis above.

4. Correlation between Flood and Atmospheric Circulation

Using averages over the Southwest China region (97.5–110.2° E, 21.2–34.2° N), we analyzed the correlations between the southwest rainstorms and floods, and the NCEP 850 hPa wind speed, 500 hPa and 100 hPa height, and the sea level pressure. Then, the years with high (low) rainstorm and flood values are defined as the years in which the anomaly of rainstorms and floods in Southwest China is larger (smaller) than two times the average number. The years with high rainstorm and flood values are 1967, 1968, 1974, 1979, 1983, 1984, 1991, 1998, 2002, 2007, 2014, 2015, 2016, 2017, 2020, and 2021; the years with low rainstorm and flood values are 1962, 1970, 1972, 1975, 1976, 1992, 1994, 1997, 2005, and 2006. The high/low rainstorm and flood years correspond well with the years with high/low values for 850 hPa wind, 500 hPa and 100 hPa height, and sea level pressure.

Rainstorms and floods have a strong correlation with the 100 hPa (Figure 6a) and 500 hPa height (Figure 6b). The correlation coefficients are 0.43 and 0.36, respectively, passing the significance levels of 99% and 95%. The correlation coefficient of rainstorms and floods with the 850 hPa wind speed (Figure 6c) and sea level pressure field (Figure 6d) is below the 90% confidence level. However, they still have a positive relationship.

4.1. Sea Level Pressure

In those years with a high frequency of rainstorms and floods (Figure 7a), the southwest region and most of the middle and east of China were mainly affected by a uniformed pressure field, and the area from the South China Sea to the Iranian Plateau was affected by low pressure, with the center value of 1004 hPa. From Balkhash Lake to most of the northwest of China, it was mainly controlled by high pressures. Two centers (1018 hPa and 1019 hPa) were located, respectively, in Balkhash Lake–Northwest Xinjiang and South Xinjiang–West Tibet. Additionally, the cold air moved eastward from the plateau to the southwest of China through a northwesterly path and converged with the warm and wet air, which was conducive to the formation of rainstorms and floods. In low rainstorm and flood years, the high pressure over the Qinghai–Tibet Plateau was that in high rainstorm and flood years, and the central value was only 1006 hPa, so the influence of the cold air in the southwest was weaker.

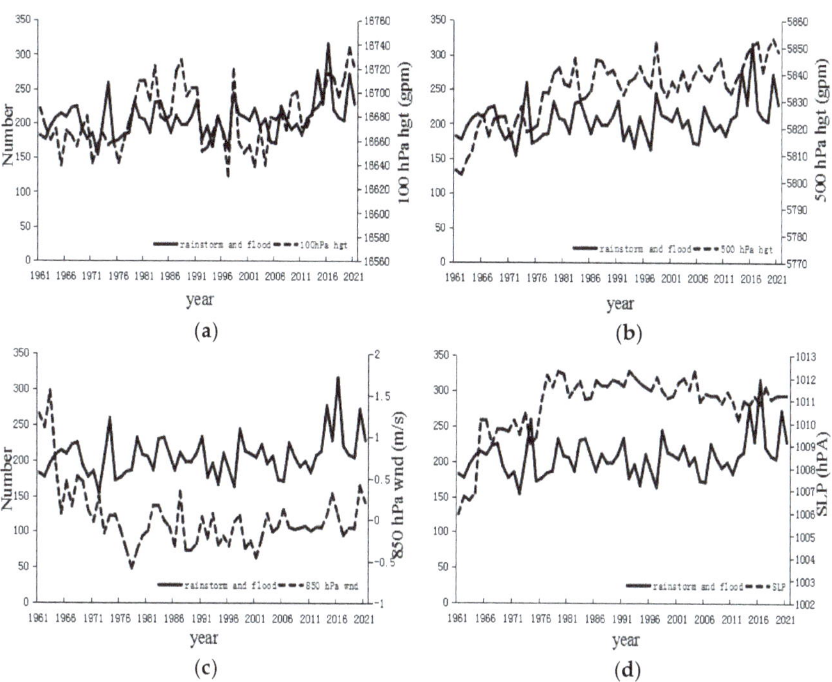

Figure 6. Observed cumulative rainstorm and flood values (solid line), 100 hPa height (**a**) and 500 hPa height (**b**), 850 hPa wind speed (**c**) and sea level pressure field (**d**), in the summer half-year of 1961–2021 (dashed lines).

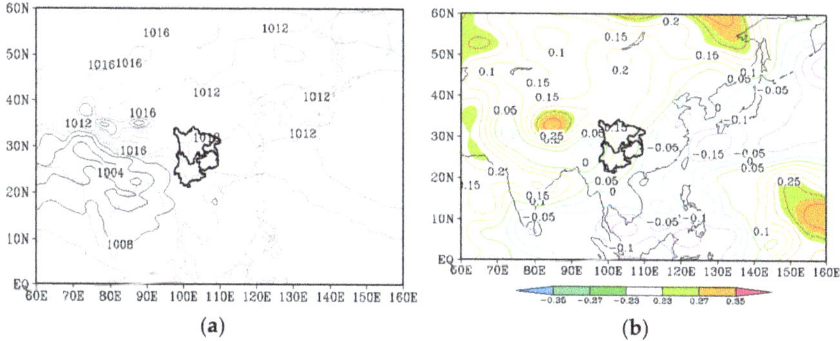

Figure 7. The sea level pressure field of rainstorms and floods in high-value years (**a**), the correlation coefficient (**b**) of rainstorms and floods and sea level pressure field in the summer half-year during 1961–2021. The shaded areas represent those where the correlation passes 90% (Green), 95% (Brown), and 99% (Rose) significance levels.

Based on the distribution of correlation coefficients between rainstorms and floods and the sea level pressure field (Figure 7b), there is a positive correlation distribution in most areas of China. By contrast, it is negative in most areas of Japan, the Philippines, and the Bay of Bengal.

The difference in sea level pressure field in high- and low-value years (figure omitted) and the distribution of correlation coefficients are consistent, indicating that when the southwest region has more rainstorms and flood disasters in the summer half-year, the pressure field at sea level is higher, and vice versa.

4.2. 850 hPa Wind Field

The weather systems in the lower atmosphere, carrying abundant water vapor, provide sufficient power and water vapor for precipitation. Figure 2b shows the change in water vapor. In years with a high number of rainstorms and flood events, the southerly airflow in Southwest China was the strongest in the country (Figure 8a). There are three main sources of water vapor. One bypasses the southern side of the Tibetan plateau. It flows through the Bay of Bengal and then moves northward in a southwesterly path, mainly affecting the central and western parts of Southwest China. One crosses the Indochina Peninsula and then moves northward, mainly affecting the central and eastern parts of Southwest China. The last one is the convergence of warm and humid air currents from the Pacific Ocean and Indochina Peninsula through the South China Sea, which mainly affects the central and eastern parts of Southwest China. At high latitudes, there is a general flat westerly airflow, with weak cold air coming down from Northern Xinjiang. The airflow meets the strong, warm, and humid airflow in Southwest China, which is conducive to precipitation. In years with a low number of rainstorms and floods, the southerly airflow affecting the southwest region is weaker than that in high-value years, especially regarding the southwesterly airflow from the Bay of Bengal, so the water vapor and energy brought by it are relatively weak.

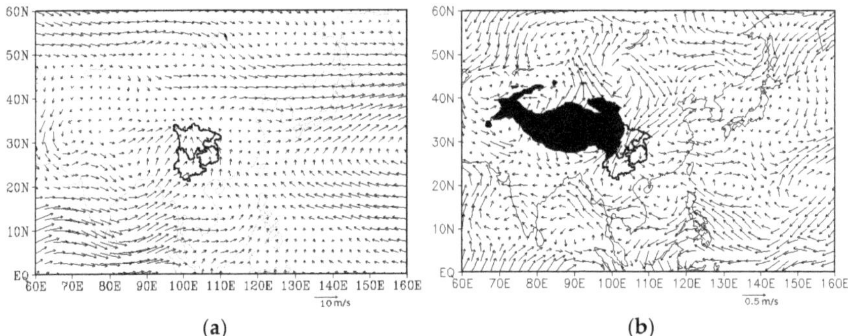

Figure 8. The 850 hPa wind field of rainstorms and floods in high-value years (**a**), the correlation coefficient (**b**) of rainstorms and floods and 850 hPa wind field in the summer half-year of 1961–2021.

From the correlation coefficients of rainstorms and floods and the 850 hPa wind speed (Figure 8b), the northwesterly air flowing southward from Eastern Mongolia and part of the easterly airflow in the front of the anticyclonic south of Lake Baikal converged in the central part of Inner Mongolia; then, the combined airflow moved southward to North China and the Hetao region. In the left branch, the northeast airflow crossed the Qinling Mountains and arrived at the Sichuan Basin and Chongqing, where it merged with the northward warm and humid airflow from the South China Sea, the Bay of Bengal, and the south side of the plateau. This was conducive to precipitation. In the middle path, the weak northward air flowed directly southward from the middle reaches of the Yangtze River to South China, which is related to the fact that the rainstorms and floods in Guizhou are weaker than those in Eastern Sichuan and Chongqing. The right branch enters the sea after passing through the East China region with the northwest airflow. We can see that the stronger the northerly airflow over the Qinling Mountains, the more the precipitation. The difference in the 850 hPa wind field in the years with a high and low number of rainstorms and floods (Figure omitted) and the distribution of correlation coefficients are consistent.

4.3. 500 hPa Height Field

In the years with a high rainstorm and flood frequency (Figure 9a), the westerly trough is located near the eastern coast of Russia to the east coast of China at a 500 hPa height. The southwest region is mainly affected by the westerly airflow in front of the plateau trough near 80° E, ranging from 584 to 586 dagpm, and the western extending point of the Western Pacific subtropical high at 586 dagpm is located near 20° N and 95° E. In addition, the ridge is located on the southern edge of the southwest region, so it is beneficial to the transport of warm and humid air from the southwest for precipitation. In low-value years, the Western Pacific subtropical high is weaker, and its 586 dagpm is located at the South China sea, with a smaller range. Therefore, the southwesterly airflow on the northwest side is weaker than that in high-value years, and thus the rainfall is weaker.

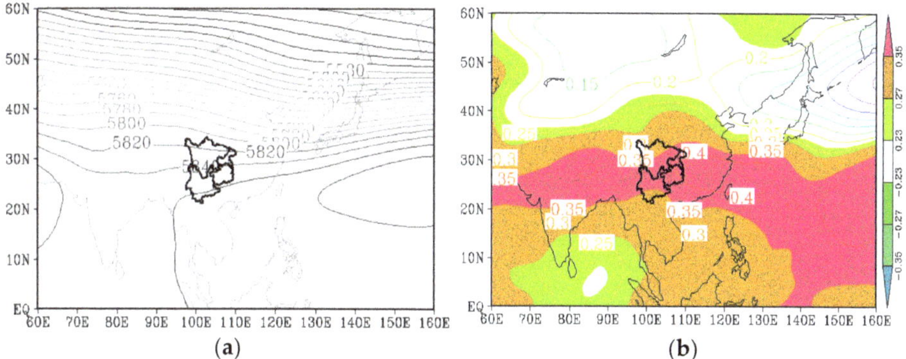

Figure 9. The 500 hPa height field of rainstorms and floods in high-value years (**a**), the correlation coefficient (**b**) of rainstorms and floods, and 500 hPa height field in the summer half-year during 1961–2021. The shaded areas represent those where the correlation passes 90% (Green), 95% (Brown), and 99% (Rose) significance levels.

From the correlation coefficients between rainstorms and floods and the 500 hPa height field (Figure 9b), all of them are positively correlated in China, with a center in the eastern southwest region. The difference in the 500 hPa height field in high- and low-value years (Figure omitted) is consistent with the correlation coefficient distribution, indicating that when the rainstorms and floods are strong, the potential height of 500 hPa significantly increases in most areas of the country, and vice versa.

4.4. 100 hPa Height Field

In the years with high rainstorm and flood frequency (Figure 10a), the South Asian high center in the upper troposphere is located around 25° N and 85° E, and its main body is slightly to the left of the western type. The subtropical long-wave trough is located in the northeastern region to the East China coast, and the southwestern region is located in the right part of the South Asian high center, with a potential height of 16,660–16,720 gpm. Combined with the east–south 588 line of 500 hPa, the rain belt is mostly in the Yangtze River basin, so the southwestern area is most affected by it. In low-value years, the central value of the southern pressure is lower (16,680 gpm), and the range is smaller. The southwestern region is not surrounded by its central line at 16,680 gpm, so the influence is weaker.

A positive correlation distribution was found in China from the correlation coefficient between rainstorms and floods and the 100 hPa height field (Figure 10b). The majority of areas between 10° N and 40° N were the centers of the positive correlation, and the southwestern area was still in the center of the positive correlation and passed the 90% reliability test. The difference in the 100 hPa height field in high- and low-value years (figure omitted) is consistent with the correlation coefficient distribution, which indicates that when the potential height field increases to 100 hPa, the South Asian high is strong, and the strong center is located in the east of the plateau, where the rainstorms and floods are more frequent, and vice versa.

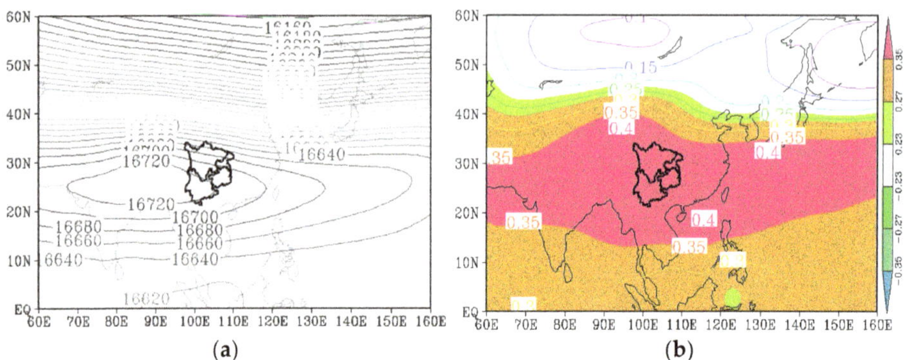

Figure 10. The 100 hPa height field of rainstorms and floods in high-value years (**a**), the correlation coefficient (**b**) of rainstorms and floods, and 100 hPa height field in the summer half-year during 1961–2021. The shaded areas represent those where the correlation passes 90% (Green), 95% (Brown), and 99% (Rose) significance levels.

5. Conclusions

This paper discussed the temporal and spatial variation in rainstorms and floods in Southwest China and its relationship with atmospheric circulation. The main results are as follows:

(1) Rainstorms and floods in Southwest China mainly occur from June to August, among which July has the highest number of rainstorms and floods, followed by August. In addition, May and September also have considerable numbers of rainstorms and floods;

(2) In the past 61 years, the number of heavy rain events and floods has shown an overall upward trend, and the interdecadal anomaly between the 1980s and the 2010s is positive, especially in the 2010s. The correlation coefficient between rainstorms and floods and the global surface temperature passes the 95% significance level. The year 2012 is the climate mutation point for rainstorms and floods, in which they change from low to high values. There are 5–10 years in the interannual cycle and 16 years and 31 years in the interdecadal cycle;

(3) The high-incidence area of rainstorms and floods in Southwest China is mainly from the east of the Sichuan Basin to the north of Guizhou, southwest of Guizhou and the southern edge of Yunnan. The low-value areas are mainly in the northwest part of the southwest region;

(4) The rich water vapor conditions are as follows: the combination of the western type of the South Asian high and the Western Pacific subtropical high; the influence of the southwesterly airflow in front of the upper trough; the transport of the warm and humid air in the middle; the lower layers; and the confluence of the cold air from the northwesterly path. Therefore, the atmospheric circulation configuration is conducive to precipitation, which can likely cause rainstorms and floods.

Author Contributions: Conceptualization, Q.X.; methodology, Q.X.; software, Q.X.; validation, Q.X.; formal analysis, Q.X.; investigation, Q.X.; resources, X.G.; data curation, G.L.; writing—original draft preparation, Q.X.; writing—review and editing, X.G.; visualization, Z.L.; supervision, Z.L.; project administration, T.T.; funding acquisition, X.G. All authors have read and agreed to the published version of the manuscript.

Funding: This work was funded by the National Key Research and Development Program of China (2018YFC1507201), Research Team of Key Technologies of Guizhou Meteorological Bureau (GGTD-202212), and National Natural Science Foundation of China "Research on Multi-model Temperature Forecast and Correction Technology in Guizhou under Complex Terrain" (42265001, Guizhou Meteorological Bureau Program (11 September 2021)).

Institutional Review Board Statement: Not applicable.

Informed Consent Statement: Not applicable.

Data Availability Statement: NCEP/NCAR reanalysis I data provided by the NOAA PSL, Boulder, CO, USA, from their website at https://psl.noaa.gov (accessed on 10 March 2022). The global Land−Ocean Temperature Index provided by the NASA, CO, USA, from their website at https://climate.nasa.gov/vital-signs/global-temperature (accessed on 10 March 2022) The daily precipitation data provided by the National Data Center for Meteorological Sciences of China, from their website at http://data.cma.cn/ (accessed on 10 March 2022).

Conflicts of Interest: The authors declare no conflict of interest.

References

1. Li, N.; Reng, Y.; Gu, W.; Cheng, Y. GIS based risk evaluation for flood hazard in Shandong Province. *J. Chin. Agric. Sci. Bull.* **2010**, *26*, 313–317.
2. Feng, Q.; Tao, S.; Wang, A.; Li, J.; Xu, L. Analysis of the influence of heavy-rain and flood disaster on social economy and human life. *J. Catastrophol.* **2001**, *16*, 44–48.
3. Feng, S.; Luo, D. Division of flood-waterlogging hazard in Southwest China. *J. Mt. Res.* **1995**, *13*, 255–260.
4. Yu, X.; Ma, Y. Spatial and Temporal Analysis of Extreme Climate Events over Northeast China. *Atmosphere* **2022**, *13*, 1197. [CrossRef]
5. Gu, J.; Cui, X.; Hong, H. A Statistical-Based Model for Typhoon Rain Hazard Assessment. *Atmosphere* **2022**, *13*, 1172. [CrossRef]
6. Cheng, S.; Xie, J.; Ma, N.; Liang, S.; Guo, J.; Fu, N. Variations in Summer Precipitation According to Different Grades and Their Effects on Summer Drought/Flooding in Haihe River Basin. *Atmosphere* **2022**, *13*, 1246. [CrossRef]
7. Di Giustino, G.; Bonora, A.; Federico, K.; Reho, M.; Lucertini, G. Spatial Analysis of the Vulnerability to Flooding in the Rural Context: The Case of the Emilia Romagna Region. *Atmosphere* **2022**, *13*, 1181. [CrossRef]
8. Cheng, W.; Chen, J.; Liu, D. Review on flood risk assessment. *J. Yangtze River Sci. Res. Inst.* **2010**, *9*, 17–24.
9. Li, C. A Statistical analysis of the storm flood disasters in China. *J. Catastrophol.* **1996**, *11*, 59–63.
10. Wang, X.; Leng, C.; Feng, X.; Zhou, Y.; Ma, S. Risk analysis of flood disaster in the middle reaches of the Yangtze River. *J. Sci. Technol. Rev.* **2008**, *26*, 61–62.
11. Werritty, A. Use of Multi-proxy Flood Records to Improve Estimates of Flood Risk: Lower River Tay, Scotland. *J. Catena* **2006**, *66*, 107–119. [CrossRef]
12. Li, J.; Feng, Q.; Wang, A. *Risk Assessment of Heavy Rain and Flood Disasters in China: Typhoon and Rainstorm Disaster Weather Monitoring, Forecasting Technology Research*; China Meteorological Press: Beijing, China, 1996.
13. Huang, C.; Liu, X.; Zhou, G.; Li, X. Agricultural natural disaster risk assessment method according to the historical disaster data. *J. Nat. Disasters* **1998**, *7*, 1–8.
14. Wan, J.; Zhou, Y.; Wang, Y. Flood disaster and risk evaluation approach based on the GIS in Hubei Province. *J. Torrential Rain Disasters* **2007**, *26*, 328–333.
15. Gong, Q.; Huang, G.; Guo, M. GIS-based risk zoning of flood hazard in Guangdong Province. *J. Nat. Disasters* **2009**, *18*, 58–63.
16. Jiang, Q.; You, Z. GIS-Based Natural Hazard Regionalization in Nantong. *J. Catastrophol.* **2005**, *20*, 110–143.
17. Tian, H.; Lu, W.; Wu, B. Integrated system for meteorological disaster monitoring and loss evaluation based on GIS. *J. Sci. Meteorol. Sin.* **2002**, *22*, 482–487.
18. Ye, Z. A Study on the Flood Disaster of Chongqing. Master's Thesis, Southwest China Normal University, Chongqing, China, 2001.
19. Lu, R. Analysis of Climate Characteristic of Rainstorm in Guizhou Province. Master's Thesis, Lanzhou University, Lanzhou, China, 2010.
20. He, H.; Sun, J. The role of upper and lower tropospheric jet in Yunnan rainstorm occurrence and their common characteristics. *J. Plateau Meteorol.* **2004**, *23*, 29–34.
21. Liu, Y.; Shegn, Y. Chuxiong "2013.8.29" heavy precipitation process analysis. *J. Yunnan Univ. Nat. Sci.* **2014**, *36*, 95–102.
22. Chen, X.; Li, H.; He, Y.; Yang, K. An analysis on the characteristics of an extreme torrential rain in Huaping and Changning on 15–16 September 2015. *J. Yunnan Univ. Nat. Sci.* **2017**, *39*, 225–234.
23. Xiong, W.; Luo, X.; Zhou, M. Comparison diagnosis and discussion of trigger mechanism of two MCC heavy rain systems in Guizhou. *J. Yunnan Univ. Nat. Sci. Ed.* **2014**, *36*, 66–78.
24. Zhao, Y.; Liu, K. Diagnostic analysis of moist potential vorticity for a heavy rain in Guizhou. *J. Yunnan Univ. Nat. Sci.* **2012**, *34*, 386–389.

25. Kalnay, E.; Kanamitsu, M.; Kistler, R.; Collins, W.; Deaven, D.; Gandin, L.; Joseph, D. The NCEP/NCAR Reanalysis 40-year Project. *Bull. Am. Meteorol. Soc.* **1996**, *77*, 437–471. [CrossRef]
26. Yu, S. *Monitoring and Forecasting of Flood in Southwest China*; China Meteorological Press: Beijing, China, 2004.
27. Wei, F. *Modern Climate Statistical Diagnosis and Prediction Technology*, 2nd ed.; China Meteorological Press: Beijing, China, 2007.

Article

Potential Predictability of Seasonal Global Precipitation Associated with ENSO and MJO

Haibo Liu [1] , Xiaogu Zheng [2,*], Jing Yuan [3] and Carsten S. Frederiksen [4,5,6]

[1] Lamont-Doherty Earth Observatory, Columbia University, Palisades, NY 10964, USA; haibo@ldeo.columbia.edu
[2] International Global Change Institute, Hamilton 3210, New Zealand
[3] International Research Institute for Climate and Society (IRI), Columbia University, Palisades, NY 10964, USA
[4] The Bureau of Meteorology, Melbourne, VIC 3001, Australia
[5] The School of Earth, Atmosphere and Environment, Monash University, Clayton, VIC 3800, Australia
[6] CSIRO, Oceans and Atmosphere, Aspendale, Melbourne, VIC 3195, Australia
* Correspondence: xiaogu.zheng@gmail.com or xiaogu.zheng@igci.org.nz

Abstract: A covariance decomposition method is applied to a monthly global precipitation dataset to decompose the interannual variability in the seasonal mean time series into an unpredictable component related to "weather noise" and to a potentially predictable component related to slowly varying boundary forcing and low-frequency internal dynamics. The "potential predictability" is then defined as the fraction of the total interannual variance accounted for by the latter component. In tropical oceans (30° E–0° W, 30° S–30° N), the consensus is that the El Nino-Southern Oscillation (ENSO, with 4–8 year cycles) is a dominant driver of the potentially predictable component, while the Madden-Julian Oscillation (MJO, with 30–90 days cycles) is a dominant driver of the unpredictable component. In this study, the consensus is verified by using the Nino3-4 SST index and a popular MJO index. It is confirmed that Nino3-4 SST does indeed explain a significant part of the potential predictable component, but only limited variability of the unpredictable component is explained by the MJO index. This raises the question of whether the MJO is dominant in the variability of the unpredictable component of the precipitation, or the current MJO indexes do not represent MJO variability well.

Keywords: predictability; global; seasonal precipitation; ENSO; MJO

1. Introduction

The seasonal mean time series of meteorological variables are widely used for analyzing interannual climate variability and predictability. Studies of potential predictability are usually based upon a decomposition of the seasonal variability into a part called the "weather noise" variability, which is fundamentally unpredictable beyond a deterministic time period, and another part assumed to be at least potentially predictable [1]. The potential predictability is measured as the fraction of the total variability accounted for by the latter part. It helps us to identify regions where the potential for making skillful climate forecasts is highest, as well as where climate noise dominates any signal. Such knowledge is useful to investigate the roles of drivers of potentially predictable variability, such as human-induced forcing, oceanic and slowly varying atmospheric states, soil moisture and snow cover conditions, and the roles of stochastic meteorological processes generated by typical weather events in modifying regional variability. They can also help provide guidance as to where and when the issuance of climate forecasts may be most reasonable.

The measure of potential predictability is clearly sensitive to how the separation of the variance is performed. Some may think that temporal filtering techniques could be employed based on the assumption that weather noise operates mainly on timescales much shorter than that of the potentially predictable variability [2]. However, weather

Citation: Liu, H.; Zheng, X.; Yuan, J.; Frederiksen, C.S. Potential Predictability of Seasonal Global Precipitation Associated with ENSO and MJO. *Atmosphere* **2023**, *14*, 695. https://doi.org/10.3390/atmos14040695

Academic Editor: Seontae Kim

Received: 17 February 2023
Revised: 2 April 2023
Accepted: 5 April 2023
Published: 7 April 2023

events include not only high-frequency day-to-day fluctuations, but also intra-seasonal fluctuations (such as MJO) which cannot be completely smoothed out by a seasonal mean filter. The residuals through the seasonal mean filter give rise to chaotic, unpredictable fluctuations in the seasonal mean time series. Therefore, it is not possible to completely isolate the potentially predictable variability through temporal filtering.

In this paper, the methodology proposed by Zheng and Frederiksen [3] is applied to analyze the potential predictability of a monthly global precipitation data. It was found that for rainfall in tropical oceans, the main driver of predictable variability is ENSO, while the unpredictable variability is related to MJO, but not as dominant as ENSO to the predictable variability. Regions with significant potential predictability are also identified. The methodology applied in this paper and other methodologies are reviewed.

The current paper begins with a description of the data and the estimation method of the potential predictability in Section 2. Section 3 is devoted to applications to monthly global precipitation data. The methodology applied in this paper and other methodologies are reviewed in Section 4. A summary is presented in Section 5.

2. Data and Methods

2.1. Data

The data used in this study include monthly precipitation data (on a $2.5° \times 2.5°$ grid) from the Global Precipitation Climatology Project (GPCP), version 2.3 [4], for the period $1979-2020$, which is a widely used global (land and ocean) precipitation dataset derived from a mix of satellite estimates over oceans and land, and rain gauge measurements from land and atolls. The GPCP is often used to study variations in precipitation at global and regional scales (https://climatedataguide.ucar.edu/climate-data/gpcp-monthly-global-precipitation-climatology-project#:~:text=The%20GPCP%20monthly%20dataset%20is,with%20some%20delay%20for%20processing) (accessed on 6 April 2023). The Nino3-4 index is calculated from the HadISST1 SST as the average of SST anomalies over the region $5°$ N–$5°$ S and $170°$–$120°$ W. The MJO index used here is the outgoing longwave radiation-based MJO index (OMI) (https://www.psl.noaa.gov/mjo/mjoindex/) (accessed on 6 April 2023). The OMI calculation is based on an empirical orthogonal function (EOF) analysis. The two components of OMI, MJO-PC1 and MJO-PC2, are the projections of 20–96 days filtered outgoing longwave radiation onto the first and the second daily spatial EOF patterns of 30–96 days eastward filtered OLR. The OMI can successfully capture the convective component of the MJO [5].

2.2. Identifying the Seasonal Predictable and Unpredictable Components

The (co-)variance decomposition method of Zheng and Frederiksen [3] is proposed for deriving the spatial patterns of interannual variability in the seasonal mean fields related to the variability of predictable and unpredictable components, based on monthly mean data. First, the annual cycle is removed from the data by subtracting the multi-year mean for each month. This serves to reduce the climate bias in the covariance estimation. Then, a conceptual statistical model for a climate variable X_{ym} in month m ($m = 1, 2, 3$ in a specific season, e.g., June, July, August) and in year y ($y = 1, \dots, Y$, where Y is the total number of years) expressed as per the following regression form:

$$X_{ym} = \mu_y + \varepsilon_{ym} \tag{1}$$

where μ_y (interception) represents the seasonal "statistical population" mean in year y; and ε_{ym} is the residual monthly departure of X_{ym} from μ_y. Then, the seasonal mean of a climate variable in year y ($X_{yo} \equiv \frac{1}{3}\sum_{m=1}^{3} X_{ym}$) can be conceptually expressed as

$$X_{yo} = \mu_y + \varepsilon_{yo} \tag{2}$$

where $\varepsilon_{yo} \equiv \frac{1}{3}\sum_{m=1}^{3} \varepsilon_{ym}$ is the seasonal mean of the monthly (also daily) residuals for μ_y that arises from intra-seasonal variability (ε_{ym}) and is not smoothed out by the seasonal mean filter. Since the weather system is chaotic, ε_{ym} is unlikely to be predictable beyond one or two weeks, so ε_{yo} is essentially unpredictable beyond the deterministic prediction period and is referred to as the unpredictable component of X_{yo}. On the other hand, the interception μ_y is a constant through the season in y year and is more likely dominated by the external forcing and slowly varying internal dynamics. Therefore, μ_y is potentially predictable beyond the deterministic prediction period and is referred to as the potentially predictable component of X_{yo} [1]. It is very difficult to numerically separate μ_y from the seasonal mean X_{yo} using filter techniques, because X_{yo} is already the best estimation of μ_y, but subject to error ε_{yo}.

Let $\left\{ X'_{ym}, y = 1, \ldots, Y; m = 1, 2, 3 \right\}$ be another climate variable in the form as Equation (1), then, using the statistical technique based on the (co-)variance decomposition [4], the interannual covariance of the unpredictable components can be calculated with monthly data by using Equation (16) in Zheng and Frederiksen [3] as follows:

$$V\left(\varepsilon_{yo}, \varepsilon'_{yo}\right) = \hat{\sigma}^2 (3 + 4\hat{\varphi}) / 9 \tag{3}$$

where

$$\hat{\sigma}^2 = \frac{a}{2(1 - \hat{\varphi})} \tag{4}$$

$$\hat{\varphi} = \min \left\{ 0.1, \max[(a + 2b) / (2a + 2b), 0] \right\} \tag{5}$$

$$a = \frac{1}{2} \left\{ \frac{1}{Y} \sum_{y=1}^{Y} [X_{y1} - X_{y2}] \left[X'_{y1} - X'_{y2} \right] + \frac{1}{Y} \sum_{y=1}^{Y} [X_{y2} - X_{y3}] \left[X'_{y2} - X'_{y3} \right] \right\} \tag{6}$$

and

$$b = \frac{1}{2} \left\{ \frac{1}{Y} \sum_{y=1}^{Y} [X_{y1} - X_{y2}] \left[X'_{y2} - X'_{y3} \right] + \frac{1}{Y} \sum_{y=1}^{Y} [X_{y2} - X_{y3}] \left[X'_{y1} - X'_{y2} \right] \right\} \tag{7}$$

The interannual covariance of the predictable components $V\left(\mu_y, \mu'_y\right)$ can be calculated as

$$V\left(\mu_y, \mu'_y\right) = V\left(X_{yo}, X'_{yo}\right) - V\left(\varepsilon_{yo}, \varepsilon'_{yo}\right) \tag{8}$$

where $V\left(X_{yo}, X'_{yo}\right)$ represents the total interannual covariance and can be estimated directly from seasonal mean variables X_{yo} and X'_{yo}, i.e.,

$$V\left(X_{yo}, X'_{yo}\right) = \frac{1}{Y} \sum_{y=1}^{Y} X_{yo} X'_{yo} \tag{9}$$

The R-code for calculating the unpredictable covariance of Equation (3) is included in Appendix A. The χ^2 test and student's t-test, used to judge the significance of the interannual covariances of the predictable and unpredictable components, i.e., Equations (3) and (8) respectively, are documented in the Appendix of Ying et al. ([6]; their Equations (10)–(14)).

The potential predictability is defined as the ratio of the predictable variance against the total variance, i.e.,

$$V\left(\mu_y, \mu_y\right) / V\left(X_{yo}, X_{yo}\right) \tag{10}$$

The significant test procedure for the potential predictability is documented in Equation (8) of Zheng et al. [7].

Define the standardised predictable and unpredictable covariances for the X'_{yo} with X_{yo} as

$$V\left(\mu'_y, \mu_y\right) / \sqrt{V\left(\mu'_y, \mu'_y\right)} \tag{11}$$

and

$$V\left(\varepsilon'_{yo}, \varepsilon_{yo}\right) / \sqrt{V\left(\varepsilon'_{yo}, \varepsilon'_{yo}\right)} \tag{12}$$

respectively. It can be proved that the square of Equation (11) (Equation (12)) is the variance of the predictable (unpredictable) component of X_{yo} explained by the predictable (unpredictable) component of X'_{yo} (see Appendix B for proof). Therefore, the fraction of the predictable (unpredictable) component of X_{yo} explained by the predictable (unpredictable) component of X'_{yo} is the square of predictable (unpredictable) correlation.

$$\frac{V^2\left(\mu_y, \mu'_y\right)}{V\left(\mu_y, \mu_y\right) V\left(\mu'_y, \mu'_y\right)} = \left[\frac{V\left(\mu_y, \mu'_y\right)}{\sqrt{V\left(\mu_y, \mu_y\right) V\left(\mu'_y, \mu'_y\right)}}\right]^2 \triangleq cor^2(\mu_y, \mu'_y) \tag{13}$$

and

$$\frac{V^2\left(\varepsilon_{yo}, \varepsilon'_{yo}\right)}{V\left(\varepsilon_{yo}, \varepsilon_{yo}\right) V\left(\varepsilon'_{yo}, \varepsilon'_{yo}\right)} = \left[\frac{V\left(\varepsilon_{yo}, \varepsilon'_{yo}\right)}{\sqrt{V\left(\varepsilon_{yo}, \varepsilon_{yo}\right) V\left(\varepsilon'_{yo}, \varepsilon'_{yo}\right)}}\right]^2 \triangleq cor^2(\varepsilon_{yo}, \varepsilon'_{yo}) \tag{14}$$

Moreover, if another climate variable X''_{yo} is statistically independent of X'_{yo}, then the fraction of the predictable (unpredictable) component of X_{yo} explained by the predictable (unpredictable) components of X'_{yo} and X''_{yo} is the sum of those for X'_{yo} and X''_{yo}, respectively.

3. Results

In principle, predictable variability is driven by the slowly varying external forcing and slowly varying internal dynamics which are stable within a season, while unpredictable variability is driven by the internal dynamics dominated by intra-seasonal variability with cycles larger than 10 days. In this section, the impacts of ENSO and MJO on the predictable and unpredictable variabilities of global precipitation, especially on tropical oceans, are studied.

3.1. El Nino-Southern Oscillation

In this study, the Nino3-4 index is used to represent the ENSO. The potential predictability of the index is more than 0.9 for all seasons indicating that the predictable variability is dominant. Therefore, any global precipitation related to ENSO is also likely to be highly predictable.

Figure 1 shows that the variability in the predictable component of precipitation is largest in the tropics in all seasons, especially in DJF. There is a clear seasonal cycle, with the largest predictable variability in DJF and the smallest variability in JJA. This coincides with the seasonal cycle of ENSO, which is also the strongest in DJF and the weakest in JJA. In the peak season DJF, the center of the largest predictable variability is in the Nino3-4 region (5° N–5° S, 170° W–120° W). This may be due to the fact that in the La Nina phase, the cold equatorial water upwells in the Nino3-4 region, associated with a strong trade wind passing through the Nino3-4 region, resulting in it being the driest; while in the El Nino phase, the sea surface temperature in the Nino3-4 region is the warmest, resulting in a large amount of convective cloud that leads to significant positive rainfall anomalies in the Nino3-4 region. All these facts indicate that ENSO is a likely dominant driver of the predictable variability of tropical precipitation. To further verify this point, the standardized predictable covariances for the Nino3-4 index with global precipitation, in all four seasons, are shown in Figure 2. Regardless of their signs, the spatial patterns and strength of their absolute values are very similar to those of the predictable component variability (Figure 1) in tropical oceans (30° E–0° W, 30° S–30° N). This is confirmed by the fact that the pattern correlations between the predictable component standard deviations (Figure 1) and the absolute values of the standardized predictable covariances for GPCP precipitation with

Nino3-4 SST (Figure 2) in the tropical oceans are 0.79, 0.82, 0.89 and 0.92 in MAM, JJA, SON and DJF, respectively. The corresponding means of the fraction of the predictable precipitation variances explained by the predictable component of Nino3-4 SST (the square of Equation (10)) over the tropical oceans are 0.24, 0.41, 0.59, 0.56, respectively. In SON and DJF when ENSO is stronger and unlikely to change phase, the correlations and explained predictable variances are more significant than those in MAM and JJA. Although the ENSO variability in MAM is stronger than that in JJA, the correlation and the explained predictable variance are less significant than those in JJA. This coincides with the fact that ENSO is more likely to change phase in MAM (especially in April) than in JJA.

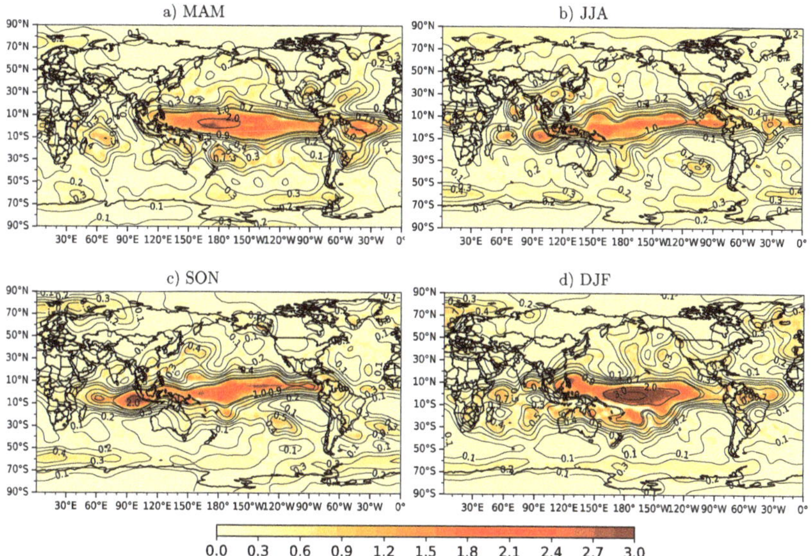

Figure 1. Predictable component standard deviation of GPCP precipitation (mm/day).

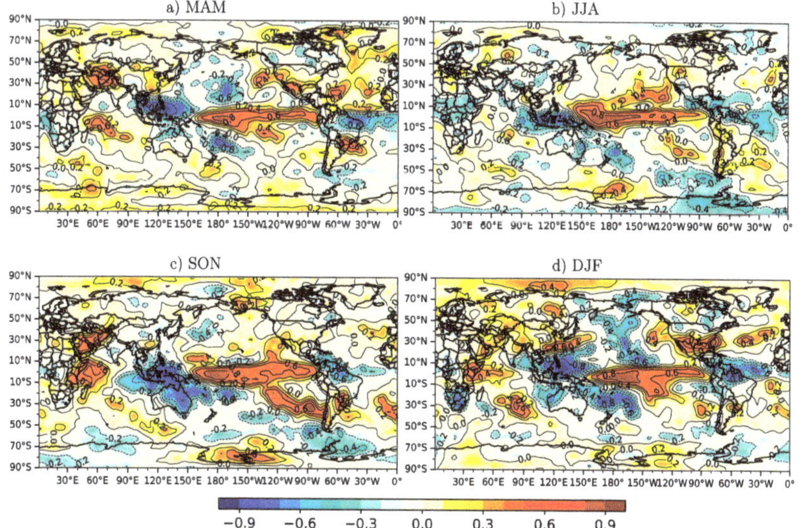

Figure 2. Standardized predictable covariances for GPCP precipitation with Nino3-4 SST.

3.2. Madden-Julian Oscillation

The Madden-Julian Oscillation (MJO) is a tropical weather system originating in the Indian Ocean that propagates eastward around the global tropics with a cycle on the order of 30–90 days [8]. The MJO indices MJO-PC1 and MJO-PC2 are the projections of 20–96 days filtered OLR [5], the seasonal potential predictability of the MJO-PC1 and MJO-PC2 are virtual zero. Thus, the MJO is more likely to be a driver of the unpredictable variability of tropical precipitation. Therefore, any global precipitation related to the two MJO indices is also likely to be highly unpredictable on a seasonal time scale.

The MJO consists of two parts; one that has strong convective rainfall (wet) and the other with suppressed rainfall (dry). The dry part of the MJO always precedes the wet part (dry in the east, wet in the west). There are less clouds in the dry part, which induces strong solar heating reaching the ocean surface underneath. The accumulation of the heating on the ocean surface causes warmer SST, which in turn causes upward movement of the air. This results in convection (wet part) moving eastward, and hence the east propagation of the MJO. The MJO has wide ranging impacts on the patterns of tropical and extratropical precipitation, atmospheric circulation, and surface temperature around the global tropics and subtropics.

We compare the seasonal cycles of the MJO described by Zhang and Dong [8] with the unpredictable variability of precipitation (Figure 3) and the standardized unpredictable covariance for the MJO index with global precipitation (Figure 4) in tropical oceans. Zhang and Dong [8] confirmed that the primary peak season is in boreal winter (December–March), during which MJO signals are mainly confined to the Indian Ocean and western Pacific Ocean, and reach their maxima in the South Pacific convergence zone. Similarly, Figure 2d shows that the primary peak season of unpredictable variability is in DJF with the most significant variability near the equatorial Indian Ocean (with standard deviation 1.5 mm/day) and western Pacific Oceans, especially in the South Pacific convergence zone (with standard deviation 2.0 mm/day). Figure 3a shows the unpredictable variability in MAM with a similar pattern to DJF, but with less variability. This coincides with the fact that March also belongs to the MJO peak season. This is also the case for the standardized unpredictable covariances for GPCP precipitation with MJO index (Figure 4d) in the tropical ocean, but with much weaker strength (with standard deviation 0.8–1.0 mm/day in the Indian Ocean and 0.6–1.0 mm/day in the South Pacific convergence zone).

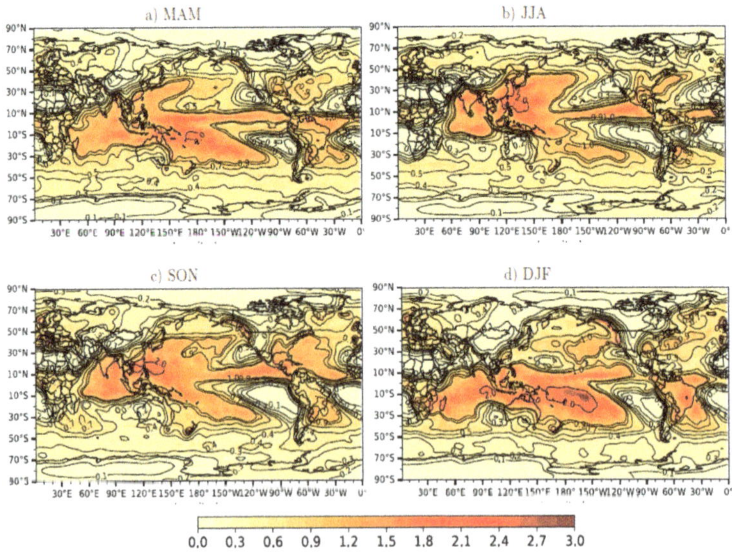

Figure 3. Unpredictable component standard deviation of GPCP precipitation (mm/day).

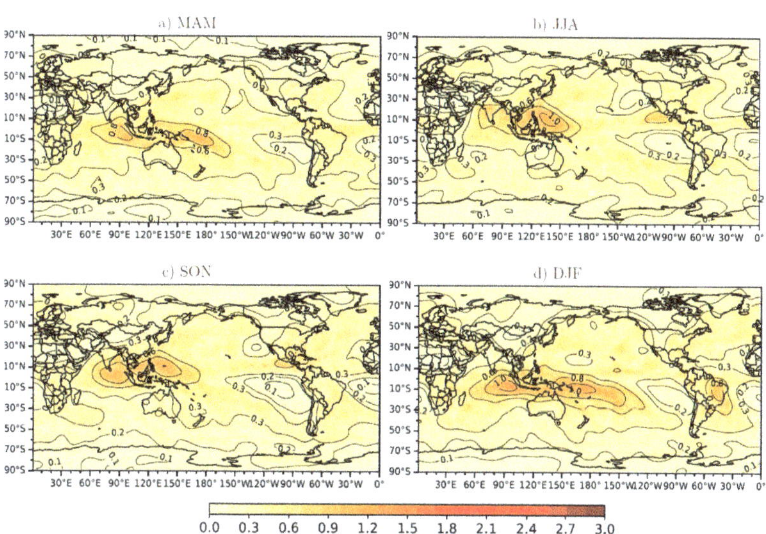

Figure 4. Standardized unpredictable covariances for GPCP precipitation with MJO index.

Zhang and Dong [8] further confirmed that the secondary peak season of the MJO is boreal summer (June–September), during which the strongest MJO occurs north of the equator from the Bay of Bengal to the South China Sea. Another separated region of strong MJO signals is located in the eastern Pacific warm pool off the Central American coast. Similarly, Figure 3b shows that the unpredictable variability in JJA also has a secondary peak north of the equator from the Bay of Bengal to the South China Sea (with standard deviation 1.5 mm/day) and a second center in the eastern Pacific warm pool off the Central American coast (with standard deviation 1.0 mm/day). This is also the case for the standardized unpredictable covariance for the global precipitations with the MJO index (Figure 4b) for which JJA also has a secondary peak north of the equator from the Bay of Bengal to the South China Sea, but with much weaker strength (with standard deviation 0.8–1.0 mm/day). There is also a second center in the eastern Pacific warm pool off the Central American coast, but with weaker strength (with standard deviation 0.8 mm/day). Figure 3c shows the unpredictable variability in SON with a similar pattern to that of JJA, but with less variability (Figure 3c), and this is also the case for the standardized unpredictable covariance for the MJO index with tropical ocean precipitations. This coincides with the fact that September also belongs to the second peak season of MJO.

Zhang and Dong [8] suggested that for the broad tropical region, the seasonality in the MJO is featured by a latitudinal migration across the equator between two peak seasons. In particular, MJO signals in the eastern Pacific (Maloney and Kiehl [9]) appear to be much stronger during boreal summer than winter. Such migration can be clearly seen for its unpredictable variability. In JJA and SON (Figure 3b,c), the 1.0 mm/day contour of the unpredictable standard deviation reaches north of Japan close to 50° N, while in DJF it draws back to north of the Philippines about 15 N. Similarly, in DJF and MAM (Figure 3a,d), the 1.0 mm/day contours reach north of New Zealand close to 30° S, while in JJA it draws back to north of Australia about 10 S. However, these are not obvious in the standardized unpredictable covariance for the MJO index with global precipitation. The pattern correlations between the unpredictable component variability (Figure 3) and the standardized unpredictable covariance of the precipitations with the MJO index (Figure 4) in tropical oceans are 0.79, 0.79, 0.77 and 0.84 in MAM, JJA, SON and DJF, respectively. The corresponding explained unpredictable variances by MJO-PC1 and MJO-PC2 are around 0.065 for all the seasons.

3.3. Potential Predictability

The spatial patterns of the potential predictability in the four seasons are shown in Figure 5. In tropical oceans, they are more similar to the predictable variability (Figure 1) than the unpredictable variability (Figure 2). In particular, the potential predictability along the tropical Pacific is lowest in JJA (Figure 5b) associated with the lowest predictable variability (Figure 1b). Potential predictability is highest in DJF (Figure 5d) associated with the highest predictable variability (Figure 4d). Predictability along the tropical Atlantic is the highest in MAM and lowest in SON (Figure 1a,c), which is consistent with the potential predictability (Figure 5a,c).

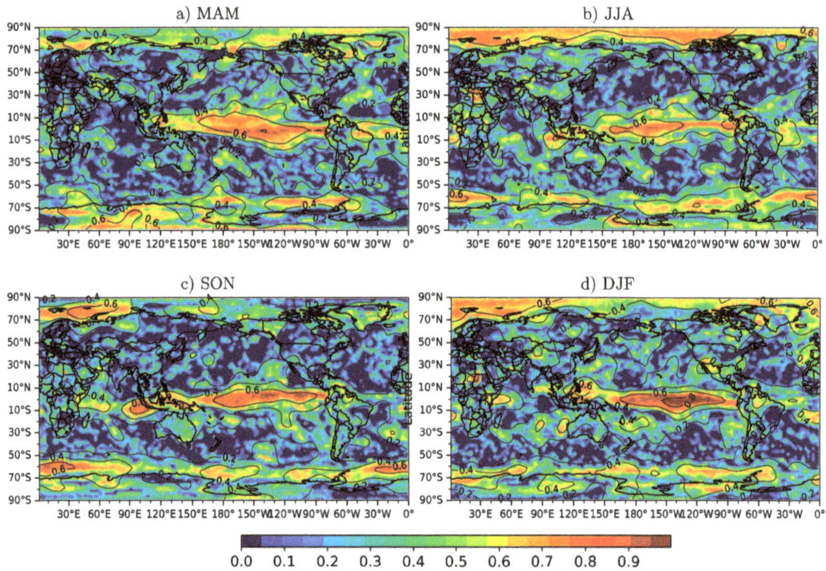

Figure 5. Seasonal potential predictability of GPCP precipitation.

The potential predictability of precipitations in the subtropics may also be related to ENSO. For example, the JJA and DJF precipitation of New Zealand has potential predictability around 0.3 (Figure 5b,d). Zheng and Frederiksen [10] further extended the decomposition methodology from the variability of single time series to the covariability of multivariate time series and established a statistical prediction scheme based on the prediction of EOFs of the covariance matrix of the predictable components. They showed that the predictability is closely related to Nino3 SST, but also related to the tropical Indian Ocean SST, the local NZ SST and the southern annular mode, with the cross-validated explained variance about 20%. For another example, the potential predictability of the eastern Asian summer monsoon rainfall [the seasonal mean rainfall in the region 5–50° N, 100–140° E in June-July-August which includes the Meiyu belt from Southern Japan through the Yangzi River catchment] is around 0.3 (Figure 5b). By using the methodology of Zheng and Frederiksen [10], Yin et al. [11] demonstrated that the predictability is not only related to the ENSO decay and developing phases, but also to seven other predictors from the Pacific, Atlantic and Indian Oceans, making the cross-validated explained variance more than 0.25, which reaches about 80% of the potential predictability.

Figure 5 also shows that there are some areas with significant potential predictability in some seasons, for example, the western Mediterranean centered in northern Egypt in JJA (Figure 5b) and northern Mexico in DJF (Figure 5d). To our knowledge, no study similar to Zheng and Frederiksen [10] has been carried out for precipitation in these regions. However, the potential predictability study in the current paper at least provides clues as to where the precipitation is more likely to be predicted and suggests that the statistical prediction scheme proposed by Zheng and Frederiksen [10] is worth attempting.

4. Discussions on Methodology

In earlier studies, daily time series were used to estimate the unpredictable weather noise component. A critical step is to separate the potentially predictable variability from daily data. For continuous metrological variables, such as temperature and pressure, Madden [1] applied a frequency domain approach where the potentially predictable variability was removed by a Fourier transform on daily time series for all seasons. Shukla [12] and Trenberth [13] applied an alternative time domain approach by assuming day-to-day weather variability is red noise, and the potentially predictability component was partially removed by a non-parametric approach in estimating the variability of the weather noise component. The red noise constraint was further relaxed as colored noise, and the influence of the potentially predictable component was completely removed by the application of the first order difference operator (Zheng [14]) or modeled as a variance parameter (Jones [15]) or multi-year mean parameters (Delsole and Feng [16]). However, daily rainfall data are intermittent and therefore the methodologies for daily continuous metrological data are no longer suitable. Madden et al. [17] applied the chain-dependent model (Katz [18]) to estimate weather noise variability, but for the daily precipitation observations, the fitted parameters are dependent on the predictable components (e.g., Katz and Zheng [19]; Zheng et al. [20]) and therefore the model is over simplified. Feng and Houser [21] modeled the daily precipitation as Gaussian auto-regressive processes described by Delsole and Feng [16], but the daily precipitation may not be Gaussian, especially if there are considerable dry days in small areas.

Zheng et al. [7] developed a method for estimating the weather noise component only, using monthly mean time series based on the likelihood estimation assuming the monthly means of the weather noise components are Gaussian. They demonstrated, by simulation, that the estimations of the weather noise component are comparable to those using daily continuous meteorological data. Zheng and Frederiksen [3] extended the methodology of Zheng et al. [7] for estimating the covariability of the potentially predictable components based on the moment estimation with a relaxed Gaussian assumption. Since the Gaussian assumption on the monthly mean of the unpredictable component is no longer required, the methodology can, in principle, be applied to precipitation data.

In this paper, the methodology of Zheng and Frederiksen [3] was applied to analyze the potential predictability of the GPCP precipitation. There are a number of advantages of the methodology applied in this study over all the previous methodologies using daily data. Monthly data are more available than daily data. This methodology can be applied to estimate the covariability of the predictable (unpredictable) components between two climate variables, which allows for the influence of one climate variable on another to be estimated, as shown in this study. Moreover, it can be used to estimate the covariance matrix of predictable (unpredictable) components of a climate field to which the singular value decomposition analysis can be applied on the estimated component covariance matrices. Such an approach was successfully applied to the seasonal prediction of New Zealand rainfall (Zheng and Frederiksen [10]) and east Asia summer monsoon rainfall (Ying et al. [11]). The successful predictions indicate that the estimation method based on monthly rainfall data is reliable. Other existing methods can only be applied to estimate the potential predictability of a single climate variable, but not the covariability between two climate variables.

By using monthly data, fewer statistical assumptions are required compared with methodologies using daily data. The only assumption is that the variance of ε_{ym} is invariant for month m, while the estimations using daily data depend on a variety of stationary assumptions and parameterizations which can lead to quite different results. For example, comparing the estimated potential predictabilities estimated in this paper (Figure 5) and those using the methodology of Delsole and Feng [16] for the GPCP data (see Figure 1 of

Feng and Houser [21]), although there is much higher predictability for the South Pole in their study, a higher predictability at the North Pole during JJA in our study is not shown in their study. Furthermore, the potential predictability of the eastern Asia summer monsoon rainfall is 30% using our method, while it is nearly zero in their study. Using our method, Ying et al. [11] developed a seasonal prediction scheme with a 0.25 cross-validated explained predictability for the eastern Asia summer monsoon rainfall.

5. Discussion and Conclusions

The covariance decomposition method proposed by Zheng and Frederiksen [3] was applied to the monthly GPCP precipitation data. The variances of the precipitation seasonal means for the four seasons were decomposed into potentially predictable components driven by slowly varying boundary forcing and low-frequency internal dynamics, and the unpredictable component related to day-to-day weather variability. It was suggested that ENSO (represented by Nino3-4 SST) contributes more than half of the predictable variability in SON and DJF when strong and unlikely to change phase. Although ENSO in JJA is weaker than ENSO in MAM, it contributes about 40% of the predictable variability in JJA compared with about 25% in MAM. This could be due to the fact that ENSO in MAM is more likely to change phase, especially in April. The seasonal distribution of the variability in the unpredictable component of GPCP precipitation is similar to that of the MJO described by Zhang and Dong [8], though less than 7% of the variability in the unpredictable component of GPCP precipitations was explained by the MJO index. This raises the question as to whether the current MJO indexes represent MJO variability well, or the MJO may not be so dominant in the variability of the unpredictable component of seasonal precipitation compared to ENSO and, therefore, this subject is worth further investigation.

Author Contributions: Conceptualization, H.L. and X.Z.; methodology, X.Z. and C.S.F.; formal analysis, H.L., X.Z. and J.Y.; data curation, J.Y.; writing—original draft preparation, X.Z and H.L.; writing—review and editing, C.S.F. All authors have read and agreed to the published version of the manuscript.

Funding: This research received no external funding.

Data Availability Statement: The data used in this study are available publicly. The GPCP precipitation is available at: https://psl.noaa.gov/data/gridded/data.gpcp.html (accessed on 6 April 2023). The MJO index is available at: https://www.psl.noaa.gov/mjo/mjoindex/ (accessed on 6 April 2023).

Acknowledgments: We sincerely thank the reviewers for their comments to improve the paper.

Conflicts of Interest: The authors declare no conflict of interest.

Appendix A

R-code for calculating the unpredictable covariance of Equation (4)

```
Cov.wmr=function(x1, x2) #x1,x2: Y by 3 matrix of the two monthly data sets
{
m <- dim(x1) [1]
X1 <- x1[, 1:2] − x1[, 2:3]
X2 <- x2[, 1:2] − x2[, 2:3]
a <- (sum(X1[, 1] * X2[, 1]) + sum(X1[, 2] * X2[, 2]))/2/m
b <- (sum(X1[, 1] * X2[, 2]) + sum(X1[, 2] * X2[, 1]))/2/m
alpha <- min(0.1, max((a + 2 * b)/(a + b)/2, 0))
sigma <- a/(1-alpha)/2
(sigma * (3 + 4 * alpha))/9
}
```

Appendix B

Proof of the variance of μ_y (ε_{yo}) explained by μ_y' (ε_{yo}').

We only prove for predictable component; proof for unpredictable component is similar.

Suppose the numerical values of μ_y and μ_y' exist, and μ_y can be regressed by μ_y' as

$$\mu_y = b\mu_y' + \xi_y$$

where b is the regression coefficient and ξ_y is the regression error. Since ξ_y and μ_y' are statistically independent, then

$$V\left(\mu_y, \mu_y'\right) = bV\left(\mu_y', \mu_y'\right)$$

so the regression coefficient b can be estimated as

$$b = V\left(\mu_y, \mu_y'\right) / V\left(\mu_y', \mu_y'\right)$$

therefore, the variance of μ_y explained by μ_y' is

$$V\left(b\mu_y', b\mu_y'\right) = b^2 V\left(\mu_y', \mu_y'\right) = V^2\left(\mu_y, \mu_y'\right) / V\left(\mu_y', \mu_y'\right)$$

Although the numerical values of μ_y and μ_y' cannot be obtained, $V(\mu_y, \mu_y')$ can be estimated using Equation (3).

References

1. Madden, R.A. Estimates of the natural variability of time averaged sea level pressure. *Mon. Wea. Rev.* **1976**, *104*, 942–952. [CrossRef]
2. Basher, R.E.; Thompson, C.S. Relationship of air temperature in New Zealand to regional anomalies in sea-surface temperature and atmospheric circulation. *Int. J. Climatol.* **1996**, *16*, 405–425. [CrossRef]
3. Zheng, X.; Frederiksen, C.S. Variability of seasonal-mean fields arising from intraseasonal variability: Part I, Methodology. *Clim. Dyn.* **2004**, *23*, 177–191. [CrossRef]
4. Adler, R.F.; Huffman, G.J.; Chang, A.; Ferraro, R.; Xie, P.P.; Janowiak, J.; Rudolf, B.; Schneider, U.; Curtis, S.; Bolvin, D.; et al. The version-2 Global Precipitation Climatology Project (GPCP) monthly precipitation analysis (1979–present). *J. Hydrometeor.* **2003**, *4*, 1147–1167. [CrossRef]
5. Kiladis, G.N.; Dias, J.; Straub, K.H.; Wheeler, M.C.; Tulich, S.N.; Kikuchi, K.; Weickmann, K.M.; Ventrice, M.J. A comparison of OLR and circulation based indices for tracking the MJO. *Mon. Wea. Rev.* **2014**, *142*, 1697–1715. [CrossRef]
6. Ying, K.; Frederiksen, C.S.; Zhao, T.; Zheng, X.; Xiong, Z.; Yi, X.; Li, C. Predictable and unpredictable modes of seasonal mean precipitation over Northeast China. *Clim. Dyn.* **2018**, *50*, 3081–3095. [CrossRef]
7. Zheng, X.; Nakamura, H.; Renwick, J.A. Potential predictability of seasonal means based on monthly time series of meteorological variables. *J. Clim.* **2000**, *13*, 2591–2604. [CrossRef]
8. Zhang, C.; Dong, M. Seasonality in the Madden–Julian Oscillation. *J. Clim.* **2004**, *17*, 1369–1380. [CrossRef]
9. Maloney, E.D.; Kiehl, J.T. MJO-related SST variations over the tropical eastern Pacific during Northern Hemisphere summer. *J. Clim.* **2002**, *15*, 675–689. [CrossRef]
10. Zheng, X.; Frederiksen, C.S. A study of predictable patterns for seasonal forecasting of New Zealand rainfalls. *J. Clim.* **2006**, *19*, 3320–3333. [CrossRef]
11. Ying, K.D.; Jiang, D.; Zheng, X.; Frederiksen, C.S.; Peng, J.; Zhao, T.; Zhong, L. Seasonal predictable source of the East Asian summer monsoon rainfall in addition to the ENSO-AO. *Clim. Dyn.* **2022**, *60*, 2459–2480. [CrossRef]
12. Shukla, J. Comments on "Natural variability and predictability". *Mon. Wea. Rev.* **1983**, *111*, 581–585. [CrossRef]
13. Trenberth, K.E. Some effects of finite sample size and persistence on meteorological statistics. Part II: Potential predictability. *Mon. Wea. Rev.* **1984**, *112*, 2369–2379. [CrossRef]
14. Zheng, X. Unbiased estimation of autocorrelations of daily meteorological variables. *J. Clim.* **1996**, *9*, 2197–2203. [CrossRef]
15. Jones, R.H. Estimating the variance of time averages. *J. Appl. Meteor.* **1975**, *14*, 159–163. [CrossRef]
16. Delsole, T.; Feng, X. The "Shukla–Gutzler" Method for estimating potential seasonal predictability. *Mon. Wea. Rev.* **2012**, *141*, 822–831. [CrossRef]
17. Madden, R.A.; Shea, D.J.; Katz, R.W.; Kidson, J.W. The potential long-range predictability of precipitation over New Zealand. *Int. J. Climatol.* **1999**, *19*, 4. [CrossRef]

18. Katz, R.W. Precipitation as a chain-dependent process. *J. Appl. Meteor.* **1977**, *16*, 671–676. [CrossRef]
19. Katz, R.W.; Zheng, X. Mixture model for overdispersion of precipitation. *J. Clim.* **1999**, *12*, 2528–2537. [CrossRef]
20. Zheng, X.; Renwick, J.A.; Clark, A. Simulation of multisite precipitation using extended chain-dependent process. *Water Resour. Res.* **2010**, *46*, W01504. [CrossRef]
21. Feng, X.; Houser, P. An examination of potential seasonal predictability in recent reanalysis. *Atmo. Sci. Lett.* **2014**, *15*, 266–274.

atmosphere

MDPI

Article

Assessment of Long-Term Rainfall Variability and Trends Using Observed and Satellite Data in Central Punjab, Pakistan

Khalil Ahmad [1,*] , Abhishek Banerjee [2,*] , Wajid Rashid [3] , Zilong Xia [4] , Shahid Karim [5] and Muhammad Asif [6]

1 School of Ecological and Environmental Sciences, East China Normal University, Shanghai 200241, China
2 State Key Laboratory of Cryospheric Sciences, Northwest Institute of Eco-Environment and Resources, Chinese Academy of Sciences, Donggang West Rd. 318, Lanzhou 730000, China
3 Department of Environmental and Conservation Sciences, University of Swat, Mingora Swat 19130, Pakistan
4 School of Geographic and Oceanographic Sciences, Nanjing University, Nanjing 210023, China
5 Department of Geography, GC University Lahore, Lahore 54000, Pakistan
6 Department of Plant Pathology, College of Plant Protection, China Agricultural University, Beijing 100193, China
* Correspondence: 52183903033@stu.ecnu.edu.cn (K.A.); abhishek@stu.ecnu.edu.cn (A.B.); Tel.: +86-1862-1101-490 (A.B.)

Abstract: This study explores the spatio-temporal distribution and trends on monthly, seasonal, and annual scales of rainfall in the central Punjab districts of Punjab province in Pakistan by using observation and satellite data products. The daily observed data was acquired from the Pakistan Metrological Department (PMD) between 1983 and 2020, along with one reanalysis, namely the Climate Hazard Infrared Group Precipitation Station (CHIRPS) and one satellite-based daily Precipitation Estimation from Remotely Sensed Information Using Artificial Neural Networks climate data record (PERSIANN-CDR) using the Google Earth Engine (GEE) web-based API platform to investigate the spatio-temporal fluctuations and inter-annual variability of rainfall in the study domain. Several statistical indices were employed to check the data similarity between observed and remotely sensed data products and applied to each district. Moreover, non-parametric techniques, i.e., Mann–Kendall (MK) and Sen's slope estimator were applied to measure the long-term spatio-temporal trends. Remotely sensed data products reveal 422.50 mm (CHIRPS) and 571.08 mm (PERSIANN-CDR) mean annual rainfall in central Punjab. Maximum mean rainfall was witnessed during the monsoon season (70.5%), followed by pre-monsoon (15.2%) and winter (10.2%). Monthly exploration divulges that maximum mean rainfall was noticed in July (26.5%), and the minimum was in November (0.84%). The district-wise rainfall estimation shows maximum rainfall in Sialkot (931.4 mm) and minimum in Pakpattan (289.2 mm). Phase-wise analysis of annual, seasonal, and monthly trends demonstrated a sharp decreasing trend in Phase-1, averaging 3.4 mm/decade and an increasing tendency in Phase-2, averaging 9.1 mm/decade. Maximum seasonal rainfall decreased in phase-1 and increased Phase-2 during monsoon season, averaging 2.1 and 4.7 mm/decade, whereas monthly investigation showed similar phase-wise tendencies in July (1.1 mm/decade) and August (2.3 mm/decade). In addition, as district-wise analyses of annual, seasonal, and monthly trends in the last four decades reveal, the maximum declined trend was in Sialkot (18.5 mm/decade), whereas other districts witnessed an overall increasing trend throughout the years. Out of them, Gujrat district experienced the maximum increasing trend in annual terns (50.81 mm/decade), and Faisalabad (25.45 mm/decade) witnessed this during the monsoon season. The uneven variability and trends have had a crucial imprint on the local environment, mainly in the primary activities.

Keywords: rainfall change; CHIRPS; PERSIANN-CDR; descriptive statistics; non-parametric trends; Google Earth Engine; central Punjab

Citation: Ahmad, K.; Banerjee, A.; Rashid, W.; Xia, Z.; Karim, S.; Asif, M. Assessment of Long-Term Rainfall Variability and Trends Using Observed and Satellite Data in Central Punjab, Pakistan. *Atmosphere* **2023**, *14*, 60. https://doi.org/10.3390/atmos14010060

Academic Editor: Haibo Liu

Received: 11 November 2022
Revised: 22 December 2022
Accepted: 23 December 2022
Published: 28 December 2022

1. Introduction

Rainfall is a key component and also a significant contributor to the global energy and water cycle for determining the overall hydro-meteorology of any region [1,2]. Rainfall has a various range of impressions on human civilization, agricultural applications [3], regional biodiversity [4], vegetation distribution and growth, local water supply, hydropower projects [5], and the inclusive geo-ecological equilibrium of a particular region [6]. In past decades, global rainfall pattern, distribution and trends have unfortunately witnessed an irregular nature, which resulted in drought [7], immense flooding [8], large-scale cloud burst, and significant damage in terms of property and infrastructure [9]. It is supposed that alterations in rainfall patterns would be linked to global temperature rises [10]. Recent studies suggested a declining annual trend over central Asia in the last few decades [2], whereas Banerjee et al. [11] reported a slightly increasing trend in annual, seasonal, and monthly rainfall over the north-west Himalayan region from 2000 to 2022. Moreover, additional studies have also reported an increasing trend of annual rainfall in the Hindu Kush Himalayan Region [12], while Zhang et al. 2014 [13] found a slightly decreasing tendency of annual and seasonal rainfall in China's Yellow River Basin. Several current studies in Pakistan found uneven rainfall patterns from 1980 to 2020, where Nawaz et al. [14] noticed an increasing rainfall trend in Punjab province using numerous remotely sensed gridded datasets. On the other hand, Nawaz et al. [15] observed an uneven distribution of trends in Punjab province with increased air temperature in the last 50 years. They further examined the dependency of elevation on rainfall trends and found higher elevation regions experienced decreased amounts of rainfall with increased air temperature, whereas lower elevation regions witnessed increased rainfall. Various other studies have been performed to estimate rainfall distribution and trends in Pakistan and noticed an overall increased trend at annual and seasonal scales in the last four decades [2,16,17]. Ullah et al. [2] reported a declined trend in winter and post-monsoon periods and an increased trend during monsoon periods, pre-monsoon periods, and winter, averaging 0.9 mm/decade from 1980 to 2016. Pakistan is a highly agriculture-dependent country, and Punjab province is the largest agricultural region [14], covering 100% of the nation's annual food demand. It is a well-established fact that rainfall is very crucial for agricultural productivity and irrigation. Thus, it is identically important to study the rainfall dynamics and their long-term variability in Punjab province, Pakistan. During the last four decades, this region has witnessed uneven alterations of rainfall dynamics and allied agricultural productivity. Decreased rainfall after the 1980s had critically impacted on agricultural production, which further coped with abrupt changes after 2002 [14,15]. The central Punjab region is mostly dominated by the monsoonal rainfall (>70%) [2,16,18], whereas moisture coming from the Bay of Bengal gradually transfers towards the central Punjab region because of the low-pressure zone at the Pamir region due to high temperatures [2,16]. Additionally, winter rainfall in the central Punjab region mainly occurred due to the western disturbances, the presence of jet streams, and the returning monsoon [2]. Ullah et al. [2] reported significant changes in seasonal rainfall, where abrupt decline alteration was reported during the pre-monsoon and winter season before 2000; afterward, a slight increased tendency has been reported using 53 rainfall stations. However, the availability of station-based observed data is very limited throughout Pakistan as well as in Punjab province, so relying on high-resolution remotely sensed data is the best option to cope with the station data scarcity [19,20]. Various satellite observations are obtainable, such as Multi-satellite Precipitation Analysis (TMPA) [21], Tropical Rainfall Measuring Mission (TRMM) [22], Climate Prediction Centre MORPHing product (CMORH) [23], Climate Hazard Group Infrared Precipitation with Stations (CHIRPS) [24], Multi-source Weighted-Ensemble Precipitation (MSWEP) [25], the Precipitation Estimation from Remotely Sensed Information using Artificial Neural Network (PERSIANN-CDR) [26], Global Precipitation Measurement (GPM), and the Asian Precipitation-Highly-Resolved Observational Data Integration Towards Evaluation (APHRODITE) [27]. A few recent studies were carried out in India and Pakistan on the performance checking of gridded datasets using numerous statistical applications and found that CHIRPS and PERSIANN-CDR datasets properly capture the rainfall events as compared to station data [2,6,11,14,15].

By considering the seasonality, variability, long-term trends, and distribution of rainfall in the highest food production region in the country, this exploration aims to study the spatio-temporal and inter-decadal dynamics in the last 38 years of rainfall in 19 districts of central Punjab province.

The objectives of this study are described as follows:

1. Collect, calculate, and analyze the CHIRPS and PERSIANN-CDR gridded rainfall products with surface observatories data from PMD of a chosen region;
2. Map seasonal, annual, and monthly rainfall variability and trends with suitable performance analysis with in situ data;
3. Determine long-term inter-decadal district-wise distribution and trends of rainfall on monthly, seasonal, and annual scales for the period of 1983–2020 by using non-parametric tests.

2. Study Area

With a total area of 205,344 sq. km, Punjab is the most populous and second largest province in Pakistan after Baluchistan, with geographical coordinates of 31.17° N and 72.70° E (Figure 1). From the total area of Punjab province, the 19 districts of central Punjab consist of 68,577 sq. km, which is 33.39% of the total. Of 110.01 million people live in Punjab province, and 63.67 million live in 19 districts of central Punjab, which is 58% of the total. Due to canal irrigation and rainfall, especially in summer, a significant part of agricultural products is produced in central Punjab [28–30]. In Punjab province, central Punjab districts produced 45% of wheat, 69% of corn, and 83% of rice in 2019–2020. Climatologically, this area is located in the subtropical dry region, having mean annual temperature ranges between 16.3 °C to 29.3 °C. Rainfall of this region varies from south–central to north-eastern, whereas it reaches its maximum during the monsoon season (50–75%) [31]. The monsoon system provides the majority of rainfall, filling yearly agricultural needs in the central Punjab region [32]. Recent increased frequency of unpredictable occurrences of rainfall impacted rice, maize, wheat, sugarcane, and cotton production along with other minor crops, such as millet, vegetables, and tree fruits, including citrus and mango in the central Punjab districts, including Nankana Sahib, Sialkot, Gujranwala, Hafizabad, Sheikhupura, and Sargodha [33]. This region is fairly vulnerable to climate change because of the large variety and intense events that occur during the summer and winter monsoon precipitation seasons [34–38]. Spatial and temporal evaluation of the GPPs (gridded satellite precipitation products) is vital to examine in-depth hydro-meteorological scenarios across the study region, where observed data are inadequate.

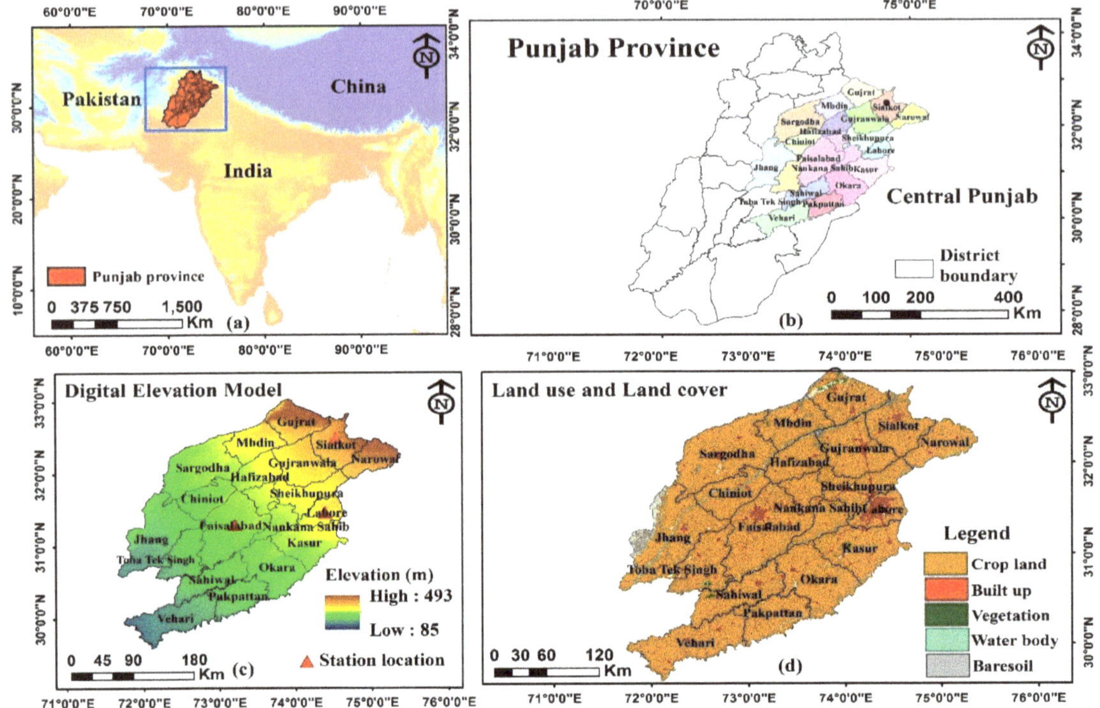

Figure 1. (**a**) Location of central Punjab, Pakistan; (**b**) Districts of central Punjab region; (**c**) Digital elevation model of central Punjab with PMD metrological observatories; (**d**) District-wise land use and land cover pattern of central Punjab.

3. Materials and Methods

3.1. Observed Data

Daily rainfall surface observatory data of three selected stations were obtained from the Climate Data Processing Centre (CPCD) of the Pakistan Meteorological Department (PMD) for the period of 38 years (1983–2020) (Table 1). Station-observed data is already quality-controlled and has been situated at the same place for the last four decades by PMD (except in Sargodha where PMD handed over the task to the Pakistan Air Force in 2016) for calculating long-term variability [2,14]. Therefore, our prime considerations to select the data were (i) recent time and long-term data availability, and (ii) fewer data gaps, as seen in Table 1.

Table 1. Data utilized to examine the rainfall distribution and trends in central Punjab.

Station-Observed Data (Collected from PMD)					
	Latitude	Longitude	Elevation (in m)	Period	Available Observation (in %)
Stations Lahore	31°33′ N	74°20′ E	214.00	1983–2020	100
Faisalabad	31°26′ N	73°08′ E	185.6	1983–2020	99.99
Sialkot	32°31′ N	74°32′ E	255.1	1983–2020	99.99
Gridded Data (Downloaded from Google Earth Engine Platform)					
Data Type	Source	Year	Spatial Resolution		
CHIRPS	UCSB/CHG	1983–2020	0.05°		
PERSIANN-CDR	NOAA/NCDC	1983–2020	0.25°		

3.2. Remotely Sensed Data

Two different sources of precipitation datasets were obtained and used; one was the CHIRPS, and the second was the PERSIANN-CDR satellite-based data. Both were extracted through the Google Earth Engine (GEE) web-based remote sensing platform for the period of 1983–2020 (Table 1). The CHIRPS data is a combination of in situ, satellite and global precipitation prediction system (https://www.chc.ucsb.edu, accessed at 10 November 2022), which was taken from the Climate Hazard Group website with the help of GEE [6]. The PERSIANN-CDR dataset is available via the NOAA National Centers for Environmental Information (NCEI) program (https://www.ncdc.noaa.gov/cdr, accessed at 10 November 2022) and was also taken from GEE via the CHRS data portal (http://chrsdata.eng.uci.edu, accessed at 10 November 2022) [39].

3.3. Data Processing and Statistical Application

Monthly, seasonal, and annual gridded rainfall products were assimilated from the GEE platform through the use of the 'ee.ImageCollection' algorithm and complete request of filter command (ee.Filter.calendarRange) for the study period (1983–2020) for this research. Subsequently, the 'clip' function was used to limit the study region and to extract the actual data as per the study domain. For matching similar spatial information with station data, the nearest point-pixel rainfall values of the gridded dataset were derived in GEE in accordance with station locations using GEE algorithms (the code can be found at Banerjee et al.) [6]. Afterward, pixel-based long-term spatio-temporal trends for gridded data sets were achieved in GEE by the use of the 'ee.Reducer.senSlope' algorithm. The gridded daily rainfall records were then aggregated to seasonal totals, namely pre-monsoon, monsoon, post-monsoon, and winter, and also on annual scale. As per recommendations by the Pakistan Meteorological Department (PMD), the season was considered in the following way: winter = January and February; pre-monsoon = March, April, and May; monsoon = June, July, August, and September; post-monsoon = October, November, and December [36,40,41]. To calculate the long-term trend, the serial auto-correlation technique has been employed to check the data similarity of a particular time-series, and then non-parametric techniques i.e., Mann–Kendall (MK) and Sen's slope estimator were applied to check the statistical significance and actual amount of alteration throughout the decades for two distinctive phases (1983–2001 and 2002–2022) at the 95% significance level [6,11], as per the status of long-term rainfall distribution and observed abrupt changes after 2001 due to the remarkable presence of the EL Nino (1998) and subsequent rapid increase in temperature and decreased rainfall pattern throughout the country [40]. Previous studies in this region also highlighted changing pattern of rainfall after 2002, so that we divide the entire study period into two phases to investigate inter-decal rainfall variability and trends (Figure S1) [14,15,40]. All of the significance tests and trend calculations at annual, seasonal, and monthly scales were performed using the GEE web-based API platform and MATLAB programming language.

The observed rainfall of three stations in central Punjab on a daily scale were obtained from the PMD archive for the period of 1983 to 2020. The missing values of possible stations were filled by taking long-term daily averages [6]. Observed rainfall records in a region, such as central Punjab, are scarce and unobtainable, because most of the rainfall data stations were established after 2004, so accessing long-term observed empirical data is still a big challenge for the research community. To ensure that high-quality satellite rainfall data concerning the station-observed data could be used for further investigation, several descriptive statistics were then applied to check data distribution and the relative performance of the gridded datasets. The statistical indices include (a) bias, (b) multiplicative bias (MBias), (c) relative bias (RBias), (d) mean error (ME), (e) mean absolute error (MAE), (f) root mean square error (RMSE), and (g) correlation coefficient (CC). To measure the uncertainty and assess the level of inaccuracy in an averaging of large-scale estimates, the following approaches were used: (i) cumulative distribution frequency (CDF) and (j) the Taylor diagram; all of the mathematical equations can be found in Banerjee et al. [6].

4. Results

4.1. The Outcome from Descriptive Statistics

Descriptive statistics revealed a blend of under- (PERSIANN-CRD) and over-estimation (CHIRPS) (Table 2) of rainfall counts as per the observed data. The ME value showed that both CHIRPS (-0.19 mm) and PERSIANN-CDR (-0.12 mm) slightly underestimate the daily rainfall at the Lahore and Sialkot stations, while at the Faisalabad station CHIRPS (-0.11 mm) closely underestimated the rainfall, and PERSIANN-CDR (0.08 mm) overestimated the rainfall (Table 2) at daily scales. The lower value of RMSE at three stations indicates a good spatial association between in situ and remotely sensed data products. The MAE value of both datasets is in the range of 1.64 mm to 2.82 mm, which indicates that gridded data is finely fitted with observed data with the least mean absolute error. A very small or negligible parentage spatial bias for three stations was observed through CHIRPS (0.23 mm to 0.28 mm) and PERSIANN-CDR (-0.1 to 0.155 mm) data. The low values of MBias (-0.61 to -1.5) indicate the less systematic errors within the datasets. The correlation coefficient (CC) and linear regression for both the gridded and station data showed a strong association (Figure 2) in terms of monthly rainfall distribution. The association of CHIRPS (r = 0.78) is relatively better as compared to PERSIANN-CDR (r = 0.76). The maximum correlation coefficient value was noticed at the Lahore station, followed by Faisalabad and Sialkot (Table 2). Additionally, the cumulative distribution frequency (CDF) graph shows the spatial distribution for average monthly rainfall (Figure 2e), which presents that more than 90% of mean monthly rainfall is evenly distributed among CHIRPS, PERSIANN-CDR, and surface observatories.

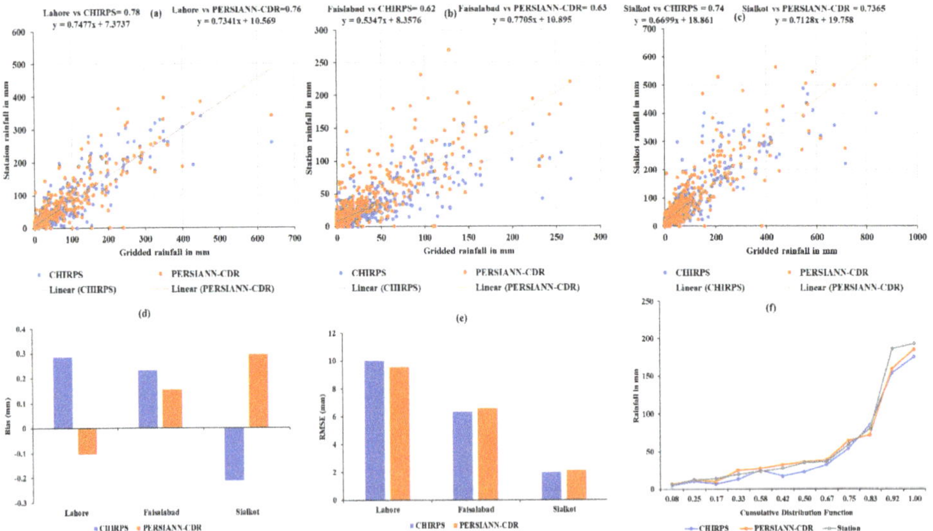

Figure 2. (**a–c**) show the relationship between the observed and gridded data (r^2), (**d**) bias, (**e**) RMSE, and (**f**) the CDF of gridded data with corresponding observatories from 1983–2020.

4.2. Phase-Wise Annual, Seasonal and Monthly Distribution of Rainfall in Central Punjab (CHIRPS, PERSIANN-CDR)

Phase-wise annual, seasonal, and monthly distribution of rainfall in central Punjab have been determined using CHIRPS and PERSIANN-CDR (Figure 3) datasets. The mean annual investigation as per the CHIRPS data shows comparatively inferior rainfall in Phase-1 (385.8 mm) compared to Phase-2 (459.2 mm), whereas PERSIANN-CDR reveals relatively higher rainfall in the last four decades (Figure S2). The seasonal average of two distinctive phases shows that the maximum rainfall is during monsoon season (70.5%), followed by pre-monsoon (12.5%), and winter (9.6%) (Figure 3a,b). In addition to this, the mean monthly investigation exhibits maximum rainfall in July, August, and September, followed by January and February (Figure 3c,d).

Table 2. Statistical indices to examine the similarity amid satellite data (CHIRPS and PERSIANN-CDR) and observed datasets.

Station	Data	ME (mm)	RMSE (mm)	MAE (mm)	Bias (mm)	MBias (mm)	RBias (%)	CC
Lahore	CHIRPS	−0.19	10.01	2.77	0.29	−1.31	0.00	0.78
	PERSIANN-CDR	−0.12	9.55	2.83	0.15	−1.57	0.00	0.76
Faisalabad	CHIRPS	−0.11	6.34	1.64	0.23	−0.62	0.00	0.62
	PERSIANN-CDR	0.08	6.57	1.88	−0.10	−1.29	−0.00	0.63
Sialkot	CHIRPS	−0.09	1.99	1.18	−0.21	0.88	0.00	0.74
	PERSIANN-CDR	0.13	2.12	2.97	0.29	1.15	0.00	0.73

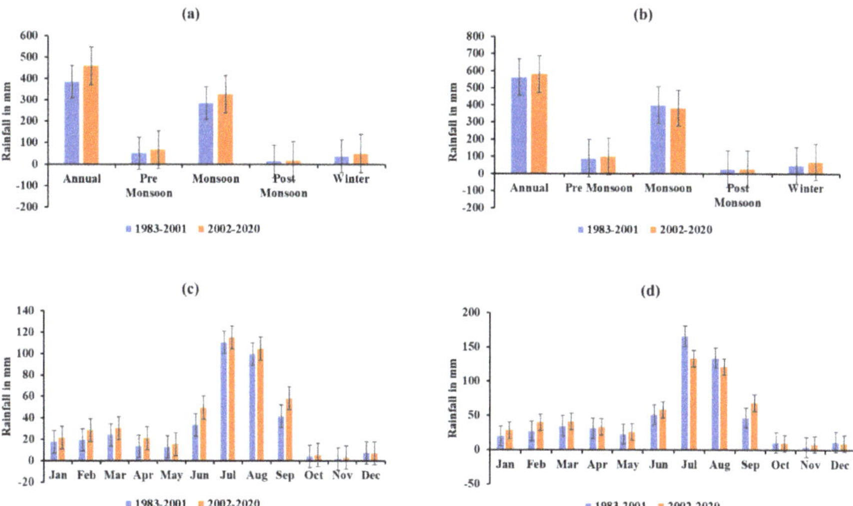

Figure 3. Phase-wise distribution of rainfall with error bars in central Punjab (from 1983–2001 and 2002–2020). (**a**) Mean annual and seasonal rainfall of CHIRPS; (**b**) mean annual and seasonal rainfall of PERSIAN-CDR; (**c**) mean monthly rainfall of CHIRPS; (**d**) mean monthly rainfall of PERSIANN-CDR.

Out of the two phases, maximum rainfall is observed in Phase-2, whereas dispersal of the PERSIANN-CDR exhibits relatively higher rainfall as compared to the CHIRPS data (Figure 3a–d).

4.3. District-Wise Distributions of Annual, Seasonal, and Monthly Rainfall (CHIRPS, PERSIANN-CDR)

District-wise distribution of mean annual, seasonal, and monthly rainfall has been carried out using CHIRPS and PERSIANN-CDR datasets for 38 years (1983–2020) in central Punjab (Figure 4). Maximum mean annual rainfall was experienced in Sialkot (904.2 mm), followed by Narowal (860.36 mm), and Gujrat (840.9 mm) as per the CHIRPS data, whereas PERSIANN-CDR data show the maximum rainfall as being in the Sialkot (958.46 mm), Narowal (912.9 mm), and Gujrat (898.78 mm) districts (Figure S2). Concurrently, the seasonal assessment reveals that maximum rainfall occurs in Sialkot (702.40 mm), followed by Narowal (656.7 mm), and Gujrat (618.00 mm), with the CHIRPS dataset during monsoon season (71.24% out of the total rainfall in central Punjab) (Table 3). Moreover, PERSIANN-CDR shows maximum rainfall during the same season in Sialkot (690.5 mm), Narowal (670.48 mm), and Gujrat (608.7 mm) districts (Figure 4a,b). Apart from the monsoon season, the pre-monsoon and winter seasons also received a significant amount of rain throughout the years; CHIRPS shows maximum rainfall in Gujrat (95.6 mm), followed by Sargodha (88.18 mm), and Narowal (87.4 mm) during the pre-monsoon season (14.74% out of the total rainfall in central Punjab), while PERSIANN-CDR highlights the maximum, albeit lower, rainfall as compared to CHIRPS in Gujrat (151.9 mm), Mandi Bahauddin (135.7 mm), and Sialkot (133.9 mm) (Figure 4c,d). Furthermore, maximum rainfall during winter season (9.96% out of the total rainfall in central Punjab) as per CHIRPS occurs in Gujrat (89.5 mm), Narowal (88.2 mm), and Sialkot (74.5 mm), where similar distribution pattern has been noticed in PERSIANN-CDR also. Furthermore, monthly exploration reveals maximum rainfall in June (43.6 mm), July (123.1 mm), August (109.9 mm), and September (56.7 mm) in CHIRPS data, while PERSIANN-CDR exhibits maximum rainfall in June (56.3 mm), July (151.9 mm), August (128.4 mm) and September (59.5 mm) (Figure 4e,f).

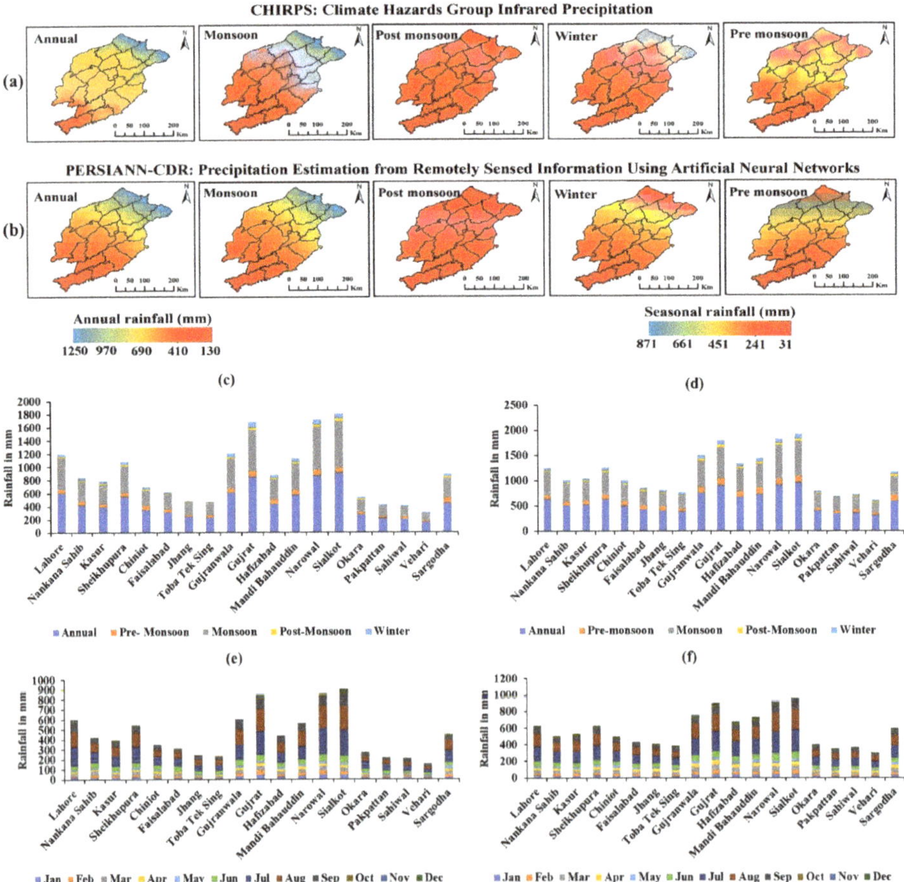

Figure 4. Spatial distribution of mean seasonal and annual rainfall. The upper panel shows (**a**) CHIRPS, and the lower panel displays (**b**) PERSIANN-CDR. Middle and lower portions show district-wise distribution of annual, seasonal, and monthly rainfall (1983–2020). (**c**) Mean annual and seasonal rainfall of CHIRPS; (**d**) mean annual and seasonal rainfall of PERSIAN-CDR; (**e**) mean monthly rainfall of CHIRPS; (**f**) mean monthly rainfall of PERSIANN-CDR.

Table 3. Phase-wise annual, seasonal, and monthly distribution of rainfall in central Punjab from gridded datasets (1983–2020).

Data	Period	January	February	March	April	May	June	July	August	September	October	November	December	Annual	Pre-Monsoon	Monsoon	Post-Monsoon	Winter
CHIRPS	1983–2001	17.61	19.20	24.14	13.37	12.73	33.45	110.68	99.64	41.48	4.23	1.81	7.51	385.83	50.23	285.24	13.55	36.81
	2002–2020	21.39	28.51	30.21	21.03	15.48	49.42	115.01	104.62	58.01	5.36	3.10	7.04	459.19	66.72	327.06	15.51	49.90
PERSIANN-CDR	1983–2001	19.98	27.56	34.98	31.44	22.89	51.22	166.20	134.20	47.15	10.99	4.37	11.77	562.74	89.31	398.77	27.12	47.54
	2002–2020	28.80	40.37	41.28	33.39	26.05	58.24	133.79	121.25	68.70	9.83	8.11	9.60	579.44	100.73	381.98	27.55	69.18

Monthly scale investigation revealed that maximum rainfall was noticed in June to September, averaging 19.5% of the total (Figure 4e,f), whereas December and January also experience remarkable rainfall throughout the years.

4.4. Phase-Wise Annual, Seasonal, and Monthly Trends of Rainfall in Central Punjab (CHIRPS, PERSIANN-CDR)

Phase-wise annual, seasonal, and monthly trends and also percentage change in rainfall in central Punjab using CHIRPS and PERSIANN-CDR datasets have been achieved (Figure 5). Both distinctive phases show a versatile temporal tendency, where Phase-1, annually detecting a statistically insignificant decreasing trend in both datasets, is an average of 5%/decade. Moreover, Phase-2 experiences a statistically significant increasing tendency in both datasets, around 19%/decade (Figure S2). Seasonal investigation exhibits an overall insignificant decreasing trend, averaging 14.3%/decade, except during post-monsoon season (-4.2%/decade) and winter (0.49 mm/decade). On the other hand, Phase-2 witnessed an overall statistically significant increasing trend in both the datasets, averaging 10.5 mm/decade (Figure 5a,b). Maximum seasonal trends occurred during monsoon season, followed by pre-monsoon and winter seasons (Figure 5a). Additionally, monthly exploration exhibits the uneven distribution of trend, where CHIRPS observed a maximum statistically significant reduction in Phase-1 in March (-28.5%/decade), followed by September, February, and December (Figure S2). Along with this, CHIRPS data during Phase-2 experienced an overall increasing trend, where the maximum statistically significant tendencies were observed in July (37.4%/decade) and April (13.3%/decade) (Figure 5e,f). However, PERSIANN-CDR witnessed relatively higher statistically significant changes in the month of April (31.4%/decade) and slightly lesser ones in June (26.8%/decade), as compared to CHIRPS data (Table S1).

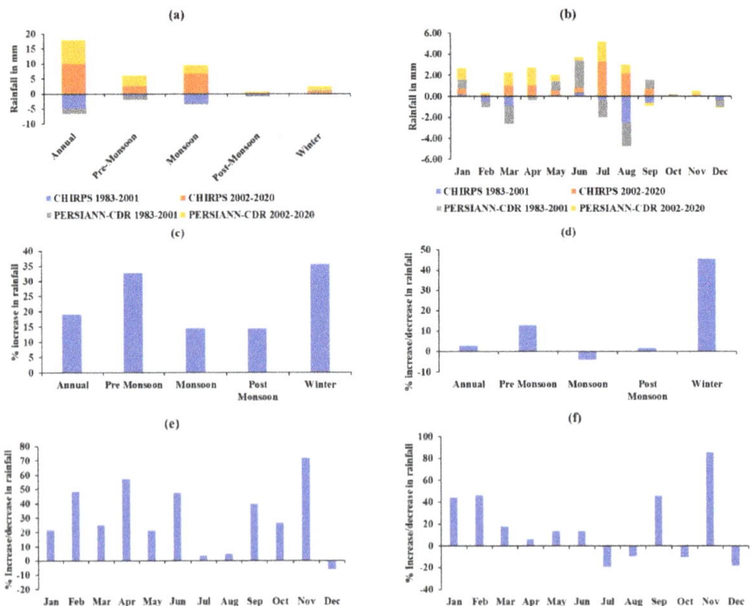

Figure 5. Phase-wise trends and mean percentage change in rainfall in central Punjab through CHIRPS and PERSIANN-CDR (1983–2001 and 2002–2020). (**a**) Trends on annual and seasonal scales; (**b**) trends on a monthly scale; (**c**) percentage mean annual and seasonal change in CHIRPS from Phase-1 to Phase-2; (**d**) percentage mean annual and seasonal change in PERSIAN-CDR from Phase-1 to Phase-2; (**e**) percentage mean monthly change in CHIRPS from Phase-1 to Phase-2; (**f**) percentage mean monthly change in PERSIANN-CDR from Phase-1 to Phase-2.

4.5. District-Wise Trend of Annual, Seasonal and Monthly Rainfall (CHIRPS, PERSIANN-CDR)

The present study further investigates the district-wise rainfall trend of two gridded datasets for the last four decades. Out of 19 districts, CHIRPS showed a maximum annual statistically significant trend in Gujrat (50.8 mm/decade), followed by Narowal (47.9 mm/decade), Mandi Bahauddin (41.7 mm), Sarghoda (39.5 mm), and Sialkot (37.2 mm/decade) (Table S2). Contrary to this, PERSIANN-CDR demonstrated a relatively low tendency, where the maximum trend was observed in Chiniot (29.5 mm/decade). Seasonal investigation illustrates a significant increase during monsoon season in both datasets, where Gujrat witnessed a higher increasing trend (29.4 mm/decade) (Figure 6a,b). However, PERSIANN-CDR showed a moderately low seasonal tendency during the monsoon period, where maximum changes were noticed in Jhang (23.5 mm/decade) (Table S3). Additionally, pre-monsoon and winter seasons also experienced a statistically significant increasing trend throughout the years. Maximum changes were observed in Chiniot (10.2 mm/decade), followed by Faisalabad (10.1 mm/decade) and Toba Tek Singh (9.6 mm/decade), during pre-monsoon season. On the other hand, Gujrat witnessed the maximum changes during the winter season (11.8 mm/decade), followed by Sialkot (9.9 mm/decade), and Narowal (9.4 mm/decade).

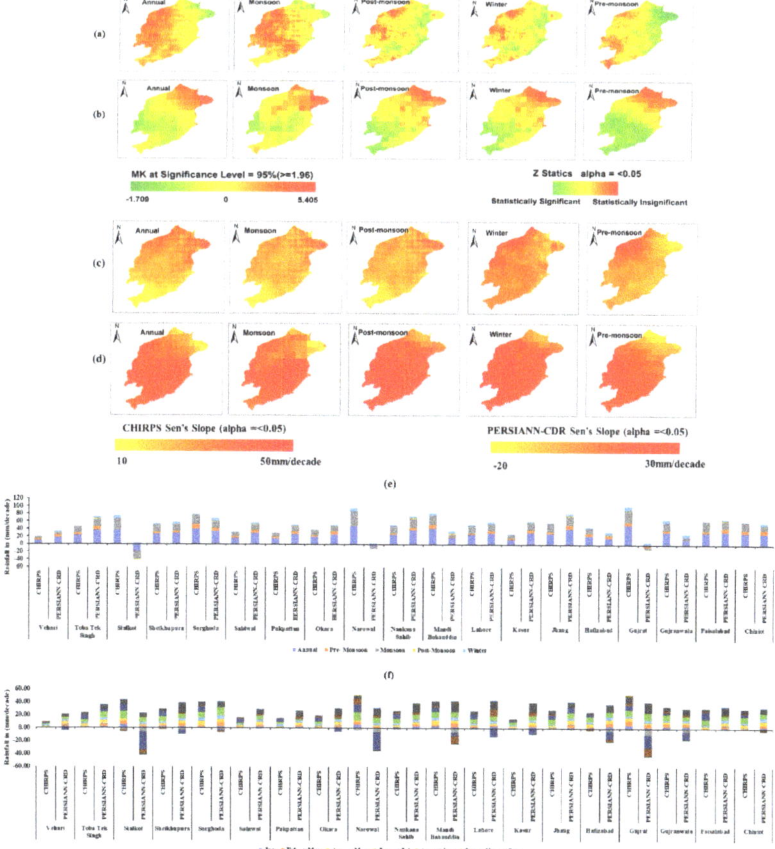

Figure 6. Spatial distribution of magnitude of trends in annual and seasonal rainfall. Top panel showing (**a**) statistical significance level of CHIRPS, (**b**) statistical significance level of PERSIANN-CDR, (**c**) magnitude of trend of CHIRPS, (**d**) magnitude of trend of PERSIANN-CDR, (**e**) annual and seasonal trends of CHIRPS and PERSIANN-CDR, and (**f**) monthly trends of CHIRPS and PERSIANN-CDR.

In addition, the study further explores long-term monthly alteration using both the gridded datasets. Maximum significant increasing changes were noticed in Gujrat district in September (9.6 mm/decade), June (8.9 mm/decade), and July (7.3 mm/decade) based on CHIRPS data. On the other hand, PERSIANN-CDR showed maximum alteration in Kasur district (13.1 mm/decade), trailed by Nankana Sahib (12.4 mm/decade), and Okara (10.6 mm/decade) in September.

5. Discussion

In places where in-situ data is scarce, the use of remotely sensed rainfall data is critical for assimilation, estimation, and forecasting, as well as enhancing model simulation performance to well understand the land–atmosphere interactions [42,43]. In this study, we intensively examined the performance of gridded datasets using two surface observatories by employing several statistical indices and found more than 90% similar rainfall counts in the last 38 years (1983–2020). Investigations of monthly, seasonal (winter, pre-monsoon, monsoon, and post-monsoon), and annual distribution, and trends of rainfall, using non-parametric statistical techniques revealed 9.59 mm/decade to 50.81 mm/decade increase in rainfall annually in central Punjab districts in the last 38 years at the 95% confidence level.

5.1. Validation of Gridded Data in Comparison to In-Situ Data (1983–2020)

There have always been concerns regarding the availability of long-term climate data, especially due to the missing surface observatory data in crucial regions, such as central Punjab in Pakistan. Due to unreachability, long-term observation stations are an exception and, hence, we relied on remotely sensed and reanalysis products to estimate uncertainties that cannot be handled with station rainfall data. Reanalysis (CHIRPS) and satellite-derived (PERSIANN-CDR) rainfall outputs can be compared and statistically assessed in terms of relative performance versus limited observational data using the GEE cloud computing capability. So far, limited research has been conducted for the validity of the CHIRPS and PERSIANN-CDR in Pakistan against the in situ data, but none so far provide deep insight at a regional level, such as central Punjab [19,44].

Because of the presence of cold clouds, ice, and snow, satellite products are inadequate in their capability to predict rainfall in mountainous locations [45]. The CHIRPS sometimes under- or overestimates in situ data in different parts of the world [46,47]. On a daily scale, CHIRPS performance was noted as being subpar; however, it was better evaluated across West Africa on a monthly, seasonal, and annual scale [48]. The dissimilarity is due to orographic features, different climate zones, and composite surface topography [47,49]. The performance of the CHIRPS was found to be satisfactory over the upper Blue Nile basin in Ethiopia and over the Central Andes of Argentina [50,51]. The CHIRPS performs relatively well in India and China as compared to other products [6,52]. In our study, however, CHIRPS slightly underestimates the rainfall in both stations, but it has ample capability to count rainfall as compared to observed data. A very low bias, Mbias, Rbias, and high CC is found. The accuracy of CHIRPS is established in Pakistan at monthly and seasonal scales in different climate zones [44]. Another study of CHIRPS in Pakistan found similar results [19].

The PERSIANN-CDR consistently overestimates the rainfall as compared to other rainfall satellite products and in situ data [18,53–55]. It underestimates the rainfall on a daily scale and overestimates on a seasonal and annual scale. The inability of infrared-based retrievals to accurately predict rainfall in PERSIANN-CDR may be the cause of overestimation [56,57]. However, the accuracy of PERSIANN-CDR is established in drought and rainfall studies with low error and bias [58–60]. Growing number of literatures have examined the effectiveness of the PERSIANN-CDR daily rainfall product in recent years to capture rainfall counts in different parts of the world and estimated its effectiveness in terms of long-term capturing and performance for hydrological modeling [61–63]. In our study, the PERSIANN-CDR underestimates the rainfall at Lahore station and overestimated it at Faisalabad station. However, the accuracy of the PERSIANN-CDR is established with high CC, and with low bias and ME. Similar findings are reported in the literature in Pakistan [44].

Our analysis also demonstrates the authenticity of two gridded data sets, with CHIRPS showing better agreement with observatory rainfall than PERSIANN-CDR. Similar temporal trends and patterns are seen in the monthly, yearly, and seasonal distributions from 1983 to 2020 from the two observatories (Figure 7).

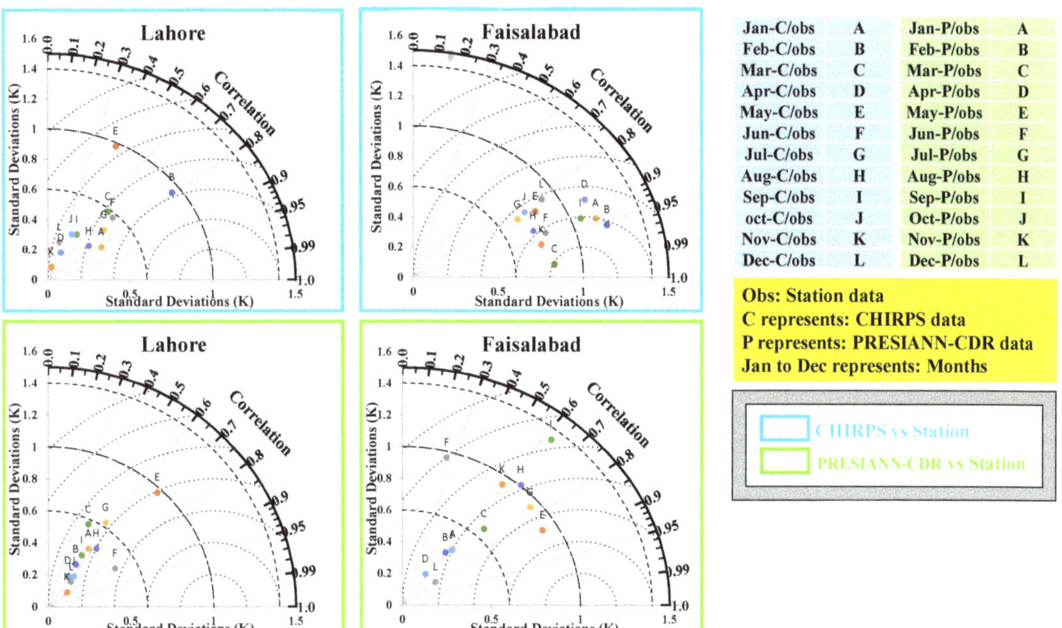

Figure 7. Taylor diagram exhibiting spatial association between station data and satellite data sets.

5.2. Variability and Trends in Rainfall

The comparison of the spatio-temporal distribution and trends of rainfall between the CHIRPS and PERSIANN-CDR (district-wise) datasets across central Punjab from 1983 to 2020 on a monthly, seasonal, and yearly scale was achieved. Both the datasets showed heavy rainfall in central Punjab during the monsoon season (70.5%), followed by pre-monsoon (15.2%) and winter (10.2%). The heavy monsoonal rainfall occurred due to uneven distribution of pressure gradients in the Pamir region and the Bay of Bengal, which brings sufficient moisture and rainfall throughout the months [11,16,64,65]. Additionally, the pre-monsoon and winter rainfall occurred due to the presence of western disturbances originating from the Mediterranean Sea and bringing more moisture towards southwest regions because of a highly intensified upper air westerly Jetstream [17,18,66]. Less rainfall during the post-monsoon season is a result of an early end of the monsoon season, but the convective systems control rainfall during that season [67,68]. The variability in connective systems can increase or decrease the rainfall in the post-monsoon season. Similar findings are reported in the literature [2,19,69]. District-wise investigation suggests that Sargodha and Faisalabad have experienced a statistically significant increasing trend in the monsoon season due to vast temperature fluctuations, land–ocean interactions, and sufficient moisture from the Arabian Sea and the Bay of Bengal.

Moreover, Narowal and Sialkot have witnessed a decreased tendency throughout the years, which can be attributed by the intense global atmospheric circulation patterns i.e., the El Niño–southern oscillation (ENSO), the Atlantic multi-decadal oscillation (AMO), inter-decadal Pacific oscillation (IPO), and urbanization and climate change [16,65,70]. Another study also found similar results in that region [70].

Additionally, anthropogenic activities, such as industrial growth, infrastructural activities, urbanization, and global increasing air temperature are also liable [38]. In Phase-2, annually, rainfall increased by 19%, which can be ascribed to the increasing tendency towards extreme events and huge temperature gradients, resulting in rapid land-ocean interactions [11,71,72].

5.3. Environmental and Socio-Economic Impacts of Rainfall Changes

Alterations in annual, seasonal, and monthly rainfall have had a critical impact on environment setup, particularly in rain-fed regions, such as central Punjab. Increased inconsistency in rainfall during the summer monsoon season, rising temperatures in arable parts, and severe water stress (such as floods and droughts) all agglomerate to have a negative impression on agricultural practices and productivity [73,74].

5.3.1. Effects of Decreasing Rainfall (Phase-I: 1983–2001)

Long-term trend analysis of central Punjab from 1983 to 2001 (Phase-1) witnessed declining rainfall (averaging 3.4 mm/decade), which resulted in a deterioration of agricultural productivity, averaging 6–18% [75]. Seasonal rainfall in this time phase also experienced a decreasing trend, averaging 0.09 mm/decade. This can be attributed to the El Niño–southern Oscillation (ENSO), the Atlantic multi-decadal oscillation (AMO), the inter-decadal Pacific oscillation (IPO), as well as urbanization and climate change [76–79].

Insufficient rainfall also resulted in less productivity of good quality cotton [80], which further caused fluctuations in the local economy [81]. Adequate rainfall during the wheat growing season also resulted from less production and growing, which troubled the country's food supply [82]. Massive reductions in livestock due to decreasing patterns of rainfall in central Punjab districts have been reported, which leads to poverty in rural areas [83]. Nevertheless, approximately 3 million people were impacted by the severe drought that occurred from 1998 to 2002, which also affected the national economy overall, and ecology and the agricultural industry in particular [84–86]. Recent studies have reported that the severe drought between 2000 and 2001 had an undesirable impact on the green sector's growth rate (-2.6%) as equated to subsequent years [87]. This declining pattern of rainfall also impacted the national water availability (a 51% reduction) and lessened river discharge from 162 billion m^3 to 109 billion m^3, which triggered a lowering GDP growth to 4.7% instead of the preliminary objective of 5.2% [88,89], as a significant proportion of population throughout the years directly or indirectly depended on water supply and allied primary activities. In rain-fed area, such as central districts of Punjab province, farm households are surrounded by uncertainty, which causes a reduction in livestock yield due to decreasing patterns of rainfall. It also increases the challenges of water availability and leads to poverty in rural areas [83]. More than 3 million people were impacted by the drought from 1998 to 2002, which also crippled the national economy [84]. Thus, it significantly affected the ecology and the agricultural industry [85,86]. In Pakistan, the severe drought between 2000 and 2001 harmed the green sector's growth rate (-2.6%), as compared to 2001 to 2002 [87]. Pakistan had a 51% increase in water shortage from the previous year, decreasing water flows in main rivers to 109 billion m^3 from 162 billion m^3, and lowering GDP growth to 4.7% instead of the preliminary objective of 5.2% [88,89]. As a result, a significant portion of the population's food security and way of life was seriously impacted, along with agricultural and animal productivity.

5.3.2. Effects of Increasing Rainfall (Phase-II: 2002–2020)

An annual increasing trend has been observed in Phase-2, according to CHIRPS data, by 10.01 mm/decade, whereas PERSIANN-CDR shows a relatively lower trend (7.94 mm/decade), which has adverse effects on the local economy, mainly in primary activities. This increased effect of rainfall meaningfully impacted on agricultural productivity in central Punjab [90]. Recent growing investigators have looked into the enlarged impacts of rainfall on harvest and yield variation and found a decreased tendency from 5% to 15% throughout September and October in rice yield in the districts of Lahore, Gujranwala, and Sheikhupura due to unnecessary heavy rainfall on mature crops [88,89]. Increased rainfall also resulted in floods in Punjab, mainly in the Dera Ghazi Khan, Rajanpur, and Rahim Yar Khan districts in 2010 and 2011, which had a massive impact on local primary production [91]. Contrary to this, a heavy increase in rainfall during the monsoon season (4.5 mm/decade in Phase-2) occasioned enormous infrastructure and property loss and also damaged about 150,000 tons of upright settled wheat crops.

6. Summary and Conclusions

The foremost intention of this study was to examine the spatio-temporal long-term phase-wise distributions and trends of annual, seasonal, and monthly rainfall in the central Punjab region, Pakistan, by using both observed and gridded (CHIRPS and PERSIANN-CDR) datasets. Several descriptive statistics and non-parametric statistical trend estimations were utilized to check the data performance and durable alterations by employing advanced web-based cloud platform computing (GEE) and MATLAB programming language. The key findings can be summarized as follows.

1. The present study initially dealt with the performance checking of remotely sensed gridded datasets (CHIRPS and PERSIANN-CDR) using station-observed data by employing numerous statistical indices and found CHIRPS has a higher spatio-temporal association with station-observed data ($r^2 = 0.76$) than PERSIANN-CDR ($r^2 = 0.63$), along with comparatively low bias and RMSE;

2. Long-term annual, seasonal, and monthly distribution of this study highlights maximum rainfall in Sialkot (904.24 mm), followed by the Narowal (860.36 mm) and Gujrat (840.99 mm) districts. In seasonal and monthly scales, monsoon season contributes 71.24% of rainfall, whereas highest rainfall observed in the month of July (26.79%). In addition, out of the 19 districts, the maximum annual statistically significant increasing trend was noticed in Gujrat (50.8 mm/decade), whereas seasonal dynamics show that the maximum increased during the monsoon season in Jhang (23.5 mm/decade). Furthermore, the maximum monthly change was observed in Gujrat district in the month of September (9.6 mm/decade);

3. The present study intensively investigated phase-wise long-term alteration in central Punjab and found a statistically decreasing trend in Phase-1 (3.5 mm/decade) and increasing in Phase-2 (7.5 mm/decade). Maximum seasonal changes were noticed during the monsoon season. Furthermore, the maximum statistically significant tendency was observed in July (3.3 mm/decade) and April (1.0 mm/decade) in Phase-2, whereas Phase-1 witnessed a statistically significant reduction in March (0.9 mm/decade);

4. This uneven nature of inter-annual long-term rainfall has had a crucial imprint on the local infrastructure and property as well as primary activities (mainly agriculture). Less rainfall in Phase-1 critically accelerated remarkable loss in agricultural productivity pf 4.7%, whereas the increased rainfall scenario in Phase-2 resulted in massive loss of mature standing crops of about 150,000 tons and damage to property due to unpredicted floods;

5. The foremost lacuna of this study is the availability of maximum ground rainfall records; however, only three observational stations were available as per the study period (1983–2020). That is why relying on remotely sensed data is identically significant in this case to exhibit overall spatial variations throughout the study domain. To achieve this aim, the applications and usefulness of a cloud computing system, such as GEE, is more efficient to gather a large quantity of datasets in a single platform with higher computational capacity.

Supplementary Materials: The following supporting information can be downloaded at: https://www.mdpi.com/article/10.3390/atmos14010060/s1, Figure S1: Annual distribution of rainfall of CHIRPS, PERSIANN-CDR and Observed datasets.; Figure S2: Seasonal distribution of rainfall of CHIRPS, PERSIANN-CDR and Observed datasets; Table S1: Distributions of mean annual, seasonal, and monthly rainfall in central Punjab districts (1983–2020); Table S2: District-wise annual, seasonal, and monthly trends of rainfall of gridded datasets (mm/decade, 1983–2020); Table S3: Phase-wise annual, seasonal, and monthly trends of rainfall in central Punjab (CHIRPS, PERSIANN-CDR).

Author Contributions: All the authors had different contribution in this research work and are mentioned here accordingly. Conceptualization, K.A. and A.B.; methodology, K.A.; software, Z.X. and M.A.; validation, K.A., S.K. and M.A.; formal analysis, K.A. and A.B.; investigation, A.B. and K.A.; resources, K.A.; data curation, K.A.; writing—original draft preparation, A.B. and K.A.; writing—review and editing, A.B., K.A. and W.R.; visualization, W.R.; supervision, A.B. and S.K.; project administration, S.K.; funding acquisition, K.A. All authors have read and agreed to the published version of the manuscript.

Funding: All the authors are equally contributed to pay the Article Processing Charge (APC) from their monthly stipend from the Shanghai Government Scholarship and Chinese Academy of Sciences.

Institutional Review Board Statement: Not applicable.

Informed Consent Statement: Not applicable.

Data Availability Statement: All the data are presented in the paper.

Acknowledgments: The authors affectionately thank the Pakistan Meteorological Department (PMD) for providing long-term station observation rainfall data. We are also thankful to the NOAA, NCDC, UCSB, and Climate Hazard Group for providing long-term high-resolution rainfall data.

Conflicts of Interest: The authors declare no conflict of interest.

References

1. Fortin, V.; Roy, G.; Stadnyk, T.; Koenig, K.; Gasset, N.; Mahidjiba, A. Ten years of science based on the Canadian precipitation analysis: A CaPA system overview and literature review. *Atmos.-Ocean* **2018**, *56*, 178–196. [CrossRef]
2. Ullah, S.; You, Q.; Ullah, W.; Ali, A. Observed changes in precipitation in China-Pakistan economic corridor during 1980–2016. *Atmos. Res.* **2018**, *210*, 1–14. [CrossRef]
3. Tian, L.; Yuan, S.; Quiring, S.M. Evaluation of six indices for monitoring agricultural drought in the south-central United States. *Agric. For. Meteorol.* **2018**, *249*, 107–119. [CrossRef]
4. Kishore, P.; Jyothi, S.; Basha, G.; Rao, S.; Rajeevan, M.; Velicogna, I.; Sutterley, T.C. Precipitation climatology over India: Validation with observations and reanalysis datasets and spatial trends. *Clim. Dyn.* **2016**, *46*, 541–556. [CrossRef]
5. Gaddam, V.K.; Kulkarni, A.V.; Gupta, A.K. Assessment of snow-glacier melt and rainfall contribution to stream runoff in Baspa Basin, Indian Himalaya. *Environ. Monit. Assess.* **2018**, *190*, 1–11. [CrossRef]
6. Banerjee, A.; Chen, R.; Meadows, M.E.; Singh, R.; Mal, S.; Sengupta, D. An analysis of long-term rainfall trends and variability in the uttarakhand himalaya using google earth engine. *Remote Sens.* **2020**, *12*, 709. [CrossRef]
7. Shukla, S.; McNally, A.; Husak, G.; Funk, C. A seasonal agricultural drought forecast system for food-insecure regions of East Africa. *Hydrol. Earth Syst. Sci.* **2014**, *18*, 3907–3921. [CrossRef]
8. Toté, C.; Patricio, D.; Boogaard, H.; Van der Wijngaart, R.; Tarnavsky, E.; Funk, C. Evaluation of satellite rainfall estimates for drought and flood monitoring in Mozambique. *Remote Sens.* **2015**, *7*, 1758–1776. [CrossRef]
9. Allen, S.K.; Rastner, P.; Arora, M.; Huggel, C.; Stoffel, M. Lake outburst and debris flow disaster at Kedarnath, June 2013: Hydrometeorological triggering and topographic predisposition. *Landslides* **2016**, *13*, 1479–1491. [CrossRef]
10. Kumar, V.; Jain, S.K. Trends in seasonal and annual rainfall and rainy days in Kashmir Valley in the last century. *Quat. Int.* **2010**, *212*, 64–69. [CrossRef]
11. Banerjee, A.; Chen, R.; Meadows, M.E.; Sengupta, D.; Pathak, S.; Xia, Z.; Mal, S. Tracking 21st century climate dynamics of the Third Pole: An analysis of topo-climate impacts on snow cover in the central Himalaya using Google Earth Engine. *Int. J. Appl. Earth Obs. Geoinf.* **2021**, *103*, 102490. [CrossRef]
12. Ren, Y.-Y.; Ren, G.-Y.; Sun, X.-B.; Shrestha, A.B.; You, Q.-L.; Zhan, Y.-J.; Rajbhandari, R.; Zhang, P.-F.; Wen, K.-M. Observed changes in surface air temperature and precipitation in the Hindu Kush Himalayan region over the last 100-plus years. *Adv. Clim. Chang. Res.* **2017**, *8*, 148–156. [CrossRef]
13. Zhang, Q.; Peng, J.; Singh, V.P.; Li, J.; Chen, Y.D. Spatio-temporal variations of precipitation in arid and semiarid regions of China: The Yellow River basin as a case study. *Glob. Planet. Chang.* **2014**, *114*, 38–49. [CrossRef]
14. Nawaz, Z.; Li, X.; Chen, Y.; Nawaz, N.; Gull, R.; Elnashar, A. Spatio-Temporal Assessment of Global Precipitation Products over the Largest Agriculture Region in Pakistan. *Remote Sens.* **2020**, *12*, 3650. [CrossRef]

15. Nawaz, Z.; Li, X.; Chen, Y.; Guo, Y.; Wang, X.; Nawaz, N. Temporal and spatial characteristics of precipitation and temperature in Punjab, Pakistan. *Water* **2019**, *11*, 1916. [CrossRef]
16. Asmat, U.; Athar, H. Run-based multi-model interannual variability assessment of precipitation and temperature over Pakistan using two IPCC AR4-based AOGCMs. *Theor. Appl. Climatol.* **2017**, *127*, 1–16. [CrossRef]
17. Dimri, A.; Niyogi, D.; Barros, A.; Ridley, J.; Mohanty, U.; Yasunari, T.; Sikka, D. Western disturbances: A review. *Rev. Geophys.* **2015**, *53*, 225–246. [CrossRef]
18. Iqbal, M.F.; Athar, H. Validation of satellite based precipitation over diverse topography of Pakistan. *Atmos. Res.* **2018**, *201*, 247–260. [CrossRef]
19. Nawaz, M.; Iqbal, M.F.; Mahmood, I. Validation of CHIRPS satellite-based precipitation dataset over Pakistan. *Atmos. Res.* **2021**, *248*, 105289. [CrossRef]
20. Arshad, M.; Ma, X.; Yin, J.; Ullah, W.; Liu, M.; Ullah, I. Performance evaluation of ERA-5, JRA-55, MERRA-2, and CFS-2 reanalysis datasets, over diverse climate regions of Pakistan. *Weather Clim. Extrem.* **2021**, *33*, 100373. [CrossRef]
21. Huffman, G.J.; Bolvin, D.T.; Nelkin, E.J.; Wolff, D.B.; Adler, R.F.; Gu, G.; Hong, Y.; Bowman, K.P.; Stocker, E.F. The TRMM multisatellite precipitation analysis (TMPA): Quasi-global, multiyear, combined-sensor precipitation estimates at fine scales. *J. Hydrometeorol.* **2007**, *8*, 38–55. [CrossRef]
22. Huffman, G.; Bolvin, D.; Braithwaite, D.; Hsu, K.; Joyce, R.; Kidd, C.; Nelkin, E.; Sorooshian, S.; Tan, J.; Xie, P. *NASA Global Precipitation Measurement (GPM) Integrated Multi-Satellite Retrievals for GPM (IMERG)*; Algorithm Theoretical Basis Document (ATBD), Version 06; National Aeronautics and Space Administration (NASA): Washington, DC, USA, 2019.
23. Joyce, R.J.; Janowiak, J.E.; Arkin, P.A.; Xie, P. CMORPH: A method that produces global precipitation estimates from passive microwave and infrared data at high spatial and temporal resolution. *J. Hydrometeorol.* **2004**, *5*, 487–503. [CrossRef]
24. Funk, C.; Peterson, P.; Landsfeld, M.; Pedreros, D.; Verdin, J.; Shukla, S.; Husak, G.; Rowland, J.; Harrison, L.; Hoell, A. The climate hazards infrared precipitation with stations—A new environmental record for monitoring extremes. *Sci. Data* **2015**, *2*, 150066. [CrossRef] [PubMed]
25. Beck, H.E.; Van Dijk, A.I.; Levizzani, V.; Schellekens, J.; Miralles, D.G.; Martens, B.; De Roo, A. MSWEP: 3-hourly 0.25 global gridded precipitation (1979–2015) by merging gauge, satellite, and reanalysis data. *Hydrol. Earth Syst. Sci.* **2017**, *21*, 589–615. [CrossRef]
26. Ashouri, H.; Hsu, K.-L.; Sorooshian, S.; Braithwaite, D.K.; Knapp, K.R.; Cecil, L.D.; Nelson, B.R.; Prat, O.P. PERSIANN-CDR: Daily precipitation climate data record from multisatellite observations for hydrological and climate studies. *Bull. Am. Meteorol. Soc.* **2015**, *96*, 69–83. [CrossRef]
27. Yatagai, A.; Kamiguchi, K.; Arakawa, O.; Hamada, A.; Yasutomi, N.; Kitoh, A. APHRODITE: Constructing a long-term daily gridded precipitation dataset for Asia based on a dense network of rain gauges. *Bull. Am. Meteorol. Soc.* **2012**, *93*, 1401–1415. [CrossRef]
28. Badar, H.; Ghafoor, A.; Adil, S.A. Factors affecting agricultural production of Punjab (Pakistan). *Pak. J. Agri. Sci.* **2007**, *44*, 506–510.
29. Statistics, S. *Statistical Yearbook*; Statistics: Copenhagen, Denmark, 2012.
30. Abid, M.; Ashfaq, M.; Khalid, I.; Ishaq, U. An economic evaluation of impact of soil quality on Bt (Bacillus thuringiensis) cotton productivity. *Soil Env.* **2011**, *30*, 78–81.
31. Shahid, M.A.; Boccardo, P.; García, W.C.; Albanese, A.; Cristofori, E. Evaluation of TRMM satellite data for mapping monthly precipitation in Pakistan by comparison with locally available data. In *Proceedings of the III CUCS Congress-Imagining Cultures of Cooperation*; Universities Working to Face the New Developemnt Challenges: Torino, Italy, 2013.
32. Cheema, S.; Hanif, M. Seasonal precipitation variation over Punjab province. *Pak. J. Meteorol.* **2013**, *10*, 61–82.
33. Bokhari, S.; Rasul, G.; Ruane, A.; Hoogenboom, G.; Ahmad, A. The past and future changes in climate of the rice-wheat cropping zone in Punjab, Pakistan. *Pak. J. Meteorol.* **2017**, *13*, 9–23.
34. Mubeen, M.; Ahmad, A.; Wajid, A.; Khaliq, T.; Bakhsh, A. Evaluating CSM-CERES-Maize Model for Irrigation Scheduling in Semi-arid Conditions of Punjab, Pakistan. *Int. J. Agric. Biol.* **2013**, *15*, 1–10.
35. Nasim, W.; Belhouchette, H.; Tariq, M.; Fahad, S.; Hammad, H.M.; Mubeen, M.; Munis, M.F.H.; Chaudhary, H.J.; Khan, I.; Mahmood, F. Correlation studies on nitrogen for sunflower crop across the agroclimatic variability. *Environ. Sci. Pollut. Res.* **2016**, *23*, 3658–3670. [CrossRef]
36. Iqbal, M.A.; Penas, A.; Cano-Ortiz, A.; Kersebaum, K.C.; Herrero, L.; del Río, S. Analysis of recent changes in maximum and minimum temperatures in Pakistan. *Atmos. Res.* **2016**, *168*, 234–249. [CrossRef]
37. Nasim, W.; Ahmad, A.; Belhouchette, H.; Fahad, S.; Hoogenboom, G. Evaluation of the OILCROP-SUN model for sunflower hybrids under different agro-meteorological conditions of Punjab—Pakistan. *Field Crops Res.* **2016**, *188*, 17–30. [CrossRef]
38. Nawaz, Z.; Li, X.; Chen, Y.; Wang, X.; Zhang, K.; Nawaz, N.; Guo, Y.; Meerzhan, A. Spatiotemporal assessment of temperature data products for the detection of warming trends and abrupt transitions over the largest irrigated area of Pakistan. *Adv. Meteorol.* **2020**, *2020*, 3584030. [CrossRef]
39. Nguyen, P.; Shearer, E.J.; Tran, H.; Ombadi, M.; Hayatbini, N.; Palacios, T.; Huynh, P.; Braithwaite, D.; Updegraff, G.; Hsu, K. The CHRS Data Portal, an easily accessible public repository for PERSIANN global satellite precipitation data. *Sci. Data* **2019**, *6*, 180296. [CrossRef]
40. Rasul, G. *Climate Data and Modelling Analysis of the Indus Region*; World Wide Fund for Nature: Gland, Switzerland, 2012.

41. Amin, A.; Nasim, W.; Mubeen, M.; Kazmi, D.H.; Lin, Z.; Wahid, A.; Sultana, S.R.; Gibbs, J.; Fahad, S. Comparison of future and base precipitation anomalies by SimCLIM statistical projection through ensemble approach in Pakistan. *Atmos. Res.* **2017**, *194*, 214–225. [CrossRef]

42. Ahmed, K.; Shahid, S.; Ismail, T.; Nawaz, N.; Wang, X.-J. Absolute homogeneity assessment of precipitation time series in an arid region of Pakistan. *Atmósfera* **2018**, *31*, 301–316. [CrossRef]

43. Ahmed, K.; Shahid, S.; Nawaz, N. Impacts of climate variability and change on seasonal drought characteristics of Pakistan. *Atmos. Res.* **2018**, *214*, 364–374. [CrossRef]

44. Ullah, W.; Wang, G.; Ali, G.; Tawia Hagan, D.F.; Bhatti, A.S.; Lou, D. Comparing multiple precipitation products against in-situ observations over different climate regions of Pakistan. *Remote Sens.* **2019**, *11*, 628. [CrossRef]

45. Hussain, Y.; Satgé, F.; Hussain, M.B.; Martinez-Carvajal, H.; Bonnet, M.-P.; Cárdenas-Soto, M.; Roig, H.L.; Akhter, G. Performance of CMORPH, TMPA, and PERSIANN rainfall datasets over plain, mountainous, and glacial regions of Pakistan. *Theor. Appl. Climatol.* **2018**, *131*, 1119–1132. [CrossRef]

46. Paredes Trejo, F.J.; Alves Barbosa, H.; Peñaloza-Murillo, M.A.; Moreno, M.A.; Farias, A. Intercomparison of improved satellite rainfall estimation with CHIRPS gridded product and rain gauge data over Venezuela. *Atmósfera* **2016**, *29*, 323–342. [CrossRef]

47. Le, A.M.; Pricope, N.G. Increasing the accuracy of runoff and streamflow simulation in the Nzoia Basin, Western Kenya, through the incorporation of satellite-derived CHIRPS data. *Water* **2017**, *9*, 114. [CrossRef]

48. Dembélé, M.; Zwart, S.J. Evaluation and comparison of satellite-based rainfall products in Burkina Faso, West Africa. *Int. J. Remote Sens.* **2016**, *37*, 3995–4014. [CrossRef]

49. Prakash, S. Performance assessment of CHIRPS, MSWEP, SM2RAIN-CCI, and TMPA precipitation products across India. *J. Hydrol.* **2019**, *571*, 50–59. [CrossRef]

50. Ayehu, G.T.; Tadesse, T.; Gessesse, B.; Dinku, T. Validation of new satellite rainfall products over the Upper Blue Nile Basin, Ethiopia. *Atmos. Meas. Tech.* **2018**, *11*, 1921–1936. [CrossRef]

51. Rivera, J.A.; Marianetti, G.; Hinrichs, S. Validation of CHIRPS precipitation dataset along the Central Andes of Argentina. *Atmos. Res.* **2018**, *213*, 437–449. [CrossRef]

52. Bai, L.; Shi, C.; Li, L.; Yang, Y.; Wu, J. Accuracy of CHIRPS satellite-rainfall products over mainland China. *Remote Sens.* **2018**, *10*, 362. [CrossRef]

53. Katiraie-Boroujerdy, P.-S.; Asanjan, A.A.; Hsu, K.-l.; Sorooshian, S. Intercomparison of PERSIANN-CDR and TRMM-3B42V7 precipitation estimates at monthly and daily time scales. *Atmos. Res.* **2017**, *193*, 36–49. [CrossRef]

54. Shrestha, N.K.; Qamer, F.M.; Pedreros, D.; Murthy, M.; Wahid, S.M.; Shrestha, M. Evaluating the accuracy of Climate Hazard Group (CHG) satellite rainfall estimates for precipitation based drought monitoring in Koshi basin, Nepal. *J. Hydrol. Reg. Stud.* **2017**, *13*, 138–151. [CrossRef]

55. Shangguan, W.; Dai, Y.; Liu, B.; Zhu, A.; Duan, Q.; Wu, L.; Ji, D.; Ye, A.; Yuan, H.; Zhang, Q. A China data set of soil properties for land surface modeling. *J. Adv. Model. Earth Syst.* **2013**, *5*, 212–224. [CrossRef]

56. Li, X.; Zhang, Q.; Xu, C.-Y. Assessing the performance of satellite-based precipitation products and its dependence on topography over Poyang Lake basin. *Theor. Appl. Climatol.* **2014**, *115*, 713–729. [CrossRef]

57. Hirpa, F.A.; Salamon, P.; Alfieri, L.; Thielen-del Pozo, J.; Zsoter, E.; Pappenberger, F. The effect of reference climatology on global flood forecasting. *J. Hydrometeorol.* **2016**, *17*, 1131–1145. [CrossRef]

58. Lashkari, A.; Salehnia, N.; Asadi, S.; Paymard, P.; Zare, H.; Bannayan, M. Evaluation of different gridded rainfall datasets for rainfed wheat yield prediction in an arid environment. *Int. J. Biometeorol.* **2018**, *62*, 1543–1556. [CrossRef]

59. Bhardwaj, A.; Ziegler, A.D.; Wasson, R.J.; Chow, W.T. Accuracy of rainfall estimates at high altitude in the Garhwal Himalaya (India): A comparison of secondary precipitation products and station rainfall measurements. *Atmos. Res.* **2017**, *188*, 30–38. [CrossRef]

60. Santos, C.A.G.; Neto, R.M.B.; do Nascimento, T.V.M.; da Silva, R.M.; Mishra, M.; Frade, T.G. Geospatial drought severity analysis based on PERSIANN-CDR-estimated rainfall data for Odisha state in India (1983–2018). *Sci. Total Environ.* **2021**, *750*, 141258. [CrossRef]

61. Ashouri, H.; Nguyen, P.; Thorstensen, A.; Hsu, K.-l.; Sorooshian, S.; Braithwaite, D. Assessing the efficacy of high-resolution satellite-based PERSIANN-CDR precipitation product in simulating streamflow. *J. Hydrometeorol.* **2016**, *17*, 2061–2076. [CrossRef]

62. Miao, C.; Ashouri, H.; Hsu, K.-L.; Sorooshian, S.; Duan, Q. Evaluation of the PERSIANN-CDR daily rainfall estimates in capturing the behavior of extreme precipitation events over China. *J. Hydrometeorol.* **2015**, *16*, 1387–1396. [CrossRef]

63. Sun, S.; Zhou, S.; Shen, H.; Chai, R.; Chen, H.; Liu, Y.; Shi, W.; Wang, J.; Wang, G.; Zhou, Y. Dissecting performances of PERSIANN-CDR precipitation product over Huai River Basin, China. *Remote Sens.* **2019**, *11*, 1805. [CrossRef]

64. Ding, Y. The variability of the Asian summer monsoon. *J. Meteorol. Soc. Jpn. Ser. II* **2007**, *85*, 21–54. [CrossRef]

65. Latif, M.; Syed, F.; Hannachi, A. Rainfall trends in the South Asian summer monsoon and its related large-scale dynamics with focus over Pakistan. *Clim. Dyn.* **2017**, *48*, 3565–3581. [CrossRef]

66. Iqbal, M.F.; Athar, H. Variability, trends, and teleconnections of observed precipitation over Pakistan. *Theor. Appl. Climatol.* **2018**, *134*, 613–632. [CrossRef]

67. Guhathakurta, P.; Rajeevan, M. Trends in the rainfall pattern over India. *Int. J. Climatol. A J. R. Meteorol. Soc.* **2008**, *28*, 1453–1469. [CrossRef]

68. Hussain, S. Gladiolus production a successful example in the climate of Khanaspur, Ayobia district Hazara, NWF (Province) Pakistan. *J. Nat. Geogr. Soc.* **2008**, *42*, 177–181.
69. Hanif, M.; Khan, A.H.; Adnan, S. Latitudinal precipitation characteristics and trends in Pakistan. *J. Hydrol.* **2013**, *492*, 266–272. [CrossRef]
70. Ali, N.; Ahmad, I.; Chaudhry, A.; Raza, M. Trend analysis of precipitation data in Pakistan. *Sci. Int.* **2015**, *27*, 803–808.
71. Syed, F.; Giorgi, F.; Pal, J.; King, M. Effect of remote forcings on the winter precipitation of central southwest Asia part 1: Observations. *Theor. Appl. Climatol.* **2006**, *86*, 147–160. [CrossRef]
72. Baig, M.H.A.; Rasul, G. The effect of Eurasian snow cover on the Monsoon rainfall of Pakistan. *Pak. J. Metcorol* **2009**, *5*, 1–11.
73. Chandio, A.A.; Magsi, H.; Ozturk, I. Examining the effects of climate change on rice production: Case study of Pakistan. *Environ. Sci. Pollut. Res.* **2020**, *27*, 7812–7822. [CrossRef]
74. Abbas, S.; Kousar, S.; Khan, M.S. The role of climate change in food security; empirical evidence over Punjab regions, Pakistan. *Environ. Sci. Pollut. Res.* **2022**, *29*, 53718–53736. [CrossRef]
75. Ullah, S. Climate change impact on agriculture of Pakistan-A leading agent to food security. *Int. J. Environ. Sci. Nat. Resour.* **2017**, *6*, 76–79.
76. Kripalani, R.; Oh, J.; Kulkarni, A.; Sabade, S.; Chaudhari, H. South Asian summer monsoon precipitation variability: Coupled climate model simulations and projections under IPCC AR4. *Theor. Appl. Climatol.* **2007**, *90*, 133–159. [CrossRef]
77. Priya, P.; Krishnan, R.; Mujumdar, M.; Houze, R.A. Changing monsoon and midlatitude circulation interactions over the Western Himalayas and possible links to occurrences of extreme precipitation. *Clim. Dyn.* **2017**, *49*, 2351–2364. [CrossRef]
78. Wang, B.; Liu, J.; Kim, H.-J.; Webster, P.J.; Yim, S.-Y. Recent change of the global monsoon precipitation (1979–2008). *Clim. Dyn.* **2012**, *39*, 1123–1135. [CrossRef]
79. Iqbal, Z.; Shahid, S.; Ahmed, K.; Ismail, T.; Nawaz, N. Spatial distribution of the trends in precipitation and precipitation extremes in the sub-Himalayan region of Pakistan. *Theor. Appl. Climatol.* **2019**, *137*, 2755–2769. [CrossRef]
80. Abbas, Q.; Ahmad, S. Effect of Different Sowing Times and Cultivars on Cotton Fiber Quality under Stable Cotton-Wheat Cropping System in Southern Punjab, Pakistan. *Pak. J. Life Soc. Sci.* **2018**, *16*, 77–84.
81. Arshad, A.; Raza, M.A.; Zhang, Y.; Zhang, L.; Wang, X.; Ahmed, M.; Habib-ur-Rehman, M. Impact of climate warming on cotton growth and yields in China and Pakistan: A regional perspective. *Agriculture* **2021**, *11*, 97. [CrossRef]
82. Ashfaq, M.; Zulfiqar, F.; Sarwar, I.; Quddus, M.A.; Baig, I.A. Impact of climate change on wheat productivity in mixed cropping system of Punjab. *Soil Environ.* **2011**, *30*, 110–114.
83. Abid, M.; Schilling, J.; Scheffran, J.; Zulfiqar, F. Climate change vulnerability, adaptation and risk perceptions at farm level in Punjab, Pakistan. *Sci. Total Environ.* **2016**, *547*, 447–460. [CrossRef]
84. Portal, P.W. History of drought in Pakistan–in detail. 2011.
85. Barlow, M.; Zaitchik, B.; Paz, S.; Black, E.; Evans, J.; Hoell, A. A review of drought in the Middle East and southwest Asia. *J. Clim.* **2016**, *29*, 8547–8574. [CrossRef]
86. Hina, S.; Saleem, F. Historical analysis (1981–2017) of drought severity and magnitude over a predominantly arid region of Pakistan. *Clim. Res.* **2019**, *78*, 189–204. [CrossRef]
87. Dahal, P.; Shrestha, N.S.; Shrestha, M.L.; Krakauer, N.Y.; Panthi, J.; Pradhanang, S.M.; Jha, A.; Lakhankar, T. Drought risk assessment in central Nepal: Temporal and spatial analysis. *Nat. Hazards* **2016**, *80*, 1913–1932. [CrossRef]
88. Ahmad, S.; Hussain, Z.; Qureshi, A.S.; Majeed, R.; Saleem, M. *Drought Mitigation in Pakistan: Current Status and Options for Future Strategies*; IWMI: Colombo, Sri Lanka, 2004; Volume 85.
89. Miyan, M.A. Droughts in Asian Least Developed Countries: Vulnerability and sustainability. *Weather Clim. Extrem.* **2015**, *7*, 8–23. [CrossRef]
90. Hussain, M.S.; Lee, S. Long-term variability and changes of the precipitation regime in Pakistan. *Asia-Pac. J. Atmos. Sci.* **2014**, *50*, 271–282. [CrossRef]
91. Tariq, M.A.U.R.; Van De Giesen, N. Floods and flood management in Pakistan. *Phys. Chem. Earth Parts A/B/C* **2012**, *47*, 11–20. [CrossRef]

Article

The Consistent Variations of Precipitable Water and Surface Water Vapor Pressure at Interannual and Long-Term Scales: An Examination Using Reanalysis

Jiawei Hao and Er Lu *

Key Laboratory of Meteorological Disaster, Ministry of Education, Joint International Research Laboratory of Climate and Environment Change, Collaborative Innovation Center on Forecast and Evaluation of Meteorological Disasters, Nanjing University of Information Science and Technology, Nanjing 221000, China
* Correspondence: elu@nuist.edu.cn

Abstract: Water vapor (WV) is a vital basis of water and energy cycles and varies with space and time. When researching the variations of moisture in the atmosphere, it is intuitive to think about the total WV of the atmosphere column, precipitable water (PW). It is an element that needs high-altitude observations. A surface quantity, surface WV pressure (SVP), has a close relationship to PW because of the internal physical linkage between them. The stability of their linkage at climatic scales is verified using monthly mean data from 1979 to 2021, while studies before mainly focused on daily and annual cycles in local areas. The consistency of their variations is checked with three reanalysis datasets from three angles, the interannual variations, the long-term trends, and the empirical orthogonal function (EOF) modes. Results show that the interannual correlation of SVP and PW can reach a level that is quite high and are significant in most areas, and the weak correlation mainly exists over low-latitude oceans. The long-term trends, as well as the first EOF modes of these two quantities, also show that their variations are consistent, with spatial correlation coefficients between the long-term trends of two variables that are generally over 0.6, but specific differences appearing in some regions including the Tropical Indian Ocean and Middle Africa. With the correspondence of PW and SVP, the variations of total column WV can be indicated by surface elements. The correspondence is also meaningful for the analysis of the co-variation in total column vapor and temperature. For example, we could research the relations between SVP and air temperature, and they can reflect the co-variance of total column vapor and near-surface air temperature, which can avoid analyzing the relation between column-integrated moisture content and surface air temperature directly.

Keywords: precipitable water; surface water vapor pressure; consistency; interannual and long-term trend

1. Introduction

As a component in the global water and energy cycles, atmospheric WV is also an important part of climate variability and change [1–3]. WV is the material basis of precipitation, and the latent heat released by precipitation is a part of the energy cycle and can drive atmospheric circulation anomalies [4,5]. The circulation may, in turn, transport WV, and change the spatiotemporal distribution of the WV [6]. As the most abundant greenhouse gas in the atmosphere, WV also affects the atmospheric radiation processes [7].

PW, a conventional quantity of humidity, is the column-integrated WV amount in the atmosphere [8]. In a regional air column, the WV evaporated from the surface, along with the WV converged from the surroundings, is used to increase the PW. When the air saturates, at least at certain levels, precipitation may be produced. In the past, acquirement of PW data nearly always relied on radiosonde observations [9,10]. Because of the high costs of radiosonde and the sparsity of observation stations, studies about empirical equations between PW and surface elements appeared. A surface quantity, SVP, has been stressed that has a close statistical linkage to PW [11–14]. This linkage can also be explained from a physical aspect. In essence, SVP can be regarded as the force exerted by the total moisture in the air column [15], in analogy with the surface pressure to the total mass of the air column. From another point of view, WV content is maximal at the near-surface level and decreases with altitude because of the existence of gravity. Hence, the SVP can well represent the column WV content [16,17]. Due to the lack of upper-air detections, the empirical relations between PW and SVP were utilized to estimate the PW with surface observations in local areas, especially during the early stage of this methodology when there were just a few sounding observations [18–21]. Some advanced methods can currently be used to obtain PW data, such as ground-based radar observations, satellite retrievals [22,23], global position system (GPS) observations [24,25], model outputs, and the reanalysis [26,27]. Because of the high spatial and temporal resolution of remote sensing products, which could be 1 h and 30 km, they can help to fill the gaps between radiosonde locations and launch times. However, the retrieval effects except for microwave-based products are often not very useful in cloudy areas [22,23]. The ground-based GPS data have a higher temporal resolution, which is about 5 min [25], but the network is not so dense, and the duration is relatively short. Except for a few elements such as precipitation, the reanalysis datasets universally have decent accuracy in broad areas, with the support of the accumulations of vast amounts of historical observations, advanced modeling, and data assimilation systems. They have been applied in studies on the hydrological fields and their effects have been verified [8].

Because of the appearance of these measuring methods, perhaps the demands of calculating PW with surface elements are not as imperious as earlier. However, analyzing the relationship between them is still meaningful. It is easy to understand that over areas with few instructions, such as oceans, the SVP can be calculated with PW retrieved by satellite remote sensing.

Numerous studies have proven the quasi-linear relation between SVP and PW from the statistical aspect, and there are empirical equations in local areas [19,21,28,29]. Due to the appearance of progressive observation methods such as satellite remote sensing, the PW data can be used to estimate the values of SVP in reverse in the areas where it is hard to establish observation stations [20]. Previous studies are mainly concentrated on daily and annual cycles in local areas when analyzing the relationship between PW and SVP. Their relationship at interannual scales was researched in a few studies, but most of them still focused on the linkage in local areas. In daily or seasonal variations, the radiation processes vary dramatically, so following the temperature, PW, and SVP both have obvious variations, and it may be easy to vary consistently for them. When analyzing the interannual or long-term variations, the amplitudes may be much smaller. The questions researched here are whether the consistency is reliable at climatical scales, and where the consistency is not good. Hence, the consistency of interannual and long-term variations of monthly mean PW and SVP are examined globally.

Under global warming, atmospheric moisture is often linked to tropospheric temperature or surface air temperature through Clausius–Clapeyron Equation [30–34]. Many researchers studied the interannual and long-term variations of WV as well as its relationship with temperature, using PW or SVP [3,35–38]. Globally, the WV follows the temperature closely in line with the Clausius–Clapeyron Equation. However, it is more exhausted when researching the variation in local areas [39]. The PW in some areas has decreasing trends, such as in Australia from 1957–2016 [37]. As for the interannual variation in WV, Hao and Lu [39] studied the response of SVP to surface air temperature and found that the response of SVP to temperature is in close line with the Clausius–Clapeyron Equation at high latitudes. Shi et al. [40] studied the relationship between upper troposphere humidity and PW. Some studies also researched the linkage between PW and surface air temperature and re-established the PW during periods that lack high-altitude observations [41,42]. Wang et al. [8] found that there is a close linkage between the first EOF mode of PW and the Niño-3.4 index. As introduced before, the surface quantity which can directly reflect the PW is SVP. Because of the physical relation between SVP and surface air temperature, there is also a tight linkage between PW and the latter.

The long-term and interannual variations of total WV in the atmosphere can be reflected by SVP if the variations of the two quantities are consistent. When studying the response of WV to temperature, it is more suitable to use SVP than PW because the quantities that are all at the surface level can be used. It can avoid the problems brought by linking the column-integrated moisture with surface-level temperature or integrating the intensive quantities. With the linkage between PW and SVP, the response of SVP to the temperature studied in [39] can be extended to that of PW to a certain degree, while the response of SVP to surface temperature is studied in another part of our work.

Consulting the internal physical relation between PW and SVP, their linkage may be stable when the atmosphere is in hydrostatic equilibrium because the vertical movements can break the direct relationship between mass and pressure of the fluid. In the areas where the convection is active, the monthly averages of vertical motions are strong and precipitation amounts are large. The climatic value of PW and SVP are both larger than in other areas, but the linear correlation between monthly mean PW and SVP are probably weaker. Hence, in some areas, it could be hard to establish an efficient empirical equation between PW and SVP, and these areas could be found in the study.

The linkage between PW and SVP is analyzed from three angles concurrently. The interannual correlation and long-term trends of them are examined globally. Different from the daily or annual cycles, the interannual and long-term variations of these two physical variables may be much smaller. Hence, whether the linkage between them is still stable and where their linkage is not so good when analyzing their interannual and long-term variations are the questions that we tried to answer. To further verify the consistency of their variations, the first EOF modes of PW and SVP are shown and compared as well. To further verify the credibility of the results, three reanalysis datasets from different institutions are used. Due achieve this, we may mainly focus on the common points among the three datasets in the analysis.

The datasets together with the methods adopted are presented in Section 2. The consistency of interannual variations and long-term trends of PW and SVP are shown in Sections 3.1 and 3.2. The EOF leading modes and time series are compared in Section 3.3. Section 4 shows the summary and discussion.

2. Data and Method

2.1. Data

Except for a few elements such as precipitation, the reanalysis datasets universally own decent accuracy in broad areas, with the support of the accumulations of vast amounts of historical observations, advanced modeling, and data assimilation systems. They have been applied in studies on the hydrological fields and their effects have been verified [8]. With the aim of the global scale study and ensuring the credibility of the results, three reanalysis datasets from different institutions are used, the European Centre for Medium-Range Weather Forecasts (ECMWF) Reanalysis v5 (ERA5), Japanese 55-year Reanalysis (JRA55), and the National Centers for Environmental Prediction and the Department of Energy (NCEP-DOE) Reanalysis-2 (NCEP2).

The ERA5 [43], provided by the ECMWF, is the fifth-generation ECMWF atmospheric reanalysis of the global climate covering the period from January 1950 to the present. ERA5 is produced by the Copernicus Climate Change Service (C3S) at ECMWF. The data cover the Earth on a 30 km grid and resolve the atmosphere using 137 levels from the surface up to a height of 80 km. The NCEP2 project performs data assimilation using past data from 1979 through the previous year [44].

Spanning from 1958 to present, JRA-55 is the longest third-generation reanalysis that uses the full observing system. Compared to the previous generation Japanese Meteorological Agency (JMA) reanalysis (JRA-25), JRA-55 uses a more advanced data assimilation scheme, increased model resolution, a new variational bias correction for satellite data, and several additional observational data sources [45]. Horizontal resolutions of the three datasets are all 2.5° × 2.5° in latitude and longitude.

The quantities used include the monthly mean PW, the 2 m temperature, and the 2 m dewpoint temperature from 1979 to 2021, which are directly downloaded from the servers. The SVP and saturated SVP are calculated with the modified Tetens formula [46], then the relative humidity can be obtained. As for the quantities at pressure levels, the moisture-related quantity directly given by ERA5 and JRA55 is specific humidity and that given by NCEP2 is relative humidity. Utilizing one of them with temperature which can be downloaded from the server, the other one can be figured out. The specific processes are illustrated below. With the temperature data, the saturated WV pressure can be calculated firstly as mentioned before. Then, with the relative humidity, the WV pressure can be archived. Finally, the specific humidity q can be obtained with the approximate empirical equation $q = \varepsilon(e/p)$, where e is WV pressure and p is the pressure. The ratio of dry and wet gas constant ε is taken as 0.622 here. To analyze the profiles, the vertical velocity at pressure levels is also used and directly downloaded. There are 12 isobaric levels from 1000 hPa to 100 hPa in all three datasets used in the paper. The specific levels can be seen in the plots that show vertical profiles. Because the surface levels of the chosen land areas are general over 1000 hPa, the lowest level shown in land areas is 925 hPa. To intuitively show the variables downloaded and calculated, they are listed in Table 1.

Table 1. The variables directly downloaded from the servers. The variables that need calculation are in brackets, and the variable in each pair of brackets is calculated with the corresponding variable out of the brackets.

Datasets	Single Level Variables	Pressure Level Variables	
ERA5	2 m temperature (saturated SVP)	vertical velocity temperature (saturated SVP)	specific humidity
JRA55	2 m dewpoint temperature (SVP)		
NCEP2	precipitable water		relative humidity (specific humidity)

2.2. Method

2.2.1. Mann–Kendall Test and Sen's Slope Estimator

There are some ways to estimate the trends of hydrological and meteorological elements, including parametric and non-parametric methods [47]. Analysis of long-term trends includes the confirmation of increasing or decreasing slopes and significance testing [48]. Mann–Kendall test was formulated by Mann [49] as a non-parametric test for trend detection, and the test statistic distribution was given by Kendall [50] for testing non-linear trends and turning points. It is an excellent non-parametric method and is preferred by many researchers [51]. It is used for trend analysis, as it eliminates the effect of serial dependence on auto-correlated data which modifies the variance of datasets [52]. The MK test is used to test the significance of long-term trends of PW and SVP.

Sen's slope estimation, also a non-parametric method, gives the magnitude of the trend [53]. Another advantage of Sen's slope is that it is not affected by outliers and single data errors in the dataset [54].

The slopes of the long-term trends are calculated with the Theil–Sen slope estimator, and it is also a non-parametric method [53]. In this model, the slope values reflect the increasing and decreasing magnitudes of variables. The long-term trends of PW and SVP at each grid are calculated with this method to examine their consistency.

2.2.2. Empirical Orthogonal Function (EOF) Analysis

In climate research, EOF analysis is often used to study possible spatial modes of variability and how they change with time. In statistics, EOF analysis is known as Principal Component Analysis (PCA) [55]. As such, EOF analysis is sometimes classified as a multivariate statistical technique. A field is partitioned into mathematically orthogonal modes which can be called EOF spatial modes, or patterns. Typically, the EOFs are found by computing the eigenvalues and eigenvectors of a spatially weighted anomaly co-variance matrix of a field. The derived eigenvectors just vary with spatial modes. Each of the eigenvectors is an EOF pattern. The derived eigenvalues provide a measure of the percent variance explained by each mode. The time series (principal components) of each spatial pattern are determined by projecting the derived eigenvectors onto the spatially weighted anomalies. This will result in the amplitude of each mode over the period of record. Each pair of spatial patterns and time series can represent the spatial-temporal varying features of the variable.

By construction, the EOF patterns and the principal components are independent. Two factors inhibit physical interpretation of EOFs. One is the orthogonality constraint and the other is that the derived patterns may be domain-dependent. Physical systems are not necessarily orthogonal and if the patterns depend on the region used, they may not exist if the domain changes. Still, even with these shortcomings, classical EOF analysis has proved to be useful.

To make sure the EOF modes of PW and SVP are distinguished from the noise signal, the Monte Carlo technique [56] is used to test the significance of the first three modes.

2.2.3. The Theoretical Linkage between PW and SVP

PW is the total WV amount in the atmosphere column and can be expressed as the integration of the WV from the surface to the tropopause. Under the condition of hydrostatic equilibrium, the equation can be transferred to the isobaric coordinate form. The σ-coordinate $\sigma \equiv p/p_s$ is employed and specific humidity is expressed with WV pressure as $q = \varepsilon(e/p)$. As shown by Lu [17], the PW can be written as

$$W = \frac{\varepsilon}{g} \int_0^1 e\, d\ln\sigma \tag{1}$$

In equations, the PW is denoted as W. The vapor pressure e normally reaches its maximum at surface level, then decreases with the altitude [57]. Vapor pressure can be seen as a function of σ approximately and expressed as $e = E\sigma^m$. In the formula, SVP is denoted as E, and m is a constant in a specific area. Then, the relationship between PW and SVP can be expressed as $W = k \cdot E$, and k is expressed as $k = \varepsilon/(gm)$ in the formula, where g is gravitational acceleration and m is the constant described before. As a result, W has a close linkage with surface vapor pressure E.

From another point of view, SVP is produced by the gravity of total WV in the air column and their linkage is the same as the relationship between surface pressure and the total quality of the column air mass. Hence, if the atmosphere is in hydrostatic equilibrium, there will be a linear linkage between PW and SVP.

3. Results

3.1. The Consistency of Interannual Variations

Figure 1 shows the maps of interannual correlation coefficients of monthly mean PW and SVP calculated with the ERA5 dataset. (The long-term trends are removed before calculating the correlation coefficient. Results obtained with JRA55 and NCEP2 are shown in Figures S1 and S2 in the Supplementary Materials). Three datasets all show that the coefficients in most areas are positive and can pass the 0.05 significance level, except in July. Just a few areas cannot pass the significance test and they are almost all in low-latitude regions. As illustrated before, the stratification of the tropical atmosphere is less stable, and the more frequent convection there probably breaks the linear relationship between PW and SVP. The linear correlation is especially strong in some areas, such as the Antarctic, Australia, Eurasian Continents, and the Southern Ocean. The distributions of strong correlation areas (SCAs) are similar among the three datasets.

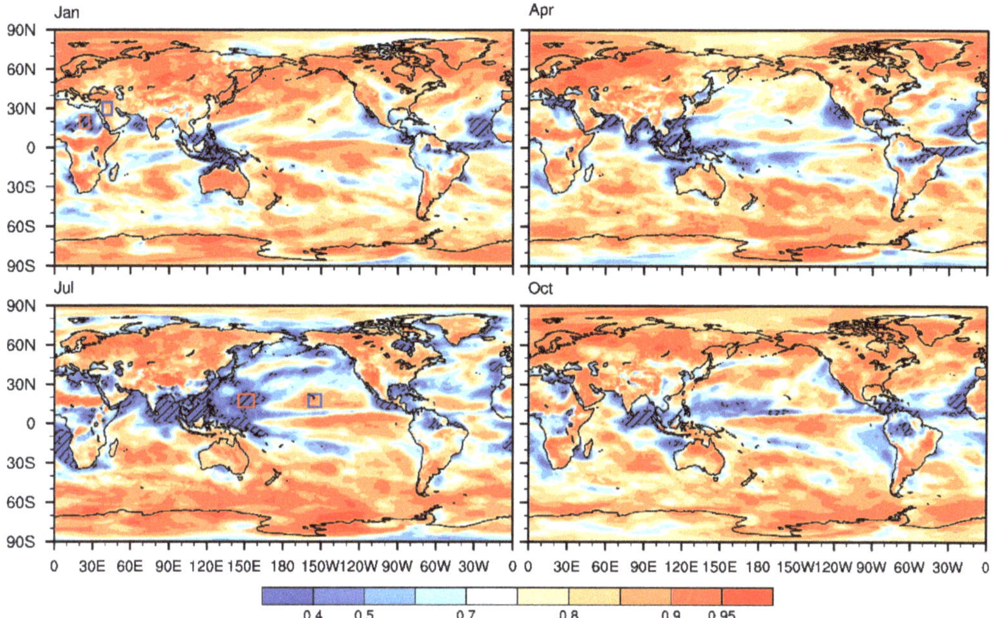

Figure 1. Maps of correlation coefficients of PW and SVP (the dataset used is ERA5). The correlation coefficients in shaded areas cannot pass the 0.05 significance level. Blue and red squares are strong and weak correlation areas, respectively, selected to show the vertical profiles.

Comparing the results calculated with different datasets, the weak correlation areas (WCAs) accordant among three datasets in all four months include the Maritime Continent, coasts of the Tropical Atlantic, the north coast of the Indian Ocean, and the east coast of the Southeast Pacific. There also exist differences among different datasets. Overall, insignificant correlation areas of NCEP2 and JRA55 are larger than those of ERA5. In JRA55, there are more WCAs in the Arctic in July. In NCEP2, there are more WCAs in the North Pacific in July and east Equatorial Pacific in July and October.

As for temporal variations, insignificant correlation areas in July are much larger than those in other months. The correlation coefficients at high latitudes in January, April, and October are large and can reach the 0.05 significance level. Differently, there exist some WCAs at mid-to-high latitudes in the Northern Hemisphere in July. It is plausible that this pattern is also relevant to stronger convection in these areas in Boreal Summer. Synthetically, the distribution of weak correlation areas, to some degree, is similar to the position and shape of the Inter Tropical Convergence Zone (ITCZ), with relatively larger areas in the West Pacific and the Indian Ocean and integrally more northward locations. The narrow belt distribution over other tropical oceans of the weak correlations is also slightly similar to the features of ITCZ.

When analyzing the monthly mean values, the areas with stronger convection could have stronger upward motion and a larger amount of precipitation. As mentioned, the linear relation between PW and SVP will be strong in hydrostatic equilibrium. In areas with strong and frequent convection, the interannual variation in vertical motion may also be stronger compared to the stable areas. The fluctuation of vertical movements will break the physical relation between quality and pressure. Hence, the weaker correlation areas in Figure 1 over oceans are probably associated with strong convection, and this is simply checked in the next section.

Results of the three datasets uniformly show that PW and SVP are strongly correlated in most extratropical areas and prove their strong linear relationship. The spatial patterns of correlation coefficients in extratropical areas are also concurrent among the three datasets, except for the weaker correlation in the Arctic in July obtained with JRA55.

The curves of normalized PW and SVP in red and blue squares in Figure 1, Figures S1 and S2 are also examined, but not shown in the main text (results are shown in Figures S3–S6 in the Supplementary Material). The variations of PW and SVP in SCAs can be quite consistent both on land and over the ocean. As for those in WCAs, on land, it seems that the variations of two variables are relatively more consistent before 1997, but this phenomenon cannot be seen over the oceans. Whether it is due to the data quality or other signals of the atmosphere variability needs further analysis.

To preliminarily analyze the reasons for the weak correlation in some areas, the vertical profiles of specific humidity and vertical velocity are exhibited. On land, the weak correlation area is in North Africa in places such as the Sahara. The vertical lapse rates are larger and more like exponential curves in SCAs (Figure 2). Due to the extreme drought in the WCAs, the WV differences between middle and low levels are relatively small compared with the SCAs. Hence, the influence of humidity at middle levels is relatively larger. Additionally, the standard deviation at middle levels is larger than that at low levels in WCAs, while the condition is reversed in the SCAs. The large ratio of fluctuations at middle levels to those at low levels can break the linear relation between PW and SVP because a large part of PW variation is at middle levels.

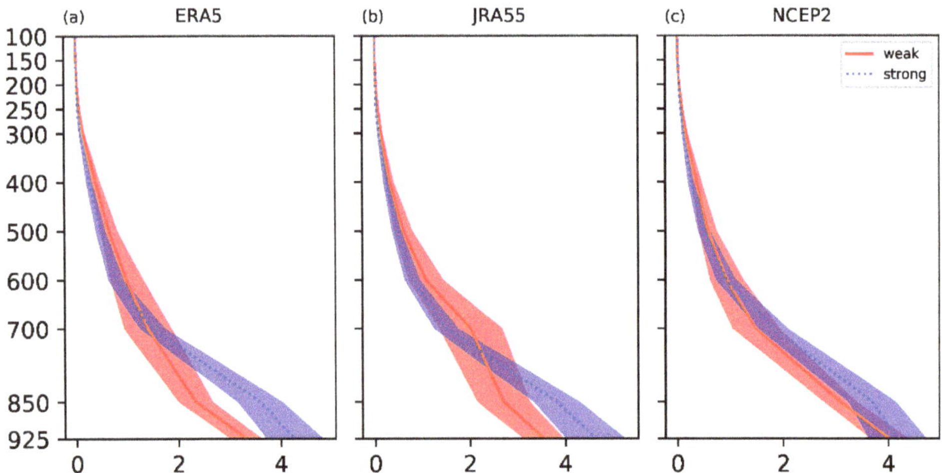

Figure 2. The vertical profiles of specific humidity in strong and weak correlation continental areas. (unit: $g \cdot kg^{-1}$). The shaded areas are the range of one standard deviation. (**a**) ERA5 datasets, (**b**) JRA55 datasets, and (**c**) NCEP2 datasets. The specific areas chosen are squared in Figure 1, Figures S1 and S2, correspondingly (Figures S1 and S2 are in the Supplementary Materials). Because the land surface pressure is around 940 hPa, the lowest level shown is 925 hPa).

Over the oceans, the specific humidity in the WCAs is more than that in the SCAs from the lower to upper troposphere (Figure 3). However, the lapse rate seems larger in the SCAs below 700 hPa. The standard deviation at middle levels is larger than that at low levels both in strong and correlation areas. The fluctuations at low levels in the WCAs are nearly zero, and it is hard for SVP, with almost no variation, to reflect the variation in PW. The small fluctuations there may be caused by the frequent convection, and the surface air is nearly saturated all the time. The convection also makes the air at middle levels own more WV.

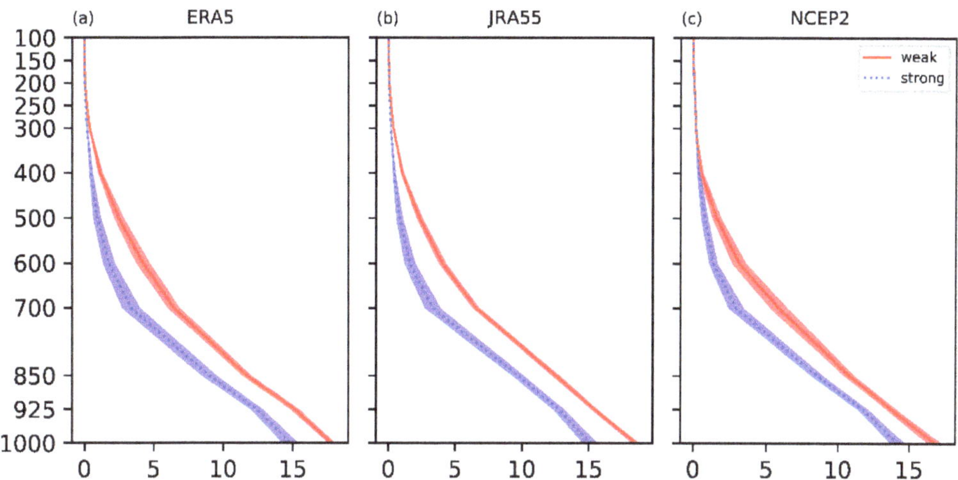

Figure 3. The contents are similar to Figure 2, but the areas squared are oceanic areas. (**a**) ERA5 datasets, (**b**) JRA55 datasets, and (**c**) NCEP2 datasets.

To further verify the convection conditions, Figures 4 and 5 exhibit the profiles of vertical velocity in the WCAs and SCAs. There are weak upward motions in the continental SCAs and relatively strong downward motions in the continental WCAs. There is not a consensus on the ratios of fluctuations in SCAs to that in WCAs among the three datasets. The stronger vertical movement is more likely to break the linear relation between PW and SVP under a drought environment over land. Because the differences between fluctuations of vertical movements in WCAs and SCAs are not so large, the fluctuations of vertical movements seem to play a less important role there.

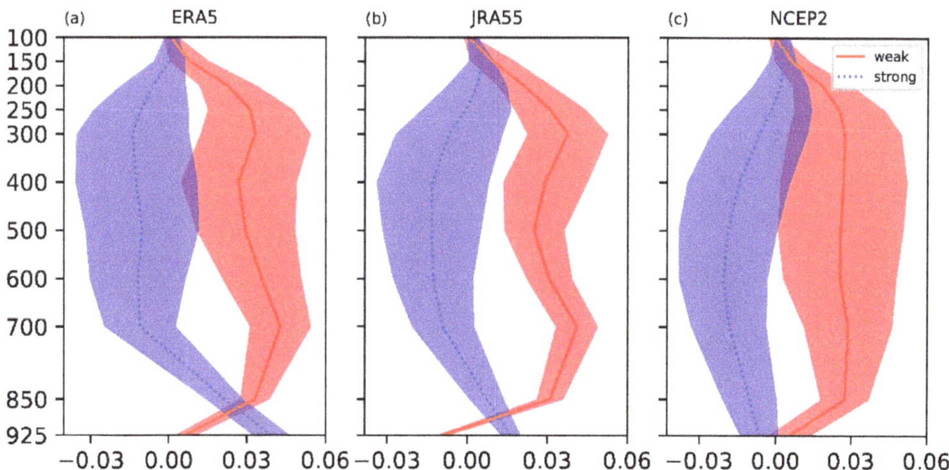

Figure 4. The vertical profiles of vertical velocity in strong and weak correlation continental areas. (unit: Pa·s^{-1}). The shaded areas are the range of one standard deviation. (**a**) ERA5 datasets, (**b**) JRA55 datasets, and (**c**) NCEP2 datasets. The specific areas chosen are squared in Figure 1, Figures S1 and S2, correspondingly. Because the land surface pressure is around 940 hPa, the lowest level shown is 925 hPa).

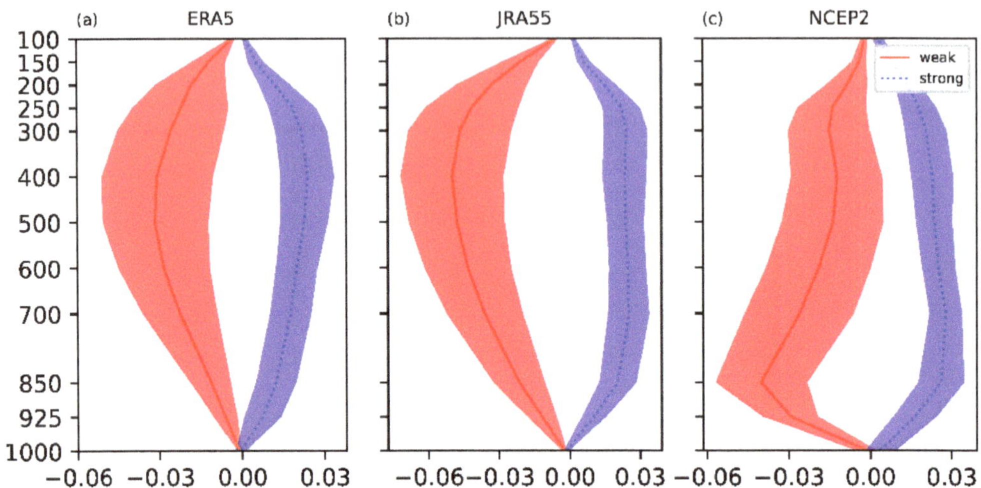

Figure 5. The contents are similar to Figure 4, but the areas squared are oceanic areas. (**a**) ERA5 datasets, (**b**) JRA55 datasets, and (**c**) NCEP2 datasets.

As for the oceanic areas, there are weak downward movements in the SCAs and relatively stronger upward movements in the WCAs. The fluctuations in the WCAs are also larger than that in the SCAs. The convection with strong upward motion is probably one of the main reasons for the weaker correlation in some oceanic areas. The unstable upward movements with large interannual variability may also have some effects.

3.2. Consistency of Long-Term Trends

Figure 6 shows the maps of long-term trends of PW and SVP for four months calculated with the ERA5 dataset (results of JRA55 and NCEP2 are shown in Figures S7 and S8 in the Supplementary Materials). In ERA5, focusing on the regions with significant trends, SVP and PW have concurrent distributions, such as the negative trends in the Southeast Pacific and the positive trends in the Northwest and Southwest Pacific. However, the trends of SVP in some oceanic areas are stronger than those of PW, including the South Pacific and North Atlantic, where the areas with significant trends of SVP are larger. In North Asia, the areas with significant trends of SVP are also larger than those of PW in April and October, while the situation is inverse in North Africa and the Northwest Pacific in July.

In the result calculated with JRA5, the trends of SVP and PW are a little less consistent than those in the result of ERA5. The phenomenon in Asia is similar, with SVP having significant trends in larger areas, and the differences between the trends of SVP and PW are larger than those of ERA5. However, in the Indian Ocean and Tropical Pacific, the trends of PW are stronger than those of SVP. The trend directions in most areas are consistent, except in some areas of Australia in January and the middle Tropical Pacific in July and October.

In NCEP2, the trends of SVP are weaker than those of PW on oceans, such as the Southeast Pacific in all four months and the Northwest Pacific in July. In North Eurasia and the Arctic, the trends of SVP are stronger. Trends of the two quantities in the North Atlantic and North America are relatively consistent, but the decreases in SVP on the east coast of South America are not so obvious as that of PW. In Africa and the middle Indian Ocean, there are contradictions between the trends of the two variables.

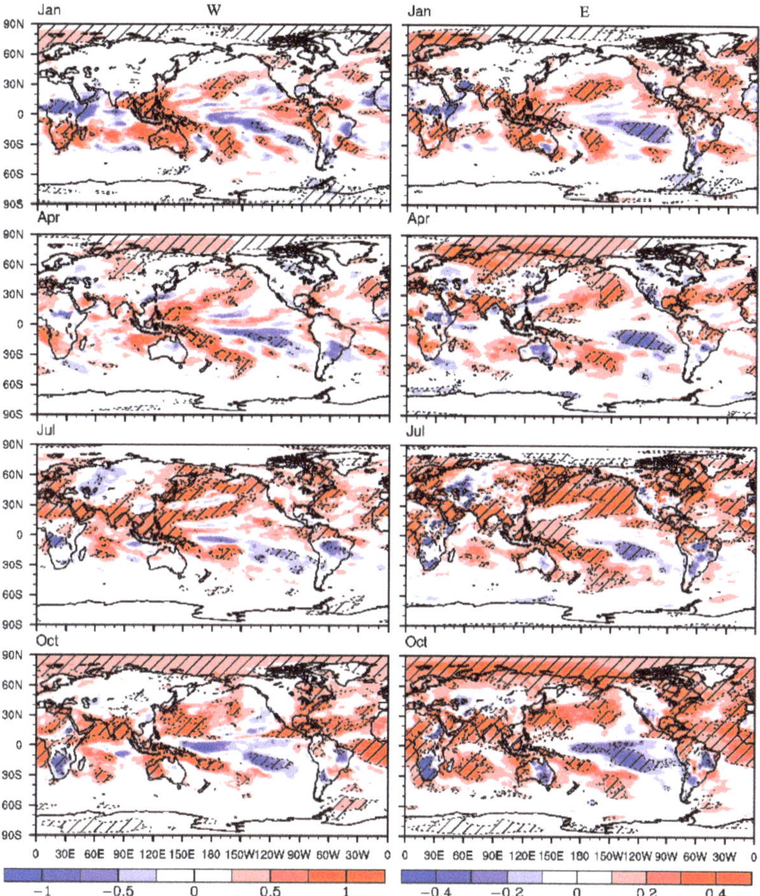

Figure 6. The long-term trends of PW (the left column, unit: kg·m^{-2}/10 yr) and SVP (the right column, unit: hPa/10 yr) in four months calculated with the ERA5 dataset. The rows from top to bottom correspond to January, April, July, and October, respectively. Trends in the shaded areas can reach the 0.05 significance level.

The ERA5 and JRA55 datasets both show that in most areas, the trend directions of PW and SVP are identical. The NCEP2 has more areas with contradictory trends of these two variables, including Middle Africa, the Indian Ocean, and the Tropical Atlantic. In all, the trend directions of PW and SVP are consistent in most areas, except in Middle Africa, the Indian Ocean, and the Tropical Atlantic. However, the trend intensities of these two variables are not so consistent.

Similar features can be seen among the three datasets. The areas with large slopes mainly exist at mid-to-low latitudes. There are also some areas with biggish discrepancies among the three datasets, such as some parts of Africa. There appears a positive-negative-positive pattern in the meridional direction. The result obtained with ERA5 exhibits a nearly reverse distribution in these areas in spring.

The distributions of trends also have temporal variations. In Australia, for example, the trends there in summer and other seasons are disparate. In some regions with the same year-round signs of slopes, the absolute values are larger in summer and weaker in winter. The absolute values of the slopes in polar areas are not large, but trends in most parts of these places can reach the 0.05 significance level, due to the climate of low air temperature and little vapor there.

To exhibit the consistency of long-term trends of PW and SVP intuitively, the scatter plots are shown in Figure 7. Every other grid is selected in meridional and zonal directions to draw the scatter plot and create the linear regression, because too many points will affect the exhibition effect if all the grids are used. The trends of the two variables are strongly correlated, with the correlation coefficients over 0.6 in all four months in ERA5. The lowest correlation coefficient is 0.49, which is in January calculated with JRA55. From the slope values of the linear regression, it can be seen that the trends of PW are generally weaker than those of SVP, but there are also some grids where the trends of PW are stronger.

The curves of global areal-weighted averaged PW, SVP, and surface air temperature are shown in Figure 8 (results calculated with JRA55 and NCEP2 datasets are in Figures S9 and S10 in the Supplementary Materials). The trends of three quantities can pass the 0.01 significance level using the MK test in all four months. It can be seen that the normalized PW and SVP overlap in quite a few parts of 43 years. Even in the rest periods, the two curves are almost parallel, and the only relatively large difference exists in October in the result of NCEP2. Hence, the global mean PW and SVP have consistent long-term trends and interannual variations, which is to say, the interannual and long-term variations of global mean SVP can well reflect those of PW.

Under greenhouse warming, the global mean SVP and PW both have significant increasing trends. From this point, results calculated with three sets of data can reach an agreement, and the trends of SVP and PW both amplify in July and October. However, when comparing the slope values of SVP and PW, there are still some differences among different datasets. Their trends are not so consistent as temperature.

Overall, from the angle of spatial distributions, three datasets all exhibit a feature that the trends of SVP are stronger than those of PW in North Eurasia. The increasing (decreasing) trends of PW can be decomposed into two parts. The first one is the increasing (decreasing) evaporation, and the other is the column-integrated vapor convergence (divergence). With the comparison of PW and SVP trends in North Asia, it can be concluded that the upward motions may be weakened there. It is plausible that the increasing trends of SVP there are mainly contributed by the increased water capacity of air and surface evaporation, but the vapor transport at higher levels may be weakened along with the weakened circulation. Eventually, the trends of PW are not so obvious as SVP in these areas.

In the Oceanic Continents and Tropical West Pacific, JRA55 and NCEP2 both show that the trends of PW are stronger than those of SVP. It can be concluded that the upward movements may be reinforced there. The stronger upward motion transports more vapor to the middle layers and the atmosphere is not in hydrostatic equilibrium. The global mean SVP and PW have consistent interannual variations and increasing trends in all three datasets.

3.3. The Consistency of the First Leading EOF Modes

To further confirm the reliability of the relationship between SVP and PW, the first EOF patterns (EOF 1) of PW and SVP are shown and compared. Figure 9 shows EOF 1 of PW and SVP obtained with the ERA5 dataset (results obtained with JRA55 and NCEP2 are shown in Figures S11 and S12 in the Supplementary Materials). With the Monte Carlo technique, the first three EOFs can be significantly distinguished from the noise. The 0.05 significance level needs the explained variances of the first three modes beyond 2.70%, 2.66%, and 2.65% respectively, with the spatial-temporal resolution (73 × 144 grids and 43 years) used here. There exist ENSO-like patterns, the inverse deviations in Northwest and Mid-to-East Pacific, in all four months, accordant with the result in [8,37]. However, what is mostly important is the similarity of patterns and time series of the two quantities. The result of ERA5 shows that the spatial patterns of the two quantities are quite similar. Relatively more differences exist in the negative value parts of the Tropical Pacific and the

Indian Ocean in April. The pattern of SVP in the Pacific is more eastern compared with those of PW in July and October.

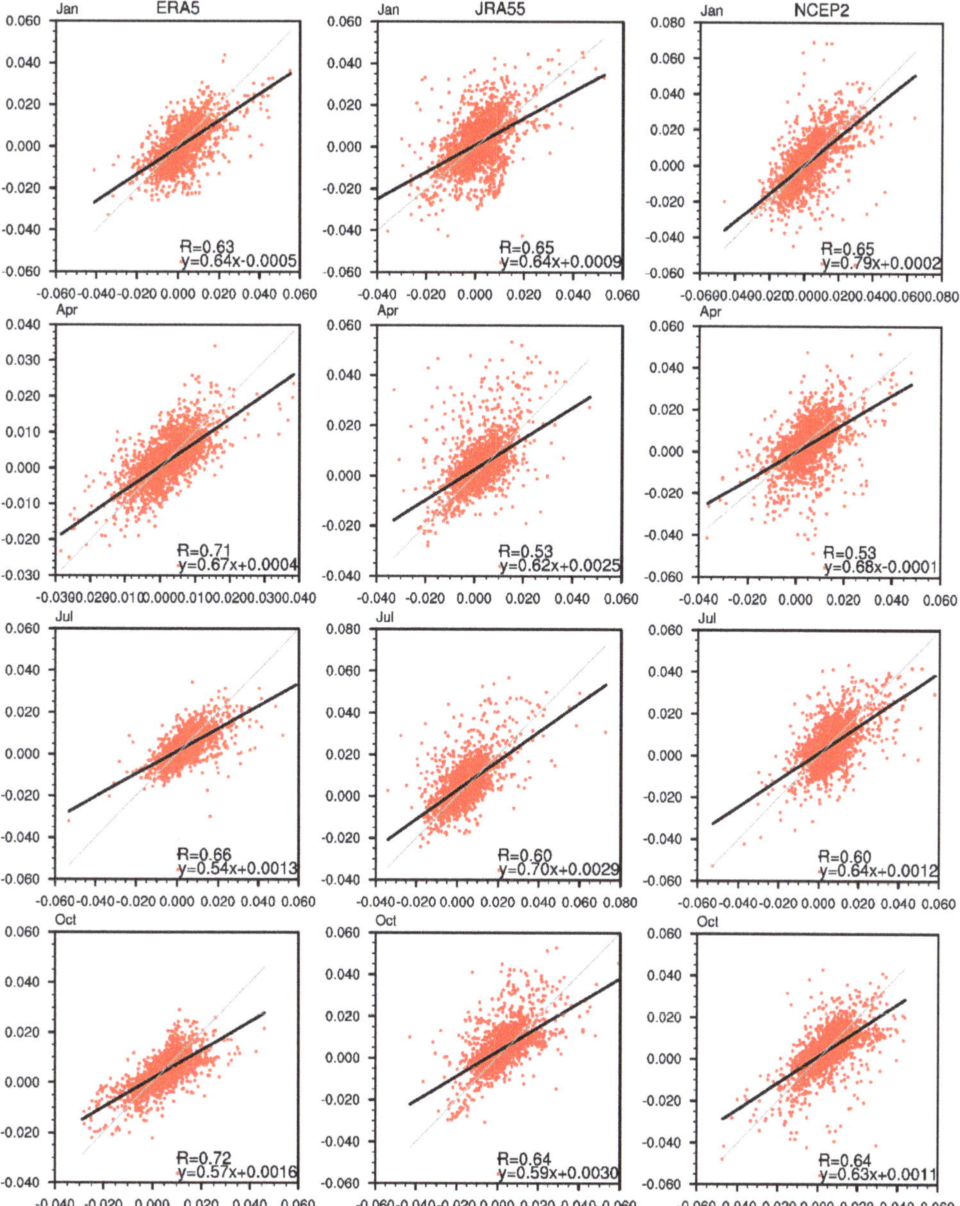

Figure 7. Scatter plots of PW and SVP trends at grids in four months. Each dot is a grid. The columns from left to right correspond to ERA5, JRA55, and NCEP2. The rows from top to bottom correspond to four months. The black line is the linear regression line of PW and SVP. The grey line is the $y = x$ line. R is the spatial correlation coefficient of PW and SVP. The equation in the right bottom corner is the regression equation.

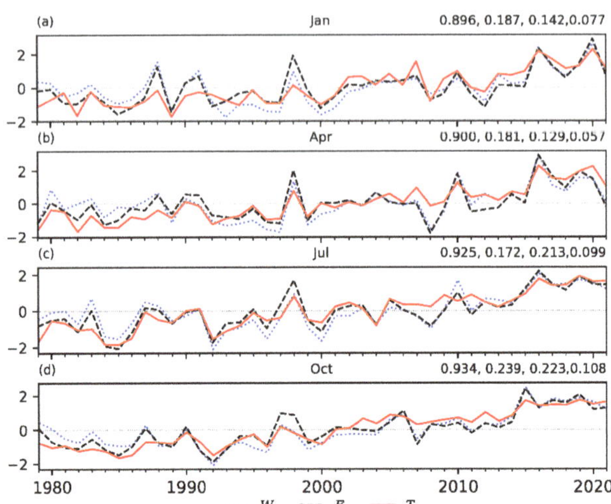

Figure 8. The time series of normalized global areal-weighted averages of 2 m temperature (*T*), PW (*W*), and SVP (*E*) in January (**a**), April (**b**), July (**c**), and October (**d**) calculated with ERA5 datasets. The four numbers at the top-right of subgraphs are the correlation coefficient between PW and SVP and Sen's slope value of three elements (*T*, *W*, and *E* in sequence; units: K/10 yr, kg·m^{-2}/10 yr, and hPa/10 yr), separately.

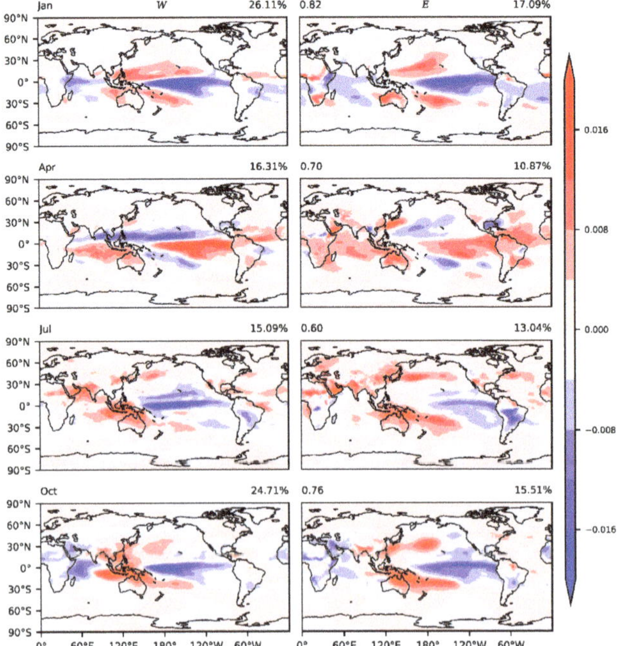

Figure 9. The first EOF (EOF1) modes of PW (**left**) and SVP (**right**) calculated with ERA5. The rows from top to bottom correspond to January, April, July, and October, respectively. The percentages at the top-right of subgraphs are fractional mode variances corresponding to PW and SVP. The numbers at the upper-left corners of graphs in the right column are spatial correlation coefficients of the first modes of PW and SVP.

The result calculated with JRA55 is similar, while the little divergences between PW and SVP are also over the Tropical Pacific in April and July. What needs attention is that there is a great distinction between the first modes of these two variables in April. After verification, it is confirmed that EOF 1 of PW corresponds well with EOF 2 of SVP. Hence, EOF 1 of SVP in April is replaced by EOF 2 in the result of JRA55.

The differences between spatial patterns of PW and SVP are relatively larger in the result of NCEP2. The distinctions of patterns are obvious in April, July, and October. The negative values of SVP on oceans are visibly weak. Especially in October, not only the is the intensity of the negative center in the Tropical Pacific not so coherent, but there are also some contradictions in the Middle East and Southeast Africa. The spatial correlation coefficient in October is just 0.37, which is much lower than those in other months.

EOF 2 and EOF 3 of the two variables are also compared, and they also correspond well (results are shown in Figures S13–S18 in the Supplementary Materials). The spatial correlation coefficients of the patterns are generally over 0.5 for EOF 2 except in October calculated with NCEP2 (the correlation coefficient is -0.34 and can pass the 0.01 significance level), and those for EOF 3 are generally over 0.4. EOF 3 of PW and SVP calculated with ERA5 is relatively less coherent in January and April, with the spatial correlation coefficient of the patterns being -0.32 and -0.29, respectively. However, because of the long sequence (73×144), the correlation coefficients can still pass the 0.01 significance level. Similar to the condition in April in the result of JRA55, EOF 3 of PW is more consistent with EOF 4 of SVP in January in ERA5. Hence, the subgraph of SVP in January in Figure S16 is EOF 4 of it.

As a whole, EOF 1 of PW and SVP obtained with ERA5 and JRA55 datasets is more coherent than that obtained with NCEP2. Even in NCEP2, spatial correlation coefficients of PW and SVP can pass the 0.01 significance level and are larger than 0.66 in January, April, and July. The first three leading modes all show good coherences between PW and SVP. Even the lowest spatial correlation coefficient between the EOF patterns of two variables can still pass the 0.01 significance level.

Corresponding to the consistency of spatial patterns, the time series of SVP and PW obtained with ERA5 almost overlaps in many periods (Figure 10; results calculated with JRA55 and NCEP2 are shown in Figures S19 and S20 in the Supplementary Materials; The time series of EOF 2 and EOF 3 calculated with three dataset are shown in Figures S21–S26 in the Supplementary Materials). The correlation coefficients of the SVP and PW time series are generally over 0.8. As for the NCEP2, the time series with the largest discrepancies exist in October, with a 0.52 correlation coefficient. In October, the time series of SVP is stable with an increasing trend, while the PW is fluctuating.

The time series of EOF 2 and EOF 3 are also examined. The absolute values of correlation coefficients of the corresponding time series are generally over 0.7, and the lowest value is 0.44, which is obtained with the time series of EOF 3 in January calculated with the ERA5 dataset. Corresponding to the relative less consistency of EOF 3 in January and April calculated with ERA5, the time series of EOF 3 in these two months is also less coherent.

Combining the spatial patterns and time series synthetically, the first three EOF leading modes and time series of PW and SVP can correspond well in ERA5 and JRA55. Most contractors exist in the Tropical Indian Ocean and the Tropical Atlantic. Most of the differences in the Pacific are intensity differences. The consistency of the EOF modes of PW and SVP in results further proves the close relationship between them in interannual and long-term variations. It is a complement to the statistical basis of using the variations of SVP to represent those of PW, or the reverse.

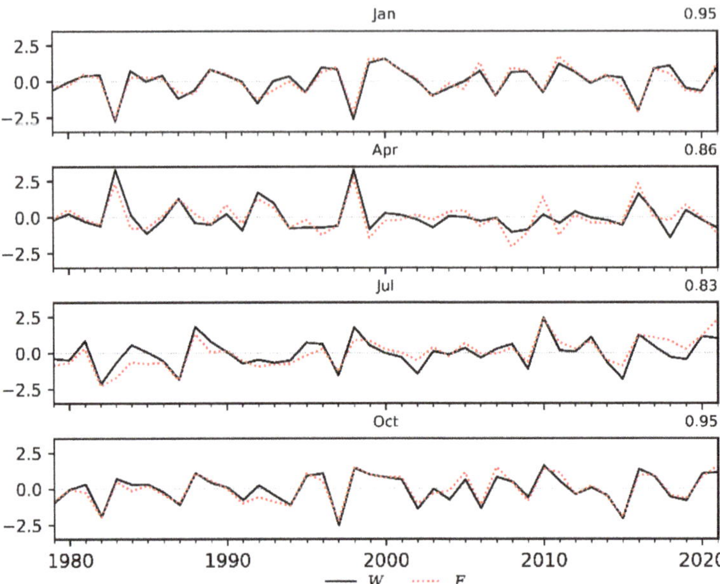

Figure 10. The principal component time series (PCs) of EOF 1 of PW and SVP was calculated with the ERA5 dataset (scaled to unit variance). The numbers at the top-right of subgraphs are correlation coefficients of two curves.

4. Summary and Discussion

Because of the deficiency of high-altitude observations and the internal physical relation of PW and SVP, the linkage between them has been researched for some decades. The time scales mainly researched are the daily and annual cycles. Whether the linkage is reliable everywhere at interannual and long-term variations is examined in this paper. With the examination, we can reveal the areas where it is difficult to obtain an efficient empirical equation between PW and SVP and the regions where the variations of SVP and PW can well reflect each other.

Results show that the interannual variations of monthly mean PW and SVP are strongly correlated in most areas, with correlation coefficients generally over 0.8. Correlations at high latitudes are stronger than those at low latitudes, and correlations in winter are stronger than those in summer. The insignificant correlations just exist over tropical oceans and some high-altitude areas in the Northern Hemisphere in Boreal Summer.

The vertical profiles of specific humidity and vertical velocity in weak and strong correlation regions are compared to figure out the reasons for the weaker correlations. Over the ocean, the weak correlation could be mainly due to the unstable and strong convection. On land, the weak correlation is associated with the extreme drought near the surface. Under the drought conditions, the downward motions in these areas probably break the linear relationship. There is only a little moisture, so the effects of vertical movements are relatively more destructive.

The long-term trends of SVP and PW are also analyzed. In a large number of areas besides some areas of the Tropical Indian Ocean and Middle Africa, the long-term trends of SVP can well reflect those of PW. Three datasets exhibit a common characteristic that the trends of SVP in North Eurasia are stronger than those of PW. The trends of SVP in high-latitude areas of Eurasia are stronger than those of PW, indicating that the upward movements are attenuated, and evaporation is enhanced along with warming. The vapor convergence at high levels may be weaker along with the weakening of circulations.

As a whole, the variations of SVP and PW are consistent in vast areas, especially in ERA5. At the interannual scale, the variation in PW can be reflected by that of SVP in most extratropical areas except in July. It is feasible to use the interannual variation in SVP to reflect that of PW in extratropical areas except in July, but researchers must be careful when performing this in the Tropical West Pacific and the Tropical Atlantic. In these areas, it should be relatively hard to establish an effective empirical equation between PW and SVP. On the long-term scale, the trend direction of PW can be well reflected by that of SVP, except in the Tropical Indian Ocean east to Africa and the Tropical Atlantic east to South America. The weaker correlations are mainly caused by the vertical movements which can cause the large fluctuation of specific humidity at middle levels and adjust the profiles of specific humidity, while it also means that the atmosphere is not in hydrostatic equilibrium.

The close linkage between SVP and PW is an extension of another study by us [39]. The response of SVP to surface temperature is studied in another part of our work. With the linkage between PW and SVP, the response characteristics of column WV content can be reflected by those of SVP. In addition, it provides a brief description of where the PW can be well regressed with SVP, or the other way around. Because the specific values of trends, interannual correlation coefficients, and EOF modes differ among different datasets, some areas with large discrepancies need more accurate observations to be verified, such as the Indian Ocean and the east coast of South America. Perhaps using SVP at several grids to estimate PW (or in reverse order) or using several surface variables to estimate PW will achieve a better effect. It is worth trying, and it may be suitable to combine some neural network methods.

Supplementary Materials: The following supporting information can be downloaded at: https: //www.mdpi.com/article/10.3390/atmos13091350/s1, Figure S1: Maps of correlation coefficients of PW and SVP (The dataset used is JRA55). The correlation coefficients in shaded areas cannot pass the 0.05 significance level. Blue and red squares are strong and weak correlation areas respectively selected to show the vertical profiles, Figure S2: Maps of correlation coefficients of PW and SVP (The dataset used is NCEP2). The correlation coefficients in shaded areas cannot pass the 0.05 significance level. Blue and red squares are strong and weak correlation areas respectively selected to show the vertical profiles, Figure S3: The curves of normalized PW and SVP in strong correlation areas on land (blue squares in January in Figure 1, Figure S1, and Figure S2), Figure S4: The curves of normalized PW and SVP in weak correlation areas on land (red squares in January in Figure 1, Figure S1, and Figure S2), Figure S5: The curves of normalized PW and SVP in strong correlation areas over the oceans (blue squares in July in Figure 1, Figure S1, and Figure S2), Figure S6: The curves of normalized PW and SVP in weak correlation areas (red squares in July in Figures 1–3) over the oceans, Figure S7: The long-term trends of PW (the left column, unit: $\mathrm{kg \cdot m^{-2}}$/10 yr) and SVP (the right column, unit: hPa/10 yr) in four months calculated with the JRA55 dataset. The rows from top to bottom respond to January, April, July, and October respectively. Trends in the shaded areas can reach the 0.05 significance level, Figure S8: The long-term trends of PW (the left column, unit: $\mathrm{kg \cdot m^{-2}}$/10 yr) and SVP (the right column, unit: hPa/10 yr) in four months calculated with the NCEP2 dataset. The rows from top to bottom respond to January, April, July, and October respectively. Trends in the shaded areas can reach the 0.05 significance level, Figure S9: The time series of normalized global areal-weighted averages of 2 m temperature (*T*), PW (*W*), and SVP (*E*) in January (**a**), April (**b**), July (**c**), and October (**d**) calculated with JRA55 datasets. The four numbers at the top-right of subgraphs are the correlation coefficient between PW and SVP and Sen's slope value of three elements (*T*, *W*, and *E* in sequence; units: K/10 yr, $\mathrm{kg \cdot m^{-2}}$/10 yr, and hPa/10 yr), separately, Figure S10: The time series of normalized global areal-weighted averages of 2 m temperature (*T*), PW (*W*), and SVP (*E*) in January (**a**), April (**b**), July (**c**), and October (**d**) calculated with NCEP2 datasets. The four numbers at the top-right of subgraphs are the correlation coefficient between PW and SVP and Sen's slope value of three elements (*T*, *W*, and *E* in sequence; units: K/10 yr, $\mathrm{kg \cdot m^{-2}}$/10 yr, and hPa/10 yr), separately, Figure S11: The first EOF modes of PW (left) and SVP (right) calculated with JRA55. (The rows from top to bottom correspond to January, April, July, and October respectively. The percentages at the top-right of subgraphs are fractional mode variances corresponding to PW and SVP. The numbers at the upper-left corners of graphs in the right column are spatial correlation coefficients of the first modes of PW and SVP), Figure S12: The first EOF modes of PW (left) and SVP (right) calculated with NCEP2. (The rows from top to bottom correspond to January, April, July, and October respectively. The percentages at the top-right of subgraphs are fractional mode variances corresponding to PW and SVP. The numbers at the upper-left corners of graphs in the right column are spatial correlation coefficients of the first modes of PW and SVP), Figure S13: The second EOF modes of PW (left) and SVP (right) calculated with ERA5. (The rows from top to bottom correspond to January, April, July, and October respectively. The percentages at the top-right of subgraphs are fractional mode variances corresponding to PW and SVP. The numbers at the upper-left corners of graphs in the right column are spatial correlation coefficients of the first modes of PW and SVP.), Figure S14: The second EOF modes of PW (left) and SVP (right) calculated with JRA55. (The rows from top to bottom correspond to January, April, July, and October respectively. The percentages at the top-right of subgraphs are fractional mode variances corresponding to PW and SVP. The numbers at the upper-left corners of graphs in the right column are spatial correlation coefficients of the first modes of PW and SVP.), Figure S15: The second EOF modes of PW (left) and SVP (right) calculated with ncep2. (The rows from top to bottom correspond to January, April, July, and October respectively. The percentages at the top-right of subgraphs are fractional mode variances corresponding to PW and SVP. The numbers at the upper-left corners of graphs in the right column are spatial correlation coefficients of the first modes of PW and SVP.), Figure S16: The third EOF modes of PW (left) and SVP (right) calculated with ERA5. (The rows from top to bottom correspond to January, April, July, and October respectively. The percentages at the top-right of subgraphs are fractional mode variances corresponding to PW and SVP. The numbers at the upper-left corners of graphs in the right column are spatial correlation coefficients of the first modes of PW and SVP.), Figure S17: The third EOF modes of PW (left) and SVP (right) calculated with JRA55. (The rows from top to bottom correspond to January, April, July, and October respectively. The percentages at the top-right of subgraphs are fractional mode variances corresponding to PW and SVP. The numbers at the upper-left corners of graphs in the right column are spatial correlation coefficients of the first modes of PW and SVP.)., Figure S18: The third EOF modes of PW (left) and SVP (right) calculated with NCEP2. (The rows from top to bottom correspond to January, April, July, and October respectively. The percentages at the top-right of subgraphs are fractional mode variances corresponding to PW and SVP. The numbers at the upper-left corners of graphs in the right column are spatial correlation coefficients of the first modes of PW and SVP.), Figure S19: The principal components of EOF 1 of PW and SVP were calculated with the JRA55 dataset (scaled to unit variance). The numbers at the top-right of subgraphs are correlation coefficients of two curves, Figure S20: The principal components of EOF 1 of PW and SVP were calculated with the NCEP2 dataset (scaled to unit variance). The numbers at the top-right of subgraphs are correlation coefficients of two curves, Figure S21: The principal components of EOF 2 of PW and SVP were calculated with the ERA5 dataset (scaled to unit variance). The numbers at the top-right of subgraphs are correlation coefficients of two curves, Figure S22: The principal components of EOF 2 of PW and SVP were calculated with the JRA55 dataset (scaled to unit variance). The numbers at the top-right of subgraphs are correlation coefficients of two curves, Figure S23: The principal components of EOF 2 of PW and SVP were calculated with the NCEP2 dataset (scaled to unit variance). The numbers at the top-right of subgraphs are correlation coefficients of two curves, Figure S24: The principal components of EOF 3 of PW and SVP were calculated with the ERA5 dataset (scaled to unit variance). The numbers at the top-right of subgraphs are correlation coefficients of two curves, Figure S25: The principal components of EOF 3 of PW and SVP were calculated with the JRA55 dataset (scaled to unit variance). The numbers at the top-right of subgraphs are correlation coefficients of two curves, Figure S26: The principal components of EOF 3 of PW and SVP were calculated with the NCEP2 dataset (scaled to unit variance). The numbers at the top-right of subgraphs are correlation coefficients of two curves.

Author Contributions: Conceptualization, J.H. and E.L.; methodology, E.L.; software, J.H.; validation, J.H.; formal analysis, J.H. and E.L.; investigation, J.H.; resources, J.H.; data curation, J.H.; writing— original draft preparation, J.H.; writing—review and editing, J.H. and E.L.; visualization, J.H.; supervision, E.L.; project administration, E.L.; funding acquisition, E.L. All authors have read and agreed to the published version of the manuscript.

Funding. This study was supported by the National Natural Science Foundation of China (Grant Number: 41991281), the National Key Research and Development Program of China (Grant Number: 2018YFC1507704), and the Priority Academic Program Development of Jiangsu Higher Education Institutions (PAPD).

Institutional Review Board Statement: Not applicable.

Informed Consent Statement: Not applicable.

Data Availability Statement: Datasets analyzed during the current study include ERA5, JRA55, and NCEP2. The ERA5 dataset is available in the Copernicus Climate Change Service Climate Date Store (https://cds.climate.copernicus.eu/#!/search?text=ERA5&type=dataset, accessed on 1 July 2022), provided by ECMWF. The JRA55 dataset is available in the Research Data Archive, managed by NCAR (https://rda.ucar.edu/datasets/ds628.1/#!access, accessed on 1 July 2022), provided by JMA. The NCEP2 dataset is available in the Physical Sciences Laboratory (PSL) (https://psl.noaa.gov/data/gridded/data.ncep.reanalysis2.html, accessed on 1 July 2022), provided by the NOAA/OAR/ESRL PSL.

Acknowledgments: The introduction of EOF analysis is cited from National Center for Atmospheric Research Staff (Eds). Last modified 22 July 2013. "The Climate Data Guide: Empirical Orthogonal Function (EOF) Analysis and Rotated EOF Analysis." retrieved from https://climatedataguide.ucar.edu/climate-data-tools-and-analysis/empirical-orthogonal-function-eof-analysis-and-rotated-eof-analysis, accessed on 1 July 2022.

Conflicts of Interest: The authors declare no conflict of interest.

References

1. Stewart, R.E.; Crawford, R.W.; Leighton, H.G.; Marsh, P.; Strong, G.S.; Moore, G.W.K.; Ritchie, H.; Rouse, W.R.; Soulis, E.D.; Kochtubajda, B. The Mackenzie GEWEX Study: The Water and Energy Cycles of a Major North American River Basin. *Bull. Am. Meteorol. Soc.* **1998**, *79*, 2665–2683. [CrossRef]
2. Semmler, T.; Jacob, D.; Schlünzen, K.H.; Podzun, R. The Water and Energy Budget of the Arctic Atmosphere. *J. Clim.* **2005**, *18*, 2515–2530. [CrossRef]
3. Trenberth, K.E.; Fasullo, J. Regional Energy and Water Cycles: Transports from Ocean to Land. *J. Clim.* **2013**, *26*, 7837–7851. [CrossRef]
4. Mathew, S.S.; Kumar, K.K. On the role of precipitation latent heating in modulating the strength and width of the Hadley circulation. *Arch. Meteorol. Geophys. Bioclimatol. Ser. B* **2018**, *136*, 661–673. [CrossRef]
5. Hao, J.; Lu, E. The quantitative comparison of contributions from vapour and temperature to midsummer precipitation in China under the influence of spring heat source over Tibet Plateau. *Int. J. Clim.* **2021**, *42*, 1754–1766. [CrossRef]
6. Zhou, T.-J.; Yu, R.-C. Atmospheric water vapor transport associated with typical anomalous summer rainfall patterns in China. *J. Geophys. Res. Atmos.* **2005**, *110*. [CrossRef]
7. Rädel, G.; Shine, K.P.; Ptashnik, I.V. Global radiative and climate effect of the water vapour continuum at visible and near-infrared wavelengths. *Q. J. R. Meteorol. Soc.* **2014**, *141*, 727–738. [CrossRef]
8. Wang, Y.; Zhang, Y.; Fu, Y.; Li, R.; Yang, Y.-J. A climatological comparison of column-integrated water vapor for the third-generation reanalysis datasets. *Sci. China Earth Sci.* **2015**, *59*, 296–306. [CrossRef]
9. Cady-Pereira, K.; Shephard, M.; Turner, D.; Mlawer, E.; Clough, S.; Wagner, T. Improved daytime column-integrated precipitable WV from Vaisala radiosonde humidity sensors. *J. Atmos. Ocean. Technol.* **2008**, *25*, 873–883. [CrossRef]
10. Durre, I.; Williams, C.N.; Yin, X.; Vose, R.S. Radiosonde-based trends in precipitable water over the Northern Hemisphere: An update. *J. Geophys. Res. Earth Surf.* **2009**, *114*. [CrossRef]
11. Wang, J.X.L.; Gaffen, D.J. Late-Twentieth-Century Climatology and Trends of Surface Humidity and Temperature in China. *J. Clim.* **2001**, *14*, 2833–2845. [CrossRef]
12. Dai, A. Recent Climatology, Variability, and Trends in Global Surface Humidity. *J. Clim.* **2006**, *19*, 3589–3606. [CrossRef]
13. Willett, K.M.; Jones, P.; Gillett, N.P.; Thorne, P. Recent Changes in Surface Humidity: Development of the HadCRUH Dataset. *J. Clim.* **2008**, *21*, 5364–5383. [CrossRef]
14. Isaac, V.; Van Wijngaarden, W. Surface WV pressure and temperature trends in North America during 1948–2010. *J. Clim.* **2012**, *25*, 3599–3609. [CrossRef]
15. Wallace, J.M.; Hobbs, P.V. *Atmospheric Science: An Introductory Survey*; Elsevier: Amsterdam, The Netherlands, 2006; 483p.
16. Lu, E.; Zeng, X. Understanding different precipitation seasonality regimes from WV and temperature fields: Case studies. *Geophys. Res. Lett.* **2005**, *32*, 1154–1158. [CrossRef]
17. Lu, E. Understanding the effects of atmospheric circulation in the relationships between water vapor and temperature through theoretical analyses. *Geophys. Res. Lett.* **2007**, *34*. [CrossRef]
18. Reber, E.E.; Swope, J.R. On the Correlation of the Total Precipitable Water in a Vertical Column and Absolute Humidity at the Surface. *J. Appl. Meteorol.* **1972**, *11*, 1322–1325. [CrossRef]
19. Viswanadham, Y. The Relationship between Total Precipitable Water and Surface Dew Point. *J. Appl. Meteorol.* **1981**, *20*, 3–8. [CrossRef]
20. Liu, W.T. Statistical Relation between Monthly Mean Precipitable Water and Surface-Level Humidity over Global Oceans. *Mon. Weather Rev.* **1986**, *114*, 1591–1602. [CrossRef]

21. Hsu, S.A.; Blanchard, B.W. The relationship between total precipitable water and surface-level humidity over the sea surface: A further evaluation. *J. Geophys. Res. Earth Surf.* **1989**, *94*, 14539–14545. [CrossRef]
22. Dostalek, J.F.; Schmit, T.J. Total Precipitable Water Measurements from GOES Sounder Derived Product Imagery. *Weather Forecast.* **2001**, *16*, 573–587. [CrossRef]
23. Deeter, M.N. A new satellite retrieval method for precipitable water vapor over land and ocean. *Geophys. Res. Lett.* **2007**, *34*. [CrossRef]
24. Radhakrishna, B.; Fabry, F.; Braun, J.J.; Van Hove, T. Precipitable Water from GPS over the Continental United States: Diurnal Cycle, Intercomparisons with NARR, and Link with Convective Initiation. *J. Clim.* **2015**, *28*, 2584–2599. [CrossRef]
25. Torri, G.; Adams, D.K.; Wang, H.; Kuang, Z. On the diurnal cycle of GPS-derived precipitable WV over Sumatra. *J. Atmos. Sci.* **2019**, *76*, 3529–3552. [CrossRef]
26. Zveryaev, I.I.; Chu, P.-S. Recent climate changes in precipitable water in the global tropics as revealed in National Centers for Environmental Prediction/National Center for Atmospheric Research reanalysis. *J. Geophys. Res. Atmos.* **2003**, *108*, ACL6.1–ACL6.10. [CrossRef]
27. Ssenyunzi, R.C.; Oruru, B.; D'ujanga, F.M.; Realini, E.; Barindelli, S.; Tagliaferro, G.; von Engeln, A.; van de Giesen, N. Performance of ERA5 data in retrieving Precipitable Water Vapour over East African tropical region. *Adv. Space Res.* **2020**, *65*, 1877–1893. [CrossRef]
28. Maghrabi, A.; Al Dajani, H. Estimation of precipitable water vapour using vapour pressure and air temperature in an arid region in central Saudi Arabia. *J. Assoc. Arab Univ. Basic Appl. Sci.* **2013**, *14*, 45–49. [CrossRef]
29. Chadwick, R.; Good, P.; Willett, K. A Simple Moisture Advection Model of Specific Humidity Change over Land in Response to SST Warming. *J. Clim.* **2016**, *29*, 7613–7632. [CrossRef]
30. Gaffen, D.J.; Elliott, W.P.; Robock, A. Relationships between tropospheric WV and surface temperature as observed by radiosondes. *Geophys. Res. Lett.* **1992**, *19*, 1839–1842. [CrossRef]
31. Sun, D.-Z.; Held, I.M. A comparison of modeled and observed relationships between interannual variations of WV and temperature. *J. Clim.* **1996**, *9*, 665–675. [CrossRef]
32. Bauer, M.; Del Genio, A.D.; Lanzante, J.R. Observed and Simulated Temperature–Humidity Relationships: Sensitivity to Sampling and Analysis. *J. Clim.* **2002**, *15*, 203–215. [CrossRef]
33. Mears, C.A.; Santer, B.D.; Wentz, F.J.; Taylor, K.E.; Wehner, M.F. Relationship between temperature and precipitable water changes over tropical oceans. *Geophys. Res. Lett.* **2007**, *34*. [CrossRef]
34. Laine, A.; Nakamura, H.; Nishii, K.; Miyasaka, T. A diagnostic study of future evaporation changes projected in CMIP5 climate models. *Clim. Dyn.* **2014**, *42*, 2745–2761. [CrossRef]
35. Trenberth, K.E.; Fasullo, J.; Smith, L. Trends and variability in column-integrated atmospheric WV. *Clim. Dyn.* **2005**, *24*, 741–758. [CrossRef]
36. Chung, E.-S.; Soden, B.; Sohn, B.J.; Shi, L. Upper-tropospheric moistening in response to anthropogenic warming. *Proc. Natl. Acad. Sci. USA* **2014**, *111*, 11636–11641. [CrossRef]
37. Zhang, Y.; Xu, J.; Yang, N.; Lan, P. Variability and Trends in Global Precipitable Water Vapor Retrieved from COSMIC Radio Occultation and Radiosonde Observations. *Atmosphere* **2018**, *9*, 174. [CrossRef]
38. Schröder, M.; Lockhoff, M.; Shi, L.; August, T.; Bennartz, R.; Brogniez, H.; Calbet, X.; Fell, F.; Forsythe, J.; Gambacorta, A.; et al. The GEWEX Water Vapor Assessment: Overview and Introduction to Results and Recommendations. *Remote Sens.* **2019**, *11*, 251. [CrossRef]
39. Hao, J.; Lu, E. Variation of Relative Humidity as Seen through Linking Water Vapor to Air Temperature: An Assessment of Interannual Variations in the Near-Surface Atmosphere. *Atmosphere* **2022**, *13*, 1171. [CrossRef]
40. Shi, L.; Schreck, C.; Schröder, M. Assessing the Pattern Differences between Satellite-Observed Upper Tropospheric Humidity and Total Column Water Vapor during Major El Niño Events. *Remote Sens.* **2018**, *10*, 1188. [CrossRef]
41. Smith, T.M.; Arkin, P.A. Improved Historical Analysis of Oceanic Total Precipitable Water. *J. Clim.* **2015**, *28*, 3099–3121. [CrossRef]
42. Wang, J.; Dai, A.; Mears, C. Global Water Vapor Trend from 1988 to 2011 and Its Diurnal Asymmetry Based on GPS, Radiosonde, and Microwave Satellite Measurements. *J. Clim.* **2016**, *29*, 5205–5222. [CrossRef]
43. Hersbach, H.; Bell, B.; Berrisford, P.; Hirahara, S.; Horanyi, A.; Muñoz-Sabater, J.; Nicolas, J.; Peubey, C.; Radu, R.; Schepers, D.; et al. The ERA5 global reanalysis. *Q. J. R. Meteorol. Soc.* **2020**, *146*, 1999–2049. [CrossRef]
44. Kanamitsu, M.; Ebisuzaki, W.; Woollen, J.; Yang, S.-K.; Hnilo, J.J.; Fiorino, M.; Potter, G.L. NCEP–DOE AMIP-II Reanalysis (R-2). *Bull. Am. Meteorol. Soc.* **2002**, *83*, 1631–1644. [CrossRef]
45. Kobayashi, S.; Ota, Y.; Harada, Y.; Ebita, A.; Moriya, M.; Onoda, H.; Onogi, K.; Kamahori, H.; Kobayashi, C.; Endo, H.; et al. The JRA-55 Reanalysis: General Specifications and Basic Characteristics. *J. Meteorol. Soc. Jpn. Ser. II* **2015**, *93*, 5–48. [CrossRef]
46. Alduchov, O.A.; Eskridge, R.E. Improved Magnus Form Approximation of Saturation Vapor Pressure. *J. Appl. Meteorol.* **1996**, *35*, 601–609. [CrossRef]
47. Duhan, D.; Pandey, A. Statistical analysis of long term spatial and temporal trends of precipitation during 1901–2002 at Madhya Pradesh, India. *Atmos. Res.* **2013**, *122*, 136–149. [CrossRef]
48. Jain, S.K.; Kumar, V. Trend analysis of rainfall and temperature data for India. *Curr. Sci.* **2012**, *102*, 37–49.
49. Mann, H.B. Nonparametric tests against trend. *Econometrica* **1945**, *13*, 245–259. [CrossRef]
50. Kendall, M.G. *Rank Correlation Methods*; Charles Grifin: London, UK, 1970; 202p.

51. Mondal, A.; Kundu, S.; Mukhopadhyay, A. Rainfall trend analysis by Mann-Kendall test: A case study of north-eastern part of Cuttack district, Orissa. *Int. J. Geol. Earth Environ. Sci.* **2012**, *2*, 70–78.
52. Hamed, K.H.; Rao, A.R. A modified Mann-Kendall trend test for autocorrelated data. *J. Hydrol.* **1998**, *204*, 182–196. [CrossRef]
53. Sen, P.K. Estimates of the regression coefficient based on Kendall's tau. *J. Am. Stat. Assoc.* **1968**, *63*, 1379–1389. [CrossRef]
54. Pal, A.B.; Khare, D.; Mishra, P.K.; Singh, L. Trend analysis of rainfall, temperature and runoff data: A case study of Rangoon watershed in Nepal. *Int. J. Stud. Res. Technol. Manag.* **2017**, *5*, 21–38. [CrossRef]
55. Schwing, F.; Murphree, T.; Green, P. The Northern Oscillation Index (NOI): A new climate index for the northeast Pacific. *Prog. Oceanogr.* **2002**, *53*, 115–139. [CrossRef]
56. Overland, J.E.; Preisendorfer, R.W. A Significance Test for Principal Components Applied to a Cyclone Climatology. *Mon. Weather Rev.* **1982**, *110*, 699–706. [CrossRef]
57. Peixoto, J.P.; Oort, A.H. *Physics of Climate*; American Institute of Physics: New York, NY, USA, 1992; 520p.

Article

Trend Analysis of Hydro-Climatological Factors Using a Bayesian Ensemble Algorithm with Reasoning from Dynamic and Static Variables

Keerthana A and Archana Nair *

Department of Mathematics, Amrita Vishwa Vidyapeetham, Kochi Campus, Ernakulam 682024, India
* Correspondence: archananair@kh.amrita.edu

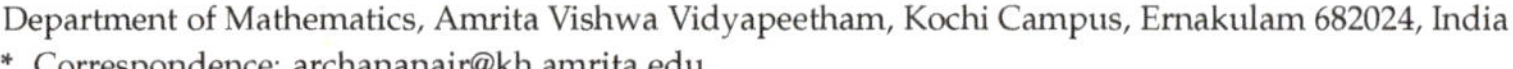

Abstract: This study examines the variations in groundwater levels from the perspectives of the dynamic layers soil moisture (SM), normalized difference vegetation index (VI), temperature (TE), and rainfall (RA), along with static layers lithology and geomorphology. Using a Bayesian Ensemble Algorithm, the trend changes are examined at 385 sites in Kerala for the years 1996 to 2016 and for the months January, April, August, and November. An inference in terms of area under the probability curve for positive, zero, and negative trend was used to deduce the changes. Positive or negative changes were noticed at 19, 32, 26, and 18 locations, in that order. These well sites will be the subject of additional dynamic and static layer investigation. According to the study, additional similar trends were seen in SM during January and April, in TE during August, and in TE and VI during November. According to the monthly order, the matching percentages were 63.2%, 59.4%, 76.9%, and 66.7%. An innovative index named SMVITERA that uses dynamic layers has been created using the aforementioned variables. The average proportion of groundwater levels that follow index trends is greater. The findings of the study can assist agronomists, hydrologists, environmentalists, and industrialists in decision making for groundwater resources.

Keywords: groundwater levels; rainfall; temperature; Mann–Kendall test; Bayesian Ensemble Algorithm

Citation: A, K.; Nair, A. Trend Analysis of Hydro-Climatological Factors Using a Bayesian Ensemble Algorithm with Reasoning from Dynamic and Static Variables. *Atmosphere* **2022**, *13*, 1961. https://doi.org/10.3390/atmos13121961

Academic Editor: Haibo Liu

Received: 20 September 2022
Accepted: 16 November 2022
Published: 24 November 2022

Publisher's Note: MDPI stays neutral with regard to jurisdictional claims in published maps and institutional affiliations.

1. Introduction

The whole life cycle of flora and fauna is impacted by hydrological and climatological changes. The combined impact is a new field of study known as hydro-climatology. All the elements of the environment are components of the water cycle, which must constantly be in balance. Life on Earth needs water to survive and be sustained, and the largest freshwater source is groundwater. In India, more than 80% of the country's needs for drinking water and 60% of its agricultural needs are met by groundwater. In the southernmost state of India, Kerala, freshwater resources are reported to be abundant. As Kerala customarily has a well for each house, the number of groundwater extraction wells has continued to grow throughout time [1]. The idea of determining the causes of oscillations in the groundwater table has arisen as a result of water shortages. Aquifers in Kerala have water depths that range from 0 to 40 mbgl. After the monsoon season, the ground water levels in most areas will be less than 5 mbgl, and subsequently, as a result of extraction, they will fall between 5 and 10 mbgl [2].

The groundwater level analysis initiates an immediate decision-making process for sustainable water management [3]. This can be done with both spatial and temporal scales [4]. Various advancements in the study of groundwater levels include groundwater storage [5,6], potential [7], resource development [8], sustainability [9], usage and management [10], trend analysis [11], and impact on agriculture and the economy [12]. The spatial variability helps the authorities understand the risks of using groundwater [13]. The trend analysis helps to identify the long-term decrease or increase in the levels. The majority of the changes seen in hydro-climatological investigations are dynamic and non-linear. In contrast, most studies have applied linear trend analysis approaches to these data sets. Mann–Kendall tests, which are also non-parametric, are widely used in the trend analysis of the groundwater level in a particular region, plain, or aquifer all over the globe. To summarize, a monotonic increasing or decreasing linear trend is applied to possibly nonlinear data. The studies related to the groundwater level trend analyses performed all over the globe include those conducted in the city of Jinan in China [14], the northwest part of Uzbekistan [15], the Tabriz plain of Iran [16], the Bacchiglione basin of Italy [17], the Yogyakarta–Sleman basin of Indonesia [18], the KwaZulu-Natal province of South Africa [19], and semi-arid Chile [20]. The trend analysis of groundwater levels carried out by researchers for different parts of India include variations in the depth of the groundwater level in the Kaithal district of Haryana state [21] and the Lower Bhavani River basin in Tamil Nadu [22]; quantification of the groundwater level trends of Gujarat [23], Punjab [24], and the semi-arid region of Telangana [25]; and groundwater variability analysis with rainfall of Jharkhand state [26]. Mann–Kendall tests are used not only for trend analysis of groundwater, but also for precipitation analysis [27,28], deposition data [29], and rainfall [30]. Modified versions of the Mann–Kendall trend test have emerged over the years. Kumar et al. [31] used four variations of Mann–Kendall in determining groundwater level trends. Recent studies have also used the innovative trend analysis method that determines the linear trend using the first and second half of the datasets [32–34].

In addition, several nonlinear approaches have been devised to overcome the limitations of the earlier linear methods. One such method is a Bayesian Ensemble algorithm called the Bayesian Estimator of Abrupt change, Seasonal change and Trend (BEAST). The primary concept is to segment the data set into seasonal signal, trend signal, and noise. The Mann–Kendall trend has been applied in a number of groundwater trend analyses, while BEAST has been applied only in a handful of studies. BEAST has several advantages over linear trend analysis methods. It forgoes the single best model concept and embraces all competing models based on Bayesian model averaging. It is a flexible tool to determine abrupt changes, cyclic variations, and nonlinear trends, and it explains how likely the detected changes are true. It is applicable to all real-valued time series datasets. These benefits make it the best method to determine the water table fluctuations. The relevant fields that have applied the BEAST method include remote sensing [35], hydrology [36], hydraulic engineering [37], atmospheric sciences [38], and climate sciences [39]. To the best of our knowledge, BEAST has not been applied to groundwater levels with the considered parameters and an index.

On a primary note, the studies so far have compared the groundwater level trends with respect to rainfall [40]. This is because rainfall acts as a natural recharge. However, in a global analysis and when analyzing a whole state as a domain, like Kerala, it is not sufficient to study just rainfall trends. Due to the rapid increase in the population, urbanization, and industrialization, along with many other developmental activities, the groundwater resources are vulnerable to depletion and quality degradation throughout the state of Kerala [41]. The other dynamic parameters under consideration are temperature, soil moisture, and normalized difference vegetation index (NDVI). The temperature is an indicator of climate change and helps to understand the water fluctuation from a climatological perspective [42]. Soil moisture plays a significant role in the hydrological cycle and tends to influence the groundwater recharge [43]. Rajesh et al. [44] used NDVI to deduce groundwater storage. Various indices are used to explain the characteristics of the groundwater level, such as the standard groundwater level index [45] and the standardized depth to water level index [46]. Presently, all the indices have been created using the levels themselves and inferences such as drought [47,48] are deduced. Studies thus far have not tried to create an index using the influential variables, which take in all the importance and contribution. With the complexity in hydraulic changes, this index can be used to determine the underlying groundwater level changes. The coefficients for the index are obtained from the Analytical Hierarchy Process (AHP). It is a commonly used multi-criteria decision-making technique that strengthens the decision-making process by verifying consistencies [49]. The applications include reinforcement of hydropower strategy [50], comparison of judgment scales [51], potential survey of photovoltaic power plants [52], and investigation of lean-green implementation practices [53]. This is first of its kind to implement AHP for creating a dynamic layer index in determining groundwater fluctuations.

The objective of the study is to analyze the Kerala groundwater level trends using BEAST in order to explain these trend changes with respect to the climatological variables of rainfall and temperature, as well as two other dynamic layers, namely soil moisture and the normalized difference vegetation index. Soil moisture is important because the rainfall needs to penetrate through the soil layer to recharge the source, while VI helps to support water retention. Apart from individual analysis, an index named SMVITERA has been developed and investigated. A final explanation is also obtained from the static layers of lithology and geomorphology. The BEAST aggregates the entire better model, and as a result, the computational part is highly mathematical. The applied studies so far have used the outputs from the model to explain trend changes. In this study, we have deduced a percentage of probability with which the trend could be positive, negative, or zero to explain a comparative level of certainty or significance. The manuscript has been arranged into 5 sections. Section 1 is the Introduction, Section 2 explains the Materials and Methods, Section 3 is the Results, Section 4 details the Discussion, and Section 5 is the Conclusions.

2. Materials and Methods

This section explains the datasets and the methodologies used to carry out this study. The datasets include the primary analysis of groundwater levels and the dynamic and static layers analyzed, whereas the methods include the AHP, Mann–Kendall, and BEAST.

2.1. Study Area and Datasets

The study is conducted in Kerala, which serves as a gateway to the Indian Monsoon. The Arabian Sea to the west, the state of Karnataka to the north, the state of Tamil Nadu to the east, and the Indian Ocean to the south form the borders of Kerala. Kerala, spread over 38,863 km^2, lies between the latitudes $8°\ 17'24''$ N to $12°\ 47'42''$ N and between the longitudes $74°\ 51'46''$ E to $77°\ 24'42''$ E. Kerala is famed for its natural beauty and backwaters, and the state is home to about 35.33 million people. The state has about 120–140 rainy days, with an average rainfall of about 2923 mm annually and a tropical climate with the seasons Summer (March–May), South–West monsoon (June–September), North–East monsoon (October–November), and Winter (December–February). The mean maximum temperature is 33 °C in March–April and 28.5 °C in July [54]. Kerala has 44 rivers, which annually yield about 70,300 mm^3 of water. Even with a copious amount of rainfall, water stress is experienced in different parts of the state during various seasons. The state is split into 14 districts: Kasaragod, Kozhikode, Kannur, Wayanad, Malappuram, Palakkad, Thrissur, Ernakulam, Alappuzha, Idukki, Kottayam, Pathanamthitta, Kollam, and Thiruvananthapuram.

The Central Groundwater Board (CGWB) maintains observation wells around the state to monitor groundwater levels. Meters below ground level (mbgl) are measured four times a year—January, April, August, and November (JAAN)—symbolizing the Winter, Pre-monsoon, Monsoon, and Post-monsoon seasons, respectively. These levels are freely accessible from the India Water Resources Information System (WRIS) (https://indiawris. gov.ins, accessed on 12 January 2022). The wells having at least 90% of the data from 1996 to 2016 during the JAAN months were chosen as suitable research locations, resulting in 385 well locations across the state (Figure 1). The studied well distributions are 41 in Malappuram, 40 in Palakkad, 37 in Thrissur, 34 in Ernakulam, 32 in Thiruvananthapuram, 30 in Kozhikode, 28 in Kollam, 23 in Kannur, 23 in Pathanamthitta, 22 in Alappuzha, 22 in Kottayam, 21 in Wayanad, 20 in Kasaragod, and 12 in Idukki.

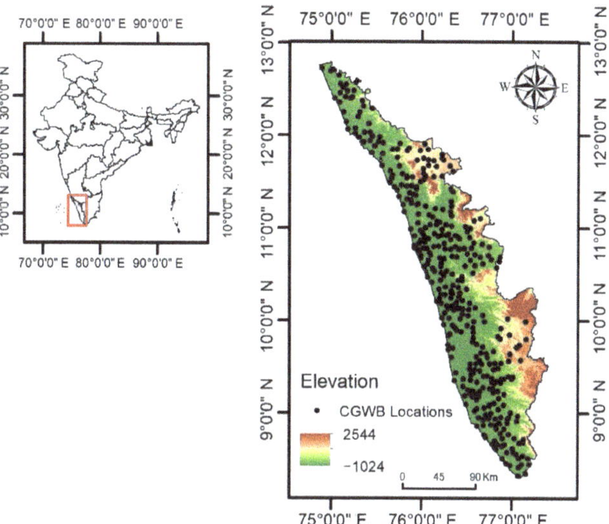

Figure 1. Kerala Map with the studied well locations on a base map of elevation.

2.2. Variable Pre-Processing

The groundwater level characteristics are analyzed with the help of the dynamic layers, namely, Soil Moisture (SM), Normalized Difference Vegetation Index (VI), Temperature (TE), and Rainfall (RA). SMVITERA, an index based on all these layers, has been developed to improvise a combined relationship.

2.2.1. Dynamic Layers

The Global Monthly Soil moisture (in millimeters) from NOAA-NCEP-CPC-GMSM is available as GeoTIFF at the IRI digital library maintained by the International Research Institute of Climate and Society, Columbia Climate School (http://iridl.ldeo.columbia.edu/SOURCES/.NOAA/.NCEP/.CPC/, accessed on 28 January 2022). Soil moisture is assessed by a one-layer hydrologic model [55], which uses observed precipitation and temperature as inputs and computes the soil moisture, evaporation, and runoff [56]. The monthly raster files are downloaded for north latitudes $8°$ N to $13°$ N and east longitudes $74°$ E to $78°$ E, having a spatial resolution of $0.5° \times 0.5°$ during the years 1996 to 2016 and in the JAAN months. In the Geographic Information System (GIS) environment, the grid point values are extracted and interpolated with the Inverse Distance Weighted (IDW) technique. IDW is calculated using the equation [57]:

$$Z_{(S_0)} = \sum_{i=1}^{N} \lambda_i Z_{(S_i)}$$

where N is the number of sampling points, λ_i is the weightage of each sample point, S_0 represents the set of sampling points in the neighborhood of S_i, $Z_{(S_0)}$ is the prediction at S_0, and $Z_{(S_i)}$ is the measured value at S_i.

The Normalized Difference Vegetation Index (VI) is an indicator of the greenness of biomes on the Earth's surface and is essential to monitoring the changes in the ecosystem. The inputs are reprocessed data from SPOT-VEGETATION and PROBA-V missions [58–61]. The formula used is:

$$NDVI = \frac{REF_{nir} - REF_{red}}{REF_{nir} + REF_{red}}$$

where REF_{nir} and REF_{red} are the spectral reflectance measured in near-infrared and red wavebands. The NetCDF NDVI files are downloaded from Copernicus Global Land Services on a scale of 10 days and 1 km resolution. Three raster images are used and averaged using Cell statistics in GIS to obtain monthly NDVI for the years from 1999 to 2016 and in the JAAN months. Focusing on the Kerala region, it is observed that there are missing values in a few regions. As a result, a model was built in ArcMap that takes these raster layers, converts them into point shapefiles, and then interpolates them using IDW interpolation. Likewise, all the raster layers are created for the considered months.

The climatological data sets on daily rainfall with a spatial resolution of $0.25^{\circ} \times 0.25^{\circ}$ [62], and maximum and minimum temperatures of $1^{\circ} \times 1^{\circ}$ spatial resolution [63], are downloaded from the India Meteorological Department (IMD) for the years 1996 to 2016 and in the JAAN months. The mean of the maximum and minimum temperatures will be the average temperature. These layers are then linearly interpolated to $0.25^{\circ} \times 0.25^{\circ}$. Using the GIS environment, all of these gridded datasets are finally interpolated using the IDW technique and the associated values for the 385 locations, the 21 years (1996 to 2016), and the JAAN months are retrieved.

2.2.2. SMVITERA Index

Each dynamic layer has its own relevance and importance regarding its water holding capacity. Moreover, their positive influence need not always result in an increase in water availability. It is difficult to analyze the variables and their influences separately, and doing so might not yield expected results as well. Here, an approach has been made to create an index based on the above-discussed dynamic layers. The index has been named after combining the layers of Soil Moisture, the normalized difference Vegetation Index, Temperature, and RAinfall to explain the effect in one instance. The formula takes the form:

$$SMIVITERA = aSM + bVI + cTE + dRA \tag{1}$$

Using the Analytical Hierarchical Process (AHP), the coefficients a, b, c and d are determined as signed normalized weights depending on the influence on water recharge. Depending on how they relate to groundwater recharge, the coefficients are given appropriate signs. The index's trend analysis aids in pinpointing the combined result and the better one more precisely.

2.2.3. Static Layers

To understand the hydro variability, which assists in determining the land setting under each analyzed place, the static layers of the lithology and geomorphology are examined. The features of a single rock unit found in the Earth's crust are the subject of the subfield of lithology in Earth sciences. Geomorphology is the study of landforms, including their formation, shapes, and underlying deposits. Bhukosh (https://bhukosh.gsi.gov.in/Bhukosh/Public, accessed on 20 February 2022), a portal and gateway to all geoscientific data of the Geological Survey of India (GSI), is where the lithology and geomorphology of the state of Kerala are found.

2.3. Methods

2.3.1. Analytical Hierarchical Process (AHP)

The Analytical Hierarchical Process (AHP) [64] is applied to obtain the index coefficients, which form a Pairwise Comparison Matrix (PCM). The PCM matrix must be normalized by dividing with the column sum. The weight vector is the average of the total, and then multiplying it by 100 gives the normalized weightage of each variable layer. The consistency of the PCM is found with the help of the Principal Eigenvalue (λ_{max}), Consistency Index (CI), and Consistency Ratio (CR). Multiply PCM by the Weight vector (W) to obtain the product matrix (M). Now, divide each of the cells in M by the row value of

W to get the Consistency Vector Matrix (CVM), whose average is the Principal Eigenvalue. The Consistency Index is obtained using the equation:

$$CI = \frac{\lambda_{max} - n}{n - 1} \tag{2}$$

The Consistency Ratio (CR) is calculated using the equation:

$$CR = \frac{CI}{RI} \tag{3}$$

where RI denotes the consistency indices for randomly produced reciprocal matrices [65]. The inconsistency is acceptable if CR is less than or equal to 0.1.

2.3.2. Mann–Kendall Trend Test

The Mann–Kendall [66,67] test is a non-parametric test to identify trends, in which all data is compared to all subsequent data. The null hypothesis states that there is no trend and the alternate hypothesis states that there exists an increasing or decreasing trend in the time series. Let $x_1, x_2, x_3, \ldots, x_n$ be a time series of length n. The Mann–Kendall test Statistic S is calculated using,

$$S = \sum_{k=1}^{n-1} \sum_{j=k+1}^{n} sign(x_j - x_k)$$

where,

$$sign(x_j - x_k) = \begin{cases} 1 & if \quad x_j - x_k > 0 \\ 0 & if \quad x_j - x_k = 0 \\ -1 & if \quad x_j - x_k < 0 \end{cases}$$

The statistic is normally distributed with mean and variance $E(S) = 0$ and $V(S) = \frac{1}{18}\left[n(n-1)(2n-5) - \sum_{p=1}^{g} t_p(t_p - 1)(2t_p + 5) \right]$, where g is the number of tied groups and t_p is the number of data points in the p th group. If there are no tied groups, the variance becomes,

$$V(S) = \frac{n(n-1)(2n-5)}{18}$$

The normalized test statistic Z in the Mann–Kendall test is,

$$Z = \begin{cases} \frac{S-1}{[VAR(S)]^{\frac{1}{2}}} & if \quad S > 0 \\ 0 & if \quad S = 0 \\ \frac{S+1}{[VAR(S)]^{\frac{1}{2}}} & if \quad S < 0 \end{cases}$$

Considering a significance level α, the null hypothesis is rejected if the absolute value of Z is greater than the table value of $Z_{1-\frac{\alpha}{2}}$. Once the null hypothesis is rejected, the time series exhibits a rise or fall.

The magnitude of the trends is obtained using the Theil–Sens approach [68–70], where the slope β is given by, $\beta = median\left[\frac{x_j - x_i}{j - i} \right]$, for all $i < j$.

2.3.3. Bayesian Estimator of Abrupt Change, Seasonal Change, and Trend (BEAST)

BEAST is the Bayesian Ensemble Algorithm used to analyze the time series and the change points [71]. The method was initially developed to study satellite time series data and can be applied to any time series that satisfies the assumptions [72]. The model presupposes that a time series can be divided into four components: the seasonal component modeled using a harmonic function, a background component modeled using a piecewise linear regression function, a certain number of possible change points for both of the components, and a certain amount of random noise. The search is for a relationship between the n points of time t and the corresponding data y, combined as time series $D = \{y_i, t_i\}$ for $i = 1, 2, \ldots, n$ via a statistical decomposition model $\hat{y}(t) = f(t)$. The model treats the time series $y_i = y(t_i)$ as a composite of the seasonal $S(.)$ and trend $T(.)$ components, abrupt changes, and the noise, formulated as follows:

$$y_i = S(t_i, \Theta_S) + T(t_i, \Theta_T) + \varepsilon_i \tag{4}$$

$S(.)$ and $T(.)$ represent the seasonal and trend signals. The noise captures the data that is not explained by these signals ε_i, which is assumed to follow Gaussian with a magnitude of σ. The general linear models are adopted to parameterize signals $S(.)$ and $T(.)$ [73,74]. The abrupt changes in the signals are implicitly encrypted in the parameters Θ_S and Θ_T, respectively.

A piecewise harmonic model is used to approximate the seasonal signal $S(t)$ with respect to p knots. These p knots divide the time series with starting time $\xi_0 = t_0$ and ending time $\xi_{p+1} = t_n$ into $p + 1$ intervals, such as $[\xi_0, \xi_1], [\xi_1, \xi_2], \ldots, [\xi_p, \xi_{p+1}]$. For each of the $p + 1$ intervals, denoted by $[\xi_k, \xi_{k+1}]$, for $k = 0, \ldots, p$, the model takes the form:

$$S(t) = \sum_{l=1}^{L_k} \left[a_{k,l} \sin\left(\frac{2\pi l t}{P}\right) + b_{k,l} \cos\left(\frac{2\pi l t}{P}\right) \right], \text{ for } \xi_k \leq t \leq \xi_{k+1}, \ k = 0, \ldots, p.$$

where P is the period of the seasonal signal, L_k is the harmonic order for the k-th segment, $a_{k,l}$ is the parameter for the sine function, and $b_{k,l}$ is the parameter for the cosine function. The seasonal harmonic curve is specified by the following set of parameters:

$$\Theta_S = \{p\} \cup \{\xi_k\}_{k=1,\ldots,p} \cup \{L_k\}_{k=0,\ldots,p} \cup \{a_{k,l}, b_{k,l}\}_{k=0,\ldots,p; l=1,\ldots,L_k}$$

A piecewise linear function is used to model the trend signal $T(t)$ with respect to m knots. These m knots divide the time series with starting time $\tau_0 = t_0$ and ending time $\tau_{m+1} = t_n$ into $m + 1$ intervals, such as $[\tau_0, \tau_1], [\tau_1, \tau_2], \ldots, [\tau_m, \tau_{m+1}]$. The trend changepoints denoted by τ_j need not be same as the seasonal changepoint ξ_k. For each of the $m + 1$ intervals, denoted by $[\tau_j, \tau_{j+1}]$, for $j = 0, \ldots, m$, the model is a line segment of the form:

$$T(t) = a_j + b_j t, \text{ for } \tau_j \leq t \leq \tau_{j+1}, \ j = 0, \ldots, m.$$

where a_j and b_j are the coefficients. The linear trend curve is specified by the following set of parameters:

$$\Theta_T = \{m\} \cup \{\tau_j\}_{j=1,\ldots,m} \cup \{a_j, b_j\}_{j=0,\ldots,m}$$

The parameters Θ_T and Θ_S are reclassified into two groups, M and β_M. The group M refers to the model structure that includes the number and timings of the seasonal and trend changepoints and the seasonal harmonic order.

$$M = \{m\} \cup \{\tau_j\}_{j=1,\ldots,m} \cup \{p\} \cup \{\xi_k\}_{k=1,\ldots,p} \cup \{L_k\}_{k=0,\ldots,p}$$

The group β_M comprises the segment specific coefficient parameters and is used to determine the exact shapes of the seasonal and trend curves.

$$\beta_M = \{a_j, b_j\}_{j=0,\ldots,m} \cup \{a_{k,l}, b_{k,l}\}_{k=0,\ldots,p; l=1,\ldots,L_k}$$

Thus, Equation (4) becomes the following:

$$y(t_i) = x_M(t_i)\beta_M + \varepsilon_i \tag{5}$$

Here, $x_M(t_i)$ and β_M are the dependent variables and associated coefficients.

In Bayesian modeling, for the time series $D = \{y_i, t_i\}$ for $i = 1, 2, \ldots, n$, the goal is to find the posterior probability distribution $p(\beta_M, \sigma^2, M | D)$. Using Baye's theorem, the posterior is the product of the likelihood and a prior model:

$$p\left(\beta_M, \sigma^2, M | D\right) \propto p\left(D | \beta_M, \sigma^2, M\right) \pi\left(\beta_M, \sigma^2, M\right) \tag{6}$$

The likelihood being Gaussian is $p(D|\beta_M, \sigma^2, M) = \prod_{i=1}^{n} N(y_i; x_M(t_i)\beta_M, \sigma^2)$ and the prior distribution is $\pi(\beta_M, \sigma^2, M) = \pi(\beta_M, \sigma^2 | M)\pi(M)$. Firstly, for $\pi(\beta_M, \sigma^2 | M)$, a normal-inverse Gamma distribution is considered and an extra vague dispersion hyper parameter ν is incorporated to reflect our vague knowledge of β_M. Secondly, for the prior $\pi(M)$, the number of changepoints are assumed to be non-negative numbers, which is equally probably a prior. Thus, the posterior model Equation (6) becomes:

$$p\left(\beta_M, \sigma^2, \nu, M | D\right) \propto \prod_{i=1}^{n} N\left(y_i; x_M(t_i)\beta_M, \sigma^2\right) \cdot \pi_\beta\left(\beta_M, \sigma^2, \nu | M\right) \cdot \pi(M)$$

The posterior distribution $p(\beta_M, \sigma^2, \nu, M | D)$ encodes all the essential information for understanding the ecosystem dynamics. As this is analytically intractable, Markov Chain Monte Carlo (MCMC) sampling is used to generate a realization of random samples for the posterior inference. A hybrid sampler that embeds a reverse jump (RJ) MCMC sampling into a Gibbs sampling framework is used. More details about the RJ-MCMC can be found in Zhao et al. [75].

2.4. Challenges and Betterness

The Mann–Kendall and Bayesian Ensemble Algorithm are assessed through various circumstances and conditions, and the following inferences are obtained:

- Mann–Kendall is a linear, nonparametric trend test often applicable to a monotonic moving dataset. As the real datasets on hydro climatological factors exhibit vibrant and non-cyclic changes, fitting a linear trend is not apt. The Bayesian algorithm overcomes this challenge by finding piecewise linearly fitted trends.
- The Mann–Kendall trend test fits a single trend line to the datasets. The interpretation of time series solely depends on the models being utilized and finding the optimal models is never easy. BEAST offers the single best model approach and adopts the concept of embracing all the models using the Bayesian model averaging scheme, which is always better.
- Sen's slope technique produces an overall slope for the Mann–Kendall test. BEAST provides the slope at each time series point of the data sets. Additionally, the Bayesian process may be used to determine how likely an event is to occur.
- The data point is removed from computations using the Mann–Kendall test if there are any missing values. Consequently, the trend analysis only considers non-NaN or non-missing variables, while the final fitted trend in BEAST will be continuous even when missing values are provided, eliminating the repercussions of missing values and discontinuity in the outcome.
- Mann–Kendall is highly sensitive to outliers, while the sensitivity is comparatively less for the Bayesian approach. BEAST studies the nature of the whole dataset rather than focusing on the end points.

2.5. Inference and Modulation

An advantage of Mann–Kendall over BEAST is that the significance level can be stated for the increasing or decreasing trend. To solve this, we used the implications from the percentage areas of positive, zero, and negative slope from the output to conclude which is greater. The method varies from traditional trend analysis in that it yields findings that take into account the likelihood of the slope. In other words, it is possible to determine the probability that the slope will be positive, zero, or negative at each time series point.

The BEAST has five major components. The first component is the piecewise linear trend along with the change point detected, the second component is the probability of the trend change point occurrence over the considered time, the third component is the order of the time-varying polynomial needed to approximate the fitted trend, the fourth component is the probabilities of the trend, and the fifth component is the error. Let the probability time series of the positive slope be denoted as $p(x)$, negative slope be denoted as $n(x)$, and zero slope be denoted as $z(x)$, for $i = 1, \ldots, N$. The area A under these probabilities is calculated using the trapezoidal method, formulated as:

$$A = \int_a^b f(x)dx \approx \frac{b-a}{2N} \sum_{n=1}^{N} (f(x_n) + f(x_{n+1}))$$

The percentage area will be $PA = \frac{A}{N-1} \times 100$.

The percentage area under each of the probability curves supplies information regarding the overall slope of the fitted trend. The greater percentage areas are taken into account to explain the nature of trend. These percentages are used instead of the significance level to convey the confidence of getting a particular trend. The piecewise linear trend helps to identify the deflection in slope at each location. Deducing an overall slope sign is explained in relation to the greatest percentage, while the slope is obtained by linearly fitting the time series trend data sets.

3. Results and Discussion

This section deals with the trend results obtained using BEAST over the datasets of groundwater levels, rainfall, temperature, soil moisture, NDVI, and SMVITERA index. The trends of these variables are compared with the groundwater level trends to obtain feasible interpretations about the influences occurring in the context of the considered years. The groundwater level trends are explained using the dynamic layers and the introduced index, as well as the static layers.

3.1. SMVITERA Index Formulation

The index acts as an integrated variable of the studied dynamics. The initial layer weights are assigned using Satty's scale, on a scale of 1 to 9. The value 1 implies equal significance, 3 indicates moderate relevance, 5 indicates strong importance, 7 indicates extremely strong importance, and 9 indicates extreme importance. The intermediate numbers 2, 4, 6, and 8 correspond to the intermediate effects. The weights are set based on the variable relevance and contribution to the water holding capacity. The vegetation index is assigned weight 8, moisture is assigned 7, rainfall is assigned 4, and temperature is assigned 2. The Pairwise Comparison Matrix (PCM) is shown in Table 1, and the normalized PCM and the final matrices are shown in Table 2.

Table 1. Pairwise comparison matrix for the variable layers.

Thematic Layer	Assigned Weight	VI	SM	RA	TE
VI	8	8/8	8/7	8/4	8/2
SM	7	7/8	7/7	7/4	7/2
RA	4	4/8	4/7	4/4	4/2
TE	2	2/8	2/7	2/4	2/2

Table 2. Normalized pairwise comparison matrix and the final matrices.

Thematic Layer	VI	SM	RA	TE	Total	Weight Vector (*W*)	Weightage	Product Matrix	Consistency Vector Matrix (CVM)
VI	8/8	8/7	8/4	8/2	1.52	0.38	38	1.52	3.998
SM	7/8	7/7	7/4	7/2	1.33	0.33	33	1.33	4.006
RA	4/8	4/7	4/4	4/2	0.76	0.19	19	0.76	3.998
TE	2/8	2/7	2/4	2/2	0.38	0.10	10	0.38	4.015

It is obtained that the normalized weights are 38 to normalized difference vegetation index, 33 to soil moisture, 19 to rainfall, and 10 to temperature. The Principal Eigenvalue is:

$$\lambda_{max} = \frac{3.998 + 4.006 + 3.998 + 4.015}{4} = \frac{16.017}{4} = 4.00425$$

The Consistency Index for $n = 4$ is calculated as follows:

$$CI = \frac{4.00425 - 4}{4 - 1} = \frac{0.00425}{3} = 0.0014167$$

For $n = 4$, RI is 0.89. The Consistency Ratio (*CR*) is calculated as:

$$CR = \frac{0.0014167}{0.89} \approx 0.001592$$

Since the obtained *CR* is much less than the threshold value, the PCM exhibits a high level of consistency.

The variables soil moisture and temperature are inversely proportional to groundwater levels. As a result, these layers are assigned negative signs in calculating the SMIVITERA index for 1999 to 2016.

$$SMIVITERA = -33SM + 38VI - 10TE + 19RA$$

3.2. General Characteristics of Variables

The analysis is on 385 well locations over Kerala state during January, April, August, and November. Groundwater levels aid in determining how much freshwater is available for consumption. It also helps in analyzing the past, tracking the present, and forecasting the future. For the years from 1996 to 2016, the average groundwater level was 0.63 to 23.48 mbgl in January, 0.82 to 24.11 mbgl in April, 0.35 to 22.03 mbgl in August, and 0.36 to 22.81 mbgl in November. The depth to water level can be used to calculate inverse water availability. For instance, the depth to water level being 2 mbgl and 5 mbgl indicates that the well with a recorded water level of 2 mbgl contains more water than the well with a water level of 5 mbgl. The variables analyzed to groundwater levels include a combination of dynamic and static layers. The dynamic variables are Soil Moisture (SM), Normalized Difference Vegetation Index (VI), Rainfall (RA), Temperature (TE), and a developed Groundwater Index (SMVITERA) based on the former dynamic layers. Lithology (LI) and geomorphology (GM) are the static factors investigated.

Rainfall and temperature are climatological factors that influence groundwater levels. Rainfall is the natural source of recharge to groundwater, but as the temperatures rises, more water is needed for all living things to consume. However, increased rainfall may not necessarily result in increased recharge since the other variables influence the site. The daily average rainfall for January, April, August, and November are evaluated. The daily rainfall values range from 0 to 0.37 mm during January, 0 to 3.986 mm during April, 0 to 26.753 mm during August, and 0 to 5.66 mm during November. It is observed that the summer or pre-monsoon season receives more average rainfall than the winter season. Temperature ranges are comparably narrower. Temperatures vary from 20.55 to 26.34 °C in January, 23.18 to 29.05 °C in April, 20.52 to 27.43 °C in August, and 20.94 to 26.53 °C in

November. Temperature ranges are highest in April due to summer and lowest in January due to winter. The average groundwater levels, rainfall, and temperature for the years from 1996 to 2016 during the JAAN months is shown in Figure 2.

The soil characteristics, such as its moisture content, help to determine the water permeability of the soil. As surface moisture increases, the water condenses on the Earth surface and does not contribute to the underlying groundwater. During the months of January, April, August, and November, it is observed that the soil moisture varies from 98.04 to 148.27 mm, 60.59 to 111.54 mm, 109.78 to 238.78 mm, and 132.33 to 203.89 mm, respectively. Owing to the monsoon, soil moisture levels are highest in August and lowest in April due to the summer season. The normalized difference vegetation index is in the range of 0 to 1. The closer the index is to 1, the greater the content of greenery in the region. The observed ranges are 0.37 to 0.83 in January, 0.29 to 0.8 in April, 0.28 to 0.75 in August, and 0.37 to 0.82 in November. With the help of all these values of dynamic layers, the SMVITERA index is calculated. The obtained index value lies in the range -5097.24 to -3368.55 in January, -3957.95 to -2216.27 in April, -8031.24 to -3426.74 in August, and -6864.86 to -4436.84 in November. The average soil moisture, NDVI, and SMVITERA index during the JAAN months are shown in Figure 3.

Over the state of Kerala, 77 divisions of lithological formations are found (Figure 4). Most of the state is covered by Acid to Intermediate Charnockite and the least by Terri Sand. The lithological formations of the studied 385 locations are of 27 categories: Acid to intermediate charnockite, Biotite gneiss, Clay, Clayey sand, Diorite, Gabbro, Garbiosill gneiss+graphite+kyanite, Garnet gneiss, Garnet biotite gneiss, Garnet sillimanite gneiss+graphite+cordierite, Granite, Granite gneiss, Grey fine sand, Hornblende biotite gneiss, Hornblende biotite syenite, Laterite, Mica Schist, Mylonite, Pyroxene granulite, Quartzite, Sand (active channel as well), Sandstone, Sericite schist, Talc tremolite actinolite, schist, and Terri Sand. A total of 18 geomorphological formations are found in the state of Kerala (Figure 5). Most of the state is covered by Pediment Pediplain Complex and Highly Dissected Hills and Valleys. Over the 385 well locations that have been studied, the formations found are Alluvial plain, Coastal plain, Dam and Reservoir, Deltaic plain, Flood plain, Low dissected plateau, Pediment pediplain complex, Waterbody—river, and also highly, moderately, and low dissected hills and valleys.

3.3. Groundwater Level Trend Analysis

The BEAST is performed over the 385 sites and individually for the JAAN months. The time series has 21 time points and spans the years 1996 to 2016. The groundwater level dataset is fitted with a piecewise linear trend even in the presence of missing values using BEAST. At each of the 21 time points, the slope of this fitted trend is also determined. The sites having the percentage areas greater in the positive or negative slope are chosen as the locations for further analysis. This positivity or negativity is also verified by linearly modelling the trend data points. Over the studied 385 locations, a combination of probabilities is obtained. During the month of January, the Well number 1 (Konnakuzhi-iii) situated in the Ernakulam district showed the greater probability of groundwater level trend to be zero (Figure 6a). That is, the green area indicating zero slope dominates in the graph. The percentage areas were 7.9% for positive slope, 58.7% for zero slope, and 33.4% for negative slope. The linear modelling showed the slope to be -0.0068, approximately zero, and the negative slope contribution is visible as the second highest percentage as well. Another example is during the month of April on the well number 170 (Kottanadu) situated in Pathanamthitta district, which showed the areas of green and blue approximately to be the same (Figure 6b). This means that there exists an approximately equal probability for the slope to be zero and negative. The percentage areas were obtained as 12.1% for positive slope, 43.3% for zero slope, and 44.6% for negative slope. We are typically interested in the well locations where the probability for negative or positive is the highest over each site. One such example is well location 20 (Pattambi) situated at Palakkad district, which has a greater percentage for positive slope during the month of April (Figure 7a), and well location 32 (Poonkulam) in Thiruvananthapuram district, which has a greater percentage for negative slope during the month of August (Figure 7b). The percentage areas are around 63.9% and 79.9%. The linearly modelled slopes are 0.0634 and −0.246.

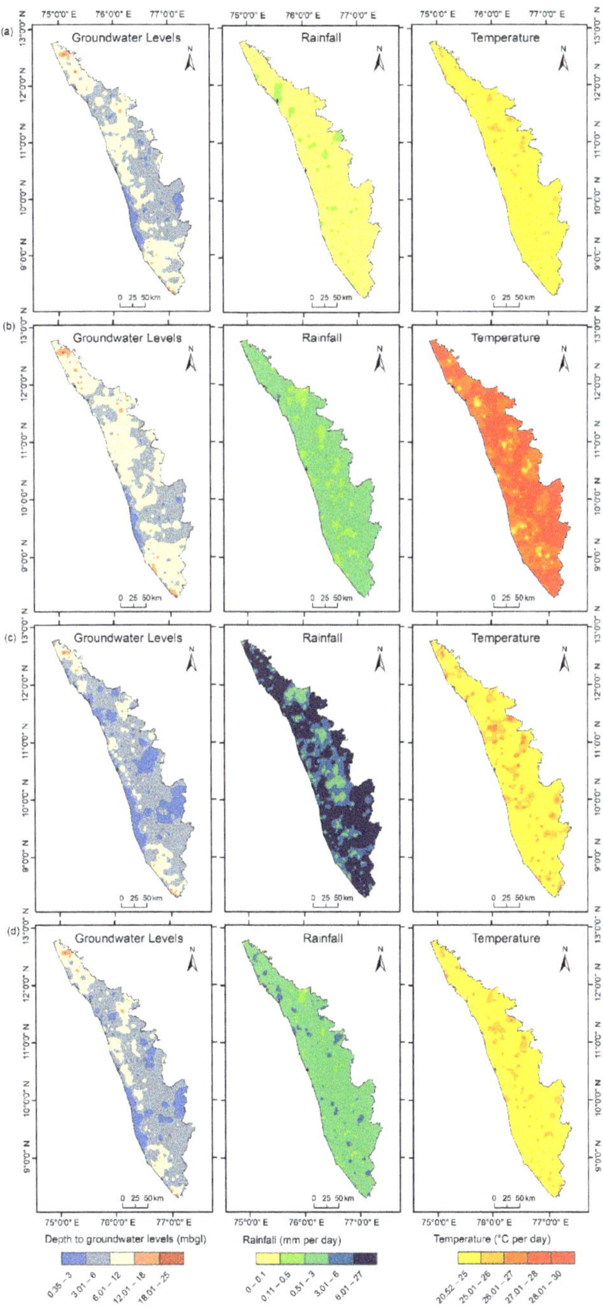

Figure 2. The average groundwater levels, rainfall, and temperature for the years 1996 to 2016 during (**a**) January, (**b**) April, (**c**) August, and (**d**) November.

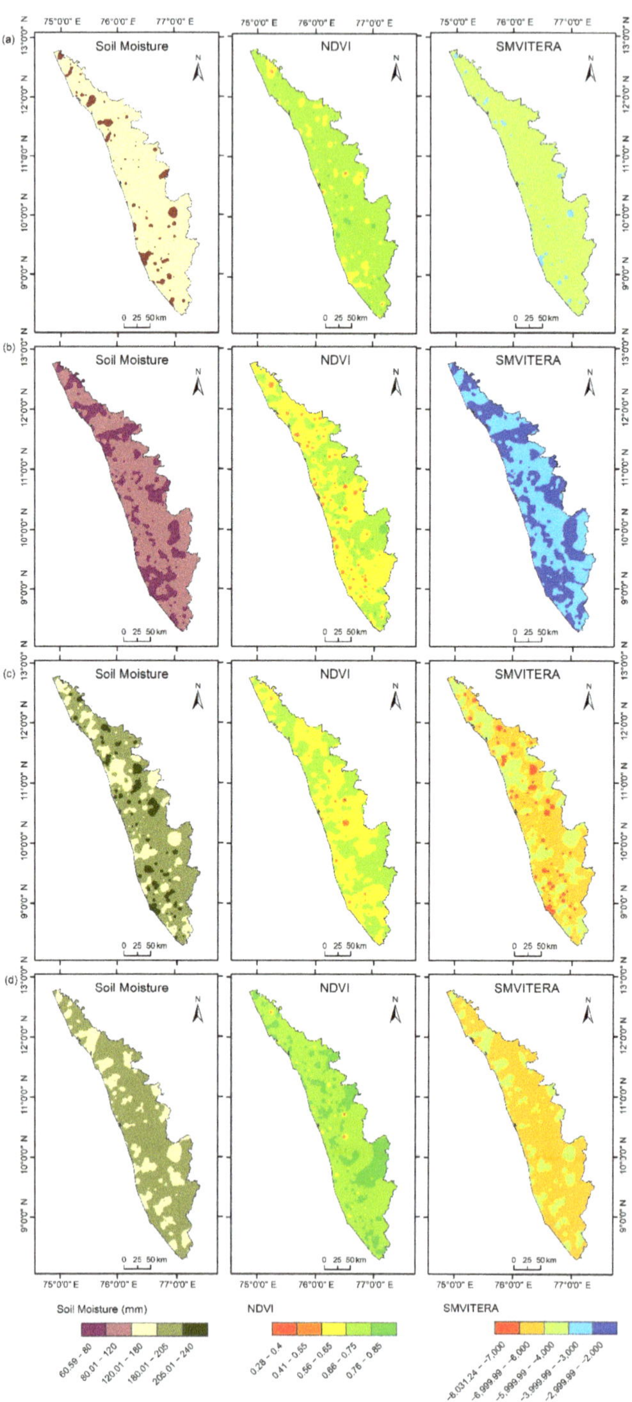

Figure 3. The average Soil Moisture, NDVI, and SMVITERA index during (**a**) January, (**b**) April, (**c**) August, and (**d**) November.

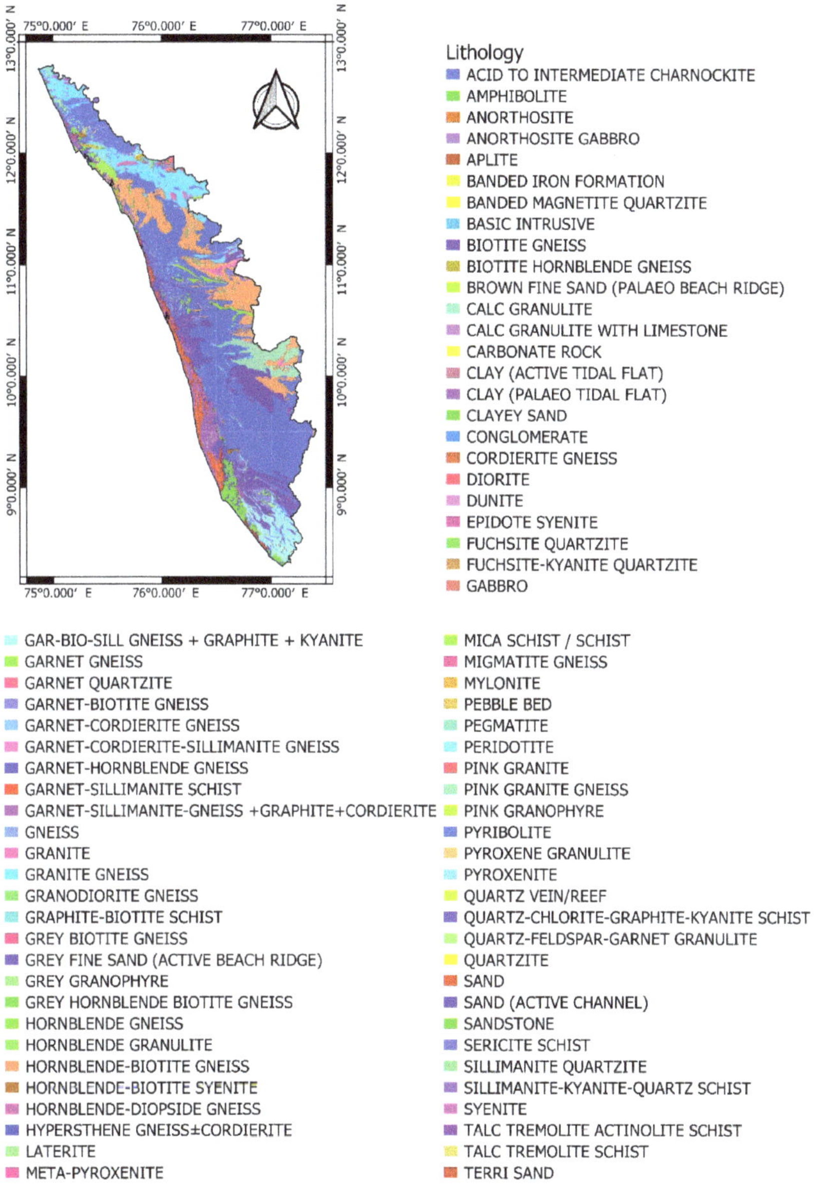

Figure 4. Kerala lithology map with legend subdivisions.

In the month of January, 19 locations resulted in comparatively positive or negative slope probability percentage areas. The locations are Poonkulam, Pambadi, Minangadi, Vayyakara, Angadimogar, Rajapuram, Cherthalai, Tachinganedam, Malipuram, Tenkara, Kuvapalli, Kayamkulam, Idukki, Parassala, Maruthamala, Vandiperiyar, Poonthura, Kuttattukulam, and Kottapuram2. The district-wise distribution is four in Thiruvananthapuram; three in Ernakulam; two each in Alappuzha, Idukki, Kasaragod, and Kottayam; and one each in Kannur, Malappuram, Palakkad, and Wayanad. Out of these, 10 locations have negative groundwater level trend, while nine wells have positive groundwater level trend.

The areas under the probabilities of positive trend vary from 51.6% to 89.5%, while the area under the probabilities of positive trend varies from 48.6% to 88.7%. The greatest percentage for positive trend is observed in the Tenkara location and the linear model showed the slope to be 0.0622, whereas the greatest percentage for negative trend is observed in the Kottapuram2 location and the slope obtained through linear modeling is −0.0512.

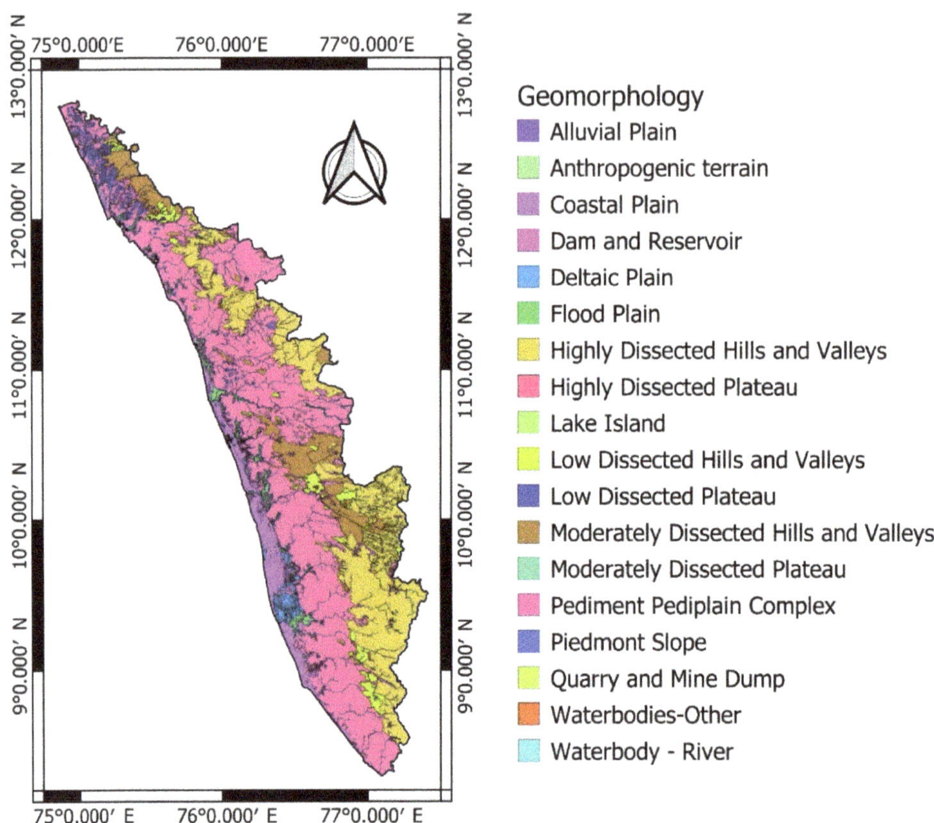

Figure 5. Kerala geomorphology map with legend subdivisions.

During the month of April, a total of 32 locations had greater percentages for slope to be positive or negative. The well location names are Chamravattom, Pattambi, Quilandy, Mavoor-ii, Mudikode, Vayyakara, Edavai, Thirunavaya, Ramantalai, Irikkur, Kottakkal, Chelakod, Thiruvallur, Cherthalai, Kottanadu, Chengamanad, Kayamkulam, Parassala, Shoranur, Mulankunnathukavu, Angamali1, Bandadka, Manattana, Pukkundu, Paral, Attingal, Palghat, Muligadde, Maruthamala, Mavoor-I, Murukumpuzha (R1), and Pathanamthitta. The district-wise well distribution is two each in Alappuzha, Ernakulam, Kasaragod, Pathanamthitta, and Thrissur; three in Palakkad; four in Kozhikode; and five each in Kannur, Malappuram, and Thiruvananthapuram. Out of these, 7 locations showed positive trend and 25 locations showed negative trends. The area under the probabilities of positive trend varies from 46.1% to 92.9%, while the area under the probabilities of negative trend varies from 42.5% to 81.3%. The greatest percentage for negative trend is observed in the Mavoor-I location and the linear model showed the slope to be 0.0622, whereas the greatest percentage for positive trend is observed in the Thirunavaya location and the slope obtained through linear modeling is −0.0512.

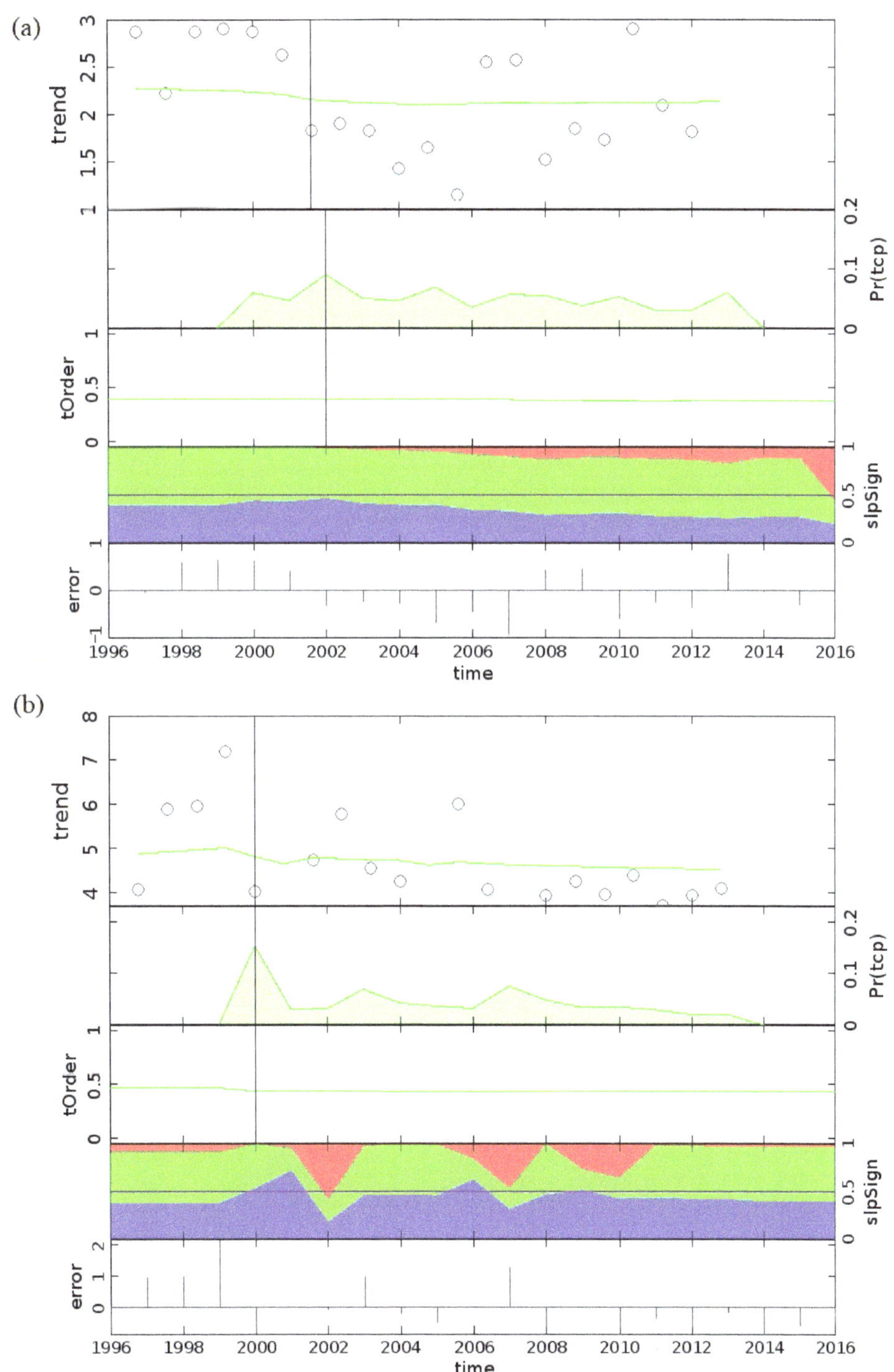

Figure 6. BEAST plot specifically for well locations (**a**) Konnakuzhi-iii and (**b**) Kottanadu.

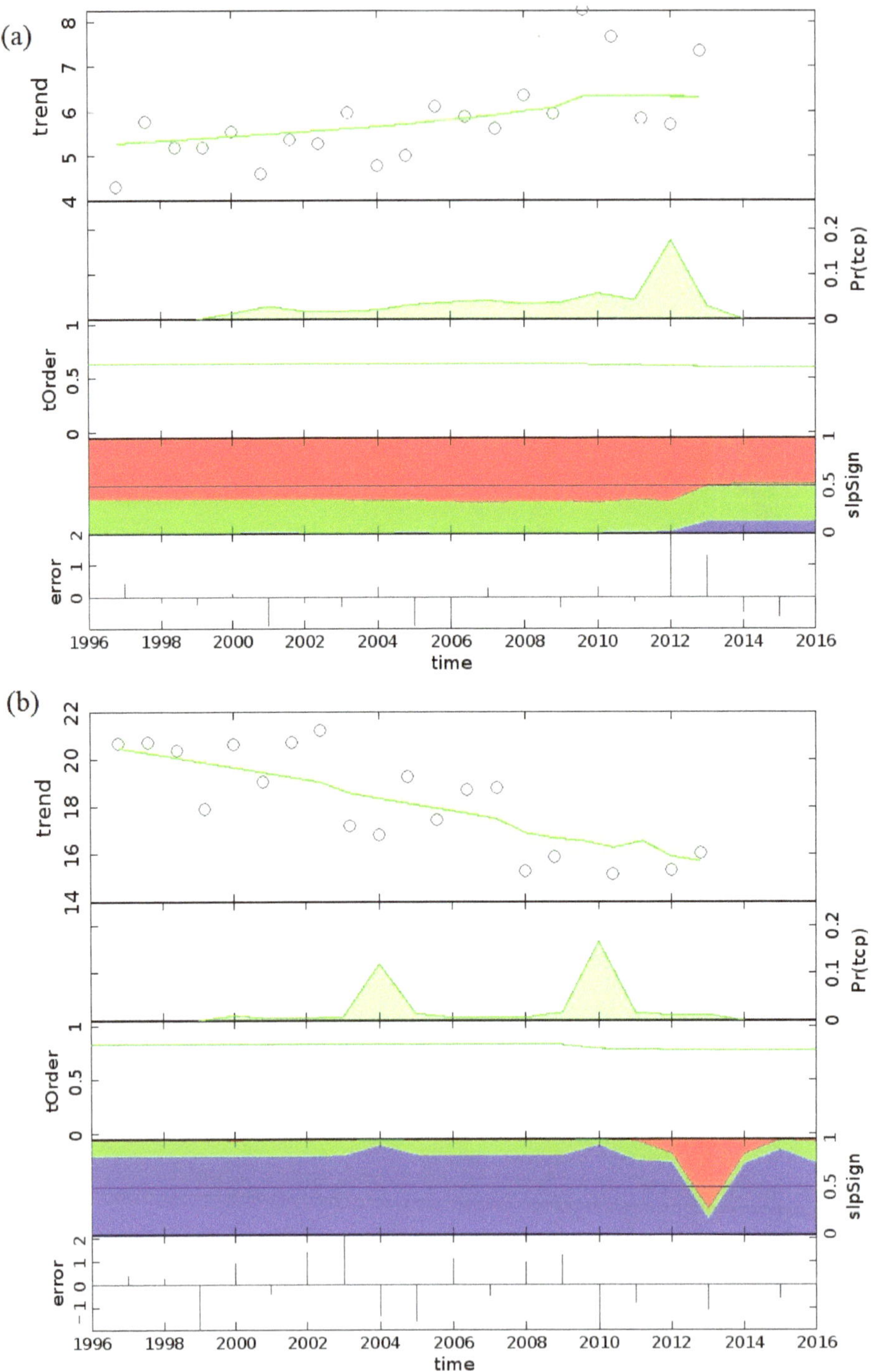

Figure 7. BEAST plot specifically for well locations (**a**) Pattambi and (**b**) Poonkulam.

In the month of August, a total of 26 wells showed trend changes. The locations are Kakkayam, Alwaye, Tholanur, Poonkulam, Pudukayam, Irikkur, Aruvikara, Mattanur, Perambra, Trichur, Akkal, Devarkoil, Lakkidi, Tenkara, Vellamunda, Kannavam, Varkala, Muliyar, Angamali1, Manattana, Cherukunnu, Attingal, Puthenchira, Elanthur, Vallom1, and Chakkarakkale. The positive trend locations count to 20 and negative trend locations count to 6. The month of August represents the monsoon season in Kerala, and the results show positive groundwater level trends in most of the well locations. The possible explanation is that the amount of water getting recharged has reduced, or over-exploitation has occurred. The percentages of area obtained lies in the range 48.9% to 83.7% for negative slope and 46.6% to 92.1% for positive slope.

The post monsoon season is indicated by the month of November. Eighteen wells showed positive or negative groundwater level trends over the studied 385 locations. The wells are Ponnani1, Tholanur, Poonkulam, Pudukayam, Arukutti, Angadimogar, Nedumudi (pupalli), Malipuram, Tenkara, Vellamunda, Kuvapalli, Manamangalam, Mavinakatta, Malampuzha, Ailara, Elappara, Murukumpuzha (R1), and Chadayamangalam1. Out of these locations, 5 wells showed negative trend and 13 wells showed positive trend. The percentages of area obtained lies in the range 47.1% to 50.7% for negative slope and 48.6% to 86.1% for positive slope. It is observed that the percentage that indicates a declining trend of depth to water level is less during the month of November.

From the positive trend, the inference is that the depth to groundwater levels is increasing and, as a result, the water availability is decreasing. Thus, a track on positive trend wells provides a balance to the water table. The focus locations 19, 32, 26, and 19 during the respective JAAN months is shown in Figure 8. The red-colored locations denote positive trend and the blue-colored location denote negative trend.

3.4. Dynamic Layers Trend Analysis

The key elements in determining climatological changes and the scenario of global warming are temperature and rainfall. Soil moisture and normalized difference vegetation index are the indicators that help to bind the water beneath the Earth surface. Using the possible groundwater level trend sites 19-32-26-18, the BEAST is run for the dynamic layer dataset received over the JAAN months.

3.4.1. Climatological Factors

Increased precipitation should potentially result in more water being available. That is, the upward trend in rainfall should be followed by a downward trend in groundwater levels. As for temperature, the upward tendency should translate into an upward trend in groundwater levels. The rainfall trend results for the 19-32-26-18 well locations are shown in Figure 9. Blue triangles indicate the increasing trend in rainfall, while red triangles indicate the decreasing trend in rainfall. The results of the rainfall trend for the month of January showed a positive trend in nine well locations, a negative trend in nine locations, and a zero trend in the Cherthalai location. Most of the increasing trends are found towards the southern part of Kerala, while the decreasing trend is scattered over the region. During the month of April, the increase in rainfall trends resulted to 14 positive trends, 17 negative trends, and a zero trend at the Kottanadu well location. It is observed that more than half of the wells had an increasing rainfall trend during April. The converse is the situation during the monsoon month of August, which can be viewed as an impact of climate change. A total of 17 negative trend and 9 positive trend rainfall locations were observed during August. Only the Ponnani1 well location showed a negative rainfall trend in November, while the remaining 17 wells showed an increase. The rainfall trends provide a rough idea of the recharge option available.

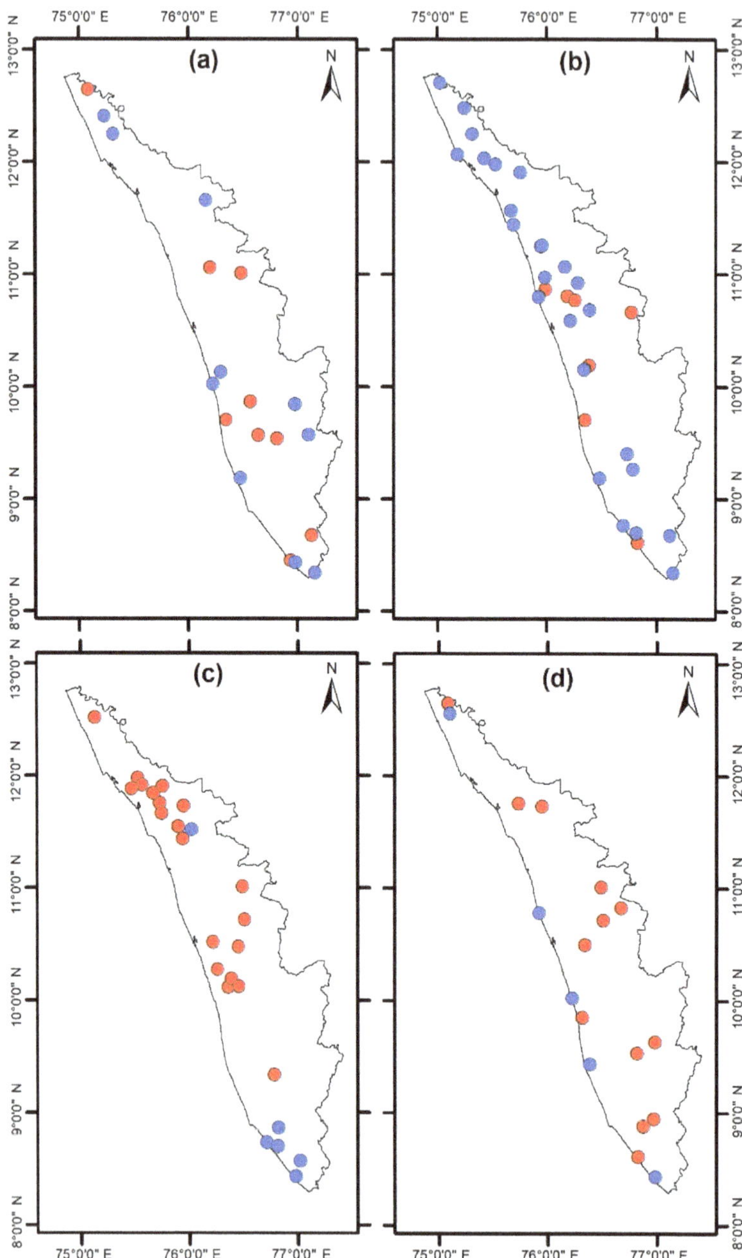

Figure 8. The positive (red) and negative (blue) groundwater level trend locations 19-32-26-18 under focus of study during (**a**) January, (**b**) April, (**c**) August, and (**d**) November.

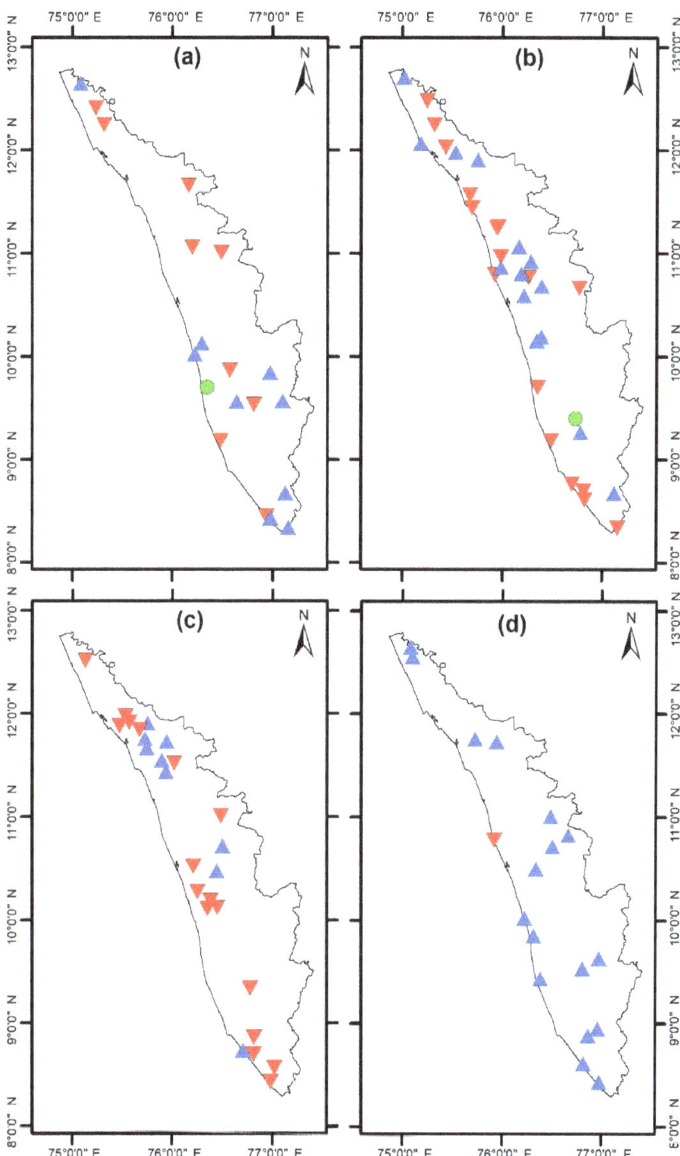

Figure 9. The positive (blue), zero (green) and negative (red) rainfall trend of locations 19-32-26-18 during (**a**) January, (**b**) April, (**c**) August, and (**d**) November.

Temperature is an influential parameter to humidity and rainfall. Over the feasible locations of 19-32-26-18, an increasing temperature trend is observed in the majority of the wells. The temperature trend results are shown in Figure 10. Red triangles denote increasing temperature and reverse influence to groundwater levels. The blue triangle represents decreasing temperature, which is helpful in increasing water quantity. Out of 18 locations during the month of November, only the Ailara well location showed a decreasing trend in temperature. The results validate the global warming scenario experienced over recent decades.

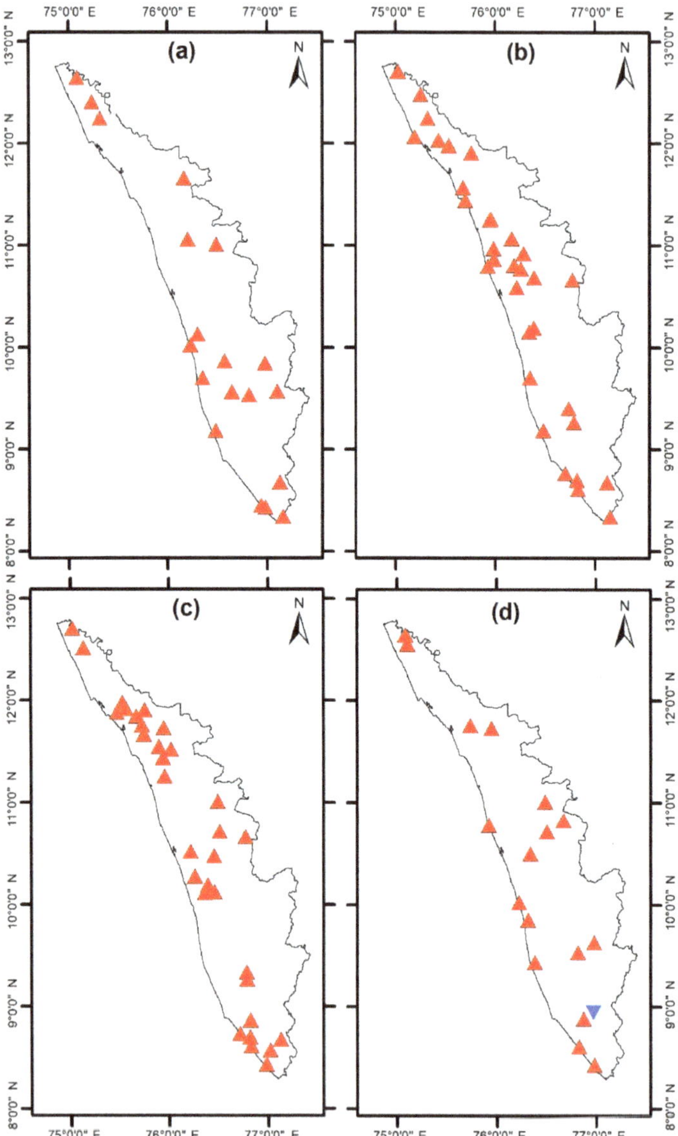

Figure 10. The positive (red) and negative (blue) temperature trend of locations 19-32-26-18 during (**a**) January, (**b**) April, (**c**) August, and (**d**) November.

3.4.2. Soil Moisture and NDVI

The soil moisture has an inverse relation with the allowance of water getting recharged below the Earth's surface. As the moisture content increases, the water gets trapped on the Earth's surface and is not able to pass through. The soil moisture trend results on 19-32-26-18 locations are shown in Figure 11. During the month of January, 3 well locations—namely Angadimogar, Tachinganedam, and Kuvapalli and Idukki—showed increasing trends, while the remaining 16 wells showed decreasing soil moisture trends. In month of April, 18 wells showed negative trends and 14 wells exhibited decreasing trends in soil moisture. A greater number of positive trends were observed in the month of August, which correspond to 15 well locations. However, in the month of November, the majority of the wells had decreasing trends, while the Kuvapalli, Malampuzha, and Elappara locations showed increasing trends. In short, a lower number of locations had negative trends in January and November.

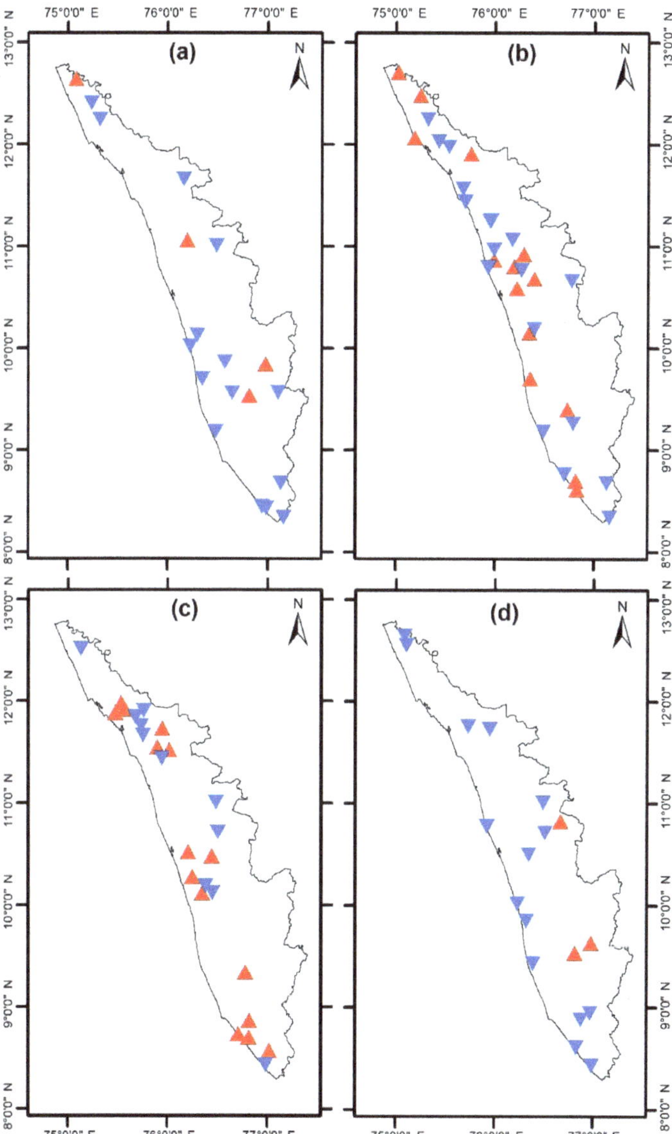

Figure 11. The positive (red) and negative (blue) soil moisture trend of locations 19-32-26-18 during (**a**) January, (**b**) April, (**c**) August, and (**d**) November.

The normalized difference vegetation index has direct influence on groundwater levels. In the month of January, three well locations—namely Poonkulam, Rajapuram, and Maruthamala—showed decreasing trends in NDVI. These wells are located in the northern and southern parts of Kerala. The remaining 16 wells have increasing trends for NDVI.

Similarly, a greater number of wells have positive trends in April and August as well. This amounts to 20 wells in April and 21 wells in August with positive NDVI trends, while 12 wells in April showed negative trends. The wells Akkal, Kannavam, Muliyar, Angamalil, and Elanthur exhibited decreasing NDVI trends during the month of August. In the month of November, nine wells each showed negative and positive trends. The NDVI trend results on 19-32-26-18 locations are shown in Figure 12. Blue triangles indicate increasing trends and thereby contributes to increasing water quantity, while red triangles indicate the decreasing NDVI trends and hence decreases in the possibility of water recharge.

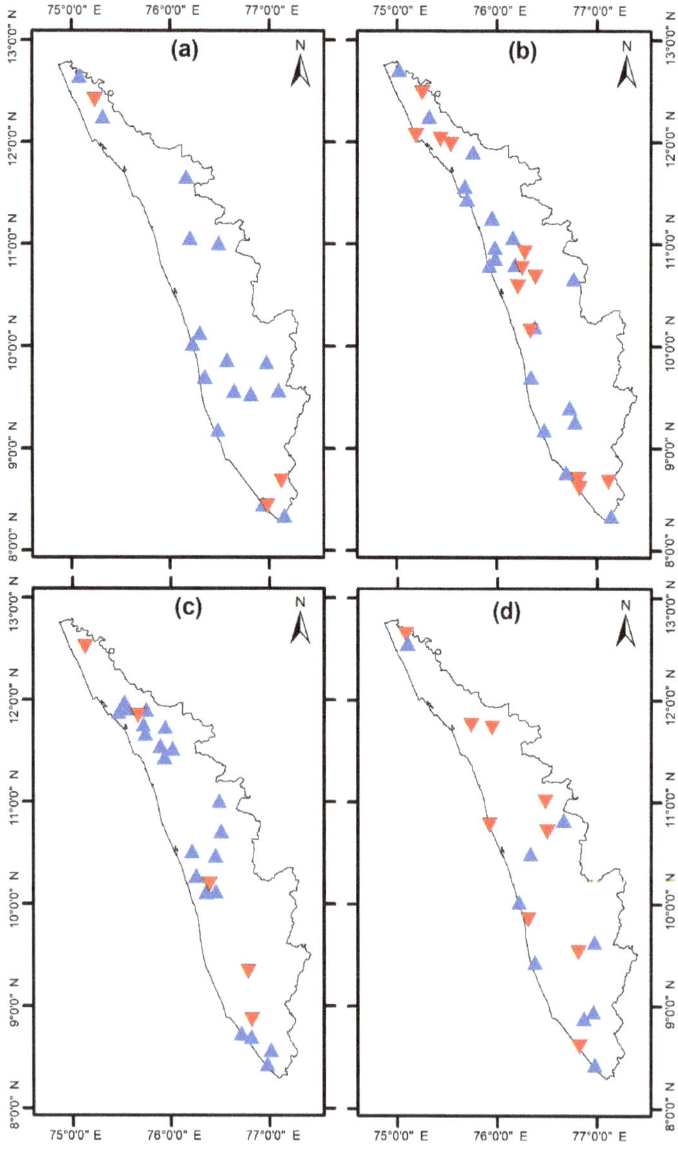

Figure 12. The positive (blue) and negative (red) normalized difference vegetation index trend of locations 19-32-26-18 during (**a**) January, (**b**) April, (**c**) August, and (**d**) November.

3.5. Dynamic Layer Trend in Accordance with Groundwater Level

Out of the dynamic variables considered, rainfall and NDVI have a direct influence on the water holding capacity, while temperature and soil moisture have an inverse relationship. That is, if there is an increasing trend of rainfall, the groundwater level in that particular location is expected to increase, considering all of the other dynamics are already apt. The case is similar for NDVI: the increasing trend should contribute to increasing water availability. In terms of positive and negative trends, a positive trend in rainfall or NDVI will help to create a negative trend in groundwater levels. When the relationship is viewed from the inversely related variables, namely temperature and soil moisture, the positive trend should result in a positive trend in the groundwater levels. If these conditions are met, then in those locations, the changes in water levels could be explained by the dynamic layers and can be termed as matching locations. If the conditions are not met, these non-matching well locations need to be studied in detail with the help of the static layers to understand the structure beneath the surface. Table 3 shows the number of matching and non-matching well locations using the trends compared. The matching locations have been further identified from increasing and decreasing trend, respectively. A plot of the groundwater level for the month of January versus each of the dynamic variables of a random well is shown in Figure 13. From the double plot it can be observed that it is difficult to conclude a single statement or inference.

Table 3. Trend matching (*M*) and non-matching (NM) sites of dynamic layers and index with respect to the groundwater level trend.

Month	RA		TE		SM		VI		Index	
	M	**NM**	**M**	**NM**	**M**	**NM**	**M**	**NM**	**M**	**NM**
Jan	6+5	8	0+9	9	3+9	7	1+8	10	9+4	6
Apr	11+4	17	0+7	25	15+4	13	15+2	15	20+1	11
Aug	1+12	13	0+20	6	0+10	16	5+4	17	1+13	12
Nov	4+0	14	0+12	6	5+3	10	4+8	6	5+2	11

The rainfall trend showed matching in 11, 15, 13, and 4 locations, respectively, in the JAAN months, which equate to approximately 57.9%, 46.9%, 50%, and 22.2%, respectively. In the case of temperature, more matching percentages were obtained in the months of August and November. The number of matching wells is 9, 7, 20, and 12, which is 47.4%, 21.9%, 76.9%, and 66.4% of the considered wells in month order. The soil moisture percentage matching is the highest with 12 wells, which is approximately 63.2% of wells in January, and the lowest with 10 wells, which is approximately 38.5% of wells in August. The month of April has 59.4% of matching wells in April and 44.4% of matching wells in November for soil moisture. The NDVI matching percentages are 17.4%, 53.1%, 34.6%, and 66.7%, in the order of months. In January and April, the greatest matching percentage is for soil moisture, while for the month of August, the temperature matching percentage is higher. The month of November has shown the greatest matching percentages using temperature and NDVI. The variable-wise average matching equates to 44.25% for rainfall, 53.225% for temperature, 51.375% for soil moisture, and 50.45% for NDVI.

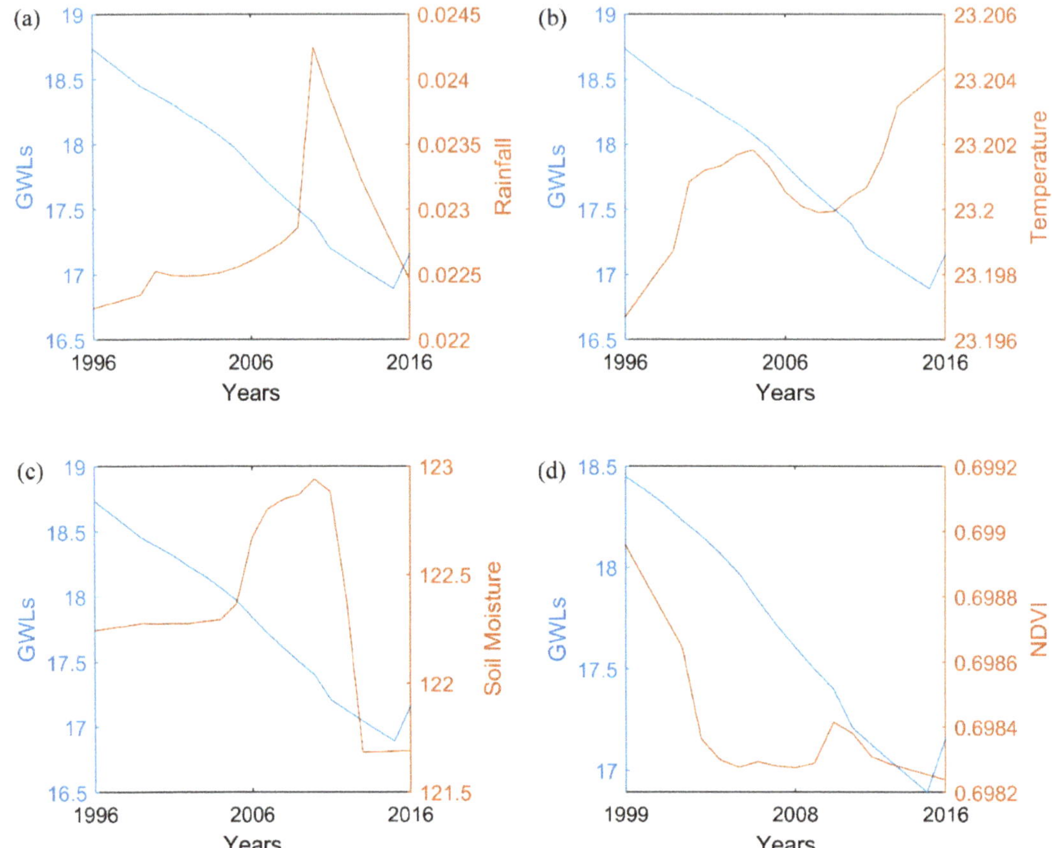

Figure 13. The January month groundwater levels of Poonkulam well location verses (**a**) Rainfall, (**b**) Temperature, (**c**) Soil Moisture, and (**d**) NDVI.

3.6. SMVITERA Index

Each of the dynamic layer trends analyzed with respect to the groundwater level trends could not yield greater matching percentages. This proves the complexity in determining the root cause for the increase or decrease in water level trends. An index has been developed that integrates all the dynamic layer advantages and disadvantages with regards to water-holding capacity. The index is then treated as a combined variable layer and BEAST is run on the obtained datasets. The trend results are shown in Figure 14, where blue triangles represent the increase in the index values and red triangles indicate the decrease in index values. In the month of January, the Angadimogar, Tachinganedam, Tenkara, Kuvapalli, and Parassala well locations showed negative trends in the index, while the remaining 13 wells showed increasing trends. Also in the month of April, 6 well locations—namely Thirunavaya, Ramantalai, Chelakod, Manattana, Paral, and Muligadde—showed decreasing trends, while 26 wells exhibited positive trends for the index. The month of August has 8 well locations with negative trends and 18 well locations with positive trends. The majority of the wells showed positive trends in November, while only two wells, Kuvapalli and Malampuzha, showed negative trends.

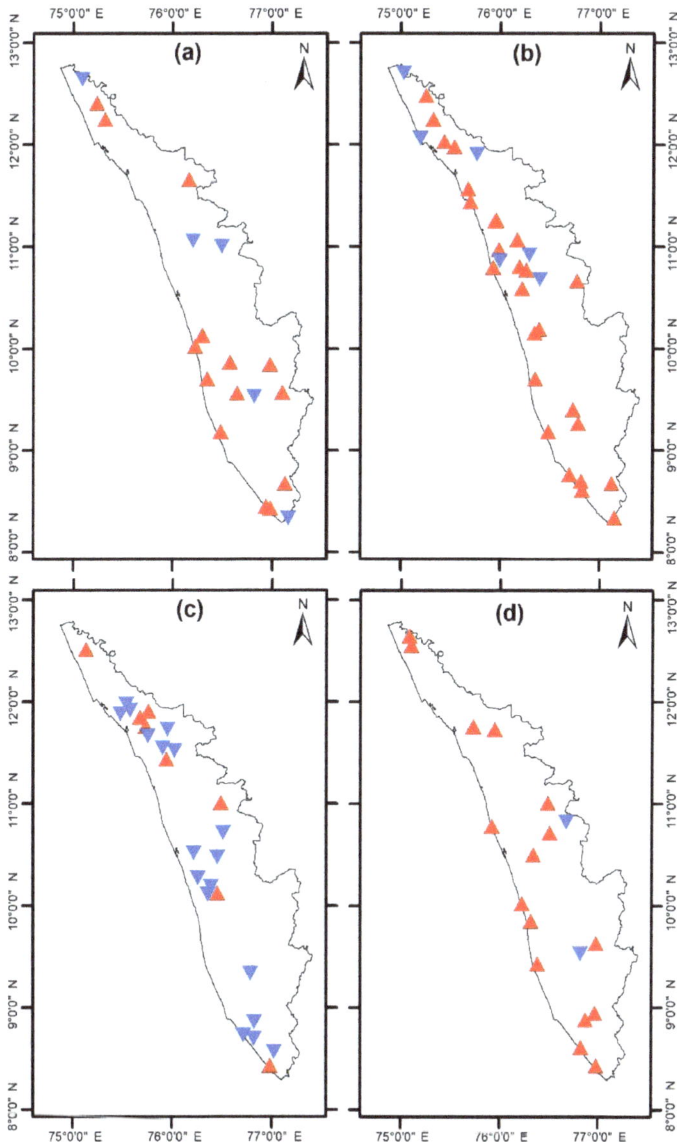

Figure 14. The positive (blue) and negative (red) SMVITERA index trend of locations 19-32-26-18 during (**a**) January, (**b**) April, (**c**) August, and (**d**) November.

The matching percentages with the groundwater levels were also verified. That is, if the index value is increasing, the water availability is increasing, or more accurately the depth to water levels is decreasing. In short, the positive index trend and negative groundwater level trend are a match, and the negative index trend and positive groundwater level trend are a match. In the order of the months, 13, 21, 14, and 7 wells revealed matching and the percentages are 68.4%, 65.6%, 53.9%, and 38.9%. The non-matching well numbers are 6, 11, 12, and 11. The average matching for the index is approximately 56.7%, which is greater than all of the dynamic layer matching percentages individually considered. This proves that index is a better indicator to explain the level changes.

3.7. Static Layers Analysis

The lithological and geomorphological structures of the non-matching wells by the SMVITERA index is extensively studied. In the month of January, six non-matching have the Quartzite, Sand, Gar-Bio-Sill Gneiss + Graphite + Kyanite, Terri Sand, and Acid to Intermediate Charnockite as the lithology formations. Quartzite is a non-foliated metamorphic rock and does not contribute much to transfer of water. Gneiss is a foliated rock and contributes to water storage and recharge, while sand is highly permeable. Charnockite, being highly non-porous, does not allow water to pass through, and as a result does not help in groundwater recharge. The geomorphological formations are Pediment Pediplain Complex and Coastal Plain, which are promising areas of water availability. Considering all of these, the locations of over-exploitation were observed as Cherthalai, Maruthamala, and Poonthura during January.

Eleven wells showed non-matching of groundwater levels with the index. The lithological formation in these wells is Laterite, Clay, Acid to Intermediate Charnockite, Sand, Granite, Garnet Gneiss, Hornblende-Biotite Gneiss, Granite Gneiss, and Terri Sand. The geomorphological formations are Pediment Pediplain Complex, Coastal Plain, and Low Dissected Plateau, which are promising areas of water availability. The identified drying wells are Cherthalai, Shoranur, Angamali1, Palghat, and Murukumpuzha (R1) during April. Similarly, Pudukayam, Perambra, Tenkara, Kannavam, Muliyar, and Vallom1 wells are found to be overly exploited during the month of August as the lithological and geomorphological formations still have sufficient permeability. The over-exploited wells during the month of November were observed to be Tholanur, Pudukayam, Tenkara, Manamangalam, Ailara, Murukumpuzha (R1), and Chadayamangalam1.

4. Discussions

It is seldom easy to explain an inference from the variations in groundwater levels. The greatest depth to groundwater level may be a result of the area's decreased water storage capacity or over-exploitation. The underlying problem must be investigated in order to identify whether a well is drying up and whether this has changed the water table. This study is carried out to determine the relationship between the groundwater levels, which also experience dynamic changes over the years and considered months, and the dynamic variables of surface factors, such as soil moisture and normalized difference vegetation index, and climatological factors, such as temperature and rainfall. This work has used the Bayesian Ensemble algorithm to determine trends rather than usual trend analysis techniques, such as the Mann–Kendall trend test and its modified variants or innovative trend analysis method. This work shows that groundwater level patterns can be very effectively described by BEAST, even though the applied method has not yet been employed to ascertain level trends.

Studies examining trends in groundwater levels over Kerala are few and far between. For the years 2008 to 2013, Sajeena and Kurien examined the groundwater hydrograph for the Kadalundi river basin, Malappuram district, and Kerala [76]. The water table generally rose from June to September and fell from October to May, according to observations. The findings showed that groundwater levels in the Kottakkal, Marakkara, and Tanur localities are unrelated to rainfall. Tachinganedam, Mudikode, and Kottakkal are the points where our study and the Kadalundi river basin converge. The depth of the groundwater at the Tachinganedam well location increased in January, while the depth of the groundwater at the latter well locations decreased in April. The results of the comparative study are consistent with the Tachinganedam well site, and it's possible that throughout the subsequent years, the features of the remaining wells have altered. The water table has risen although rainfall has been trending downward in the Kottakkal well location. This is uncorrelated as there is an obvious trend mismatch. According to the groundwater potential assessment study by Jayasankar and Babu, the districts of Kannur, Wayanad, and Idukki have critical water resource quantities for the decades 1989 to 1999 and 1999 to 2009 [77]. The intersection wells in our study are Idukki, Vandiperiyar, and Elappara from Idukki district; Minangadi, Lakkidi, and Vellamuda from Wayand district; and Vayyakara, Ramantalai, Irikkur, Pukkundu, Manattana, Mattanur, Kannavam, and Chakkarakkale from Kannur district. Groundwater levels decreased in the Idukki, Vandiperiyar, Minangadi, and Vayyakara well locations in January. The depth to groundwater levels also decreased in April at the well locations in Vayyakara, Ramantalai, Irikkur, Pukkundu, and Manattana. The Lakkidi well location showed a declining trend in August, whereas the well locations in Vellamuda, Irikkur, Manattana, Mattanur, Kannavam, and Chakkarakkale showed an upward trend. Elappara and Vellamuda displayed a growing trend in the month of November. According to the report, the majority of well locations experienced a decrease in the water table in August and November, but experienced an increase in January and April. While the situation has marginally improved during the winter and pre-monsoon seasons, the trends are still being followed in the monsoon and post-monsoon seasons. Due to the location-based methodology used in this study, there is some inconsistency in the overall district or state comparisons.

The study by Jagadeesh and Anupama over the Bharathapuzha river basin is to determine rainfall trend analysis using the Mann–Kendall test [78]. The four rain gauge stations looked at were Eruthempathy, Thrithala, and the Malampuzha dam. Except at the Eruthempathy location, the study's results showed a downward trend in rainfall during the northeast monsoon. Our study location coincides with the Malampuzha well, where we have seen increasing rainfall trends. In Kerala state, there were no discernible rainfall patterns during the annual, autumn, spring, summer, or winter according to Jaman et al. [79]. When Sai and Joseph used the Mann–Kendall trend test to assess the rainfall trends in the Pattambi region, they found negative Z values in the months of January, June, July, August, and November [80]. According to our investigation, April saw a favorable trend in rainfall at the Pattambi well location, which was consistent with previous findings. The Vamanapuram river basin's rainfall patterns from 1984 to 2013 were examined by Berma and John [81]. For the months of January, February, May, and August, negative Z values were found. The Maruthamala and Attingal wells in our study are located in the basin. Maruthamala displayed an upward trend in January, whereas Attingal displayed a descending trend in August. Attingal exhibited a declining trend in April, whereas Maruthamala showed an upward trend. The month of August coincides with the findings of our study.

Nearly all of the well locations analyzed for this study had rising trends in average temperatures. Similar results were seen in other studies that covered the state of Kerala. Anjali and Roshni [82] attempted to relate changes in Kerala's forest cover to rainfall and land surface temperature (LST) using satellite data. Between 2000 and 2019, the minimum and maximum LST significantly increased as a result of the Mann–Kendall trend. The patterns of climate change over Kerala's Bharathapuzha river basin since 1900 were examined by George and Athira [83]. Both the monthly average temperature and the monthly lowest temperature showed a consistent long-term upward trend. Using the Mann–Kendall trend test, Varughese et al. [84] examined historical patterns in climate change over the same basin from 1951 to 2013. Mean, maximum, and minimum temperature trends all indicated a noticeable rise over time. Subash and Sikka [85] looked into India's temperature and rainfall trends. The findings indicated an upward trend in the yearly maximum temperature. Over 15 basins in India, Jain and Kumar examined the temperature and rainfall patterns. The findings showed that temperatures were on the rise at the majority of the stations in southern India. The trend analysis of the mean maximum and mean minimum temperature over 13 observatories in Kerala state was conducted by Kabbilawsh et al. [86]. While the mean lowest temperature over Alappuzha indicated a declining trend, the Mann–Kendall test revealed an increase in nine places.

There aren't many studies on NDVI and soil moisture trends over Kerala and India. Bhimala et al. [87] examined the soil moisture, rainfall, and land use and cover data in India. South and northwest India both experienced an increase in soil moisture. Parida et. al. [88] investigated the greening and browning trends of the vegetation in India. The study came to the conclusion that the southern peninsula's vegetation and croplands were no longer greening. Chakraborty et al. [89] used NDVI to evaluate the changes in seasonal greenness across different forest types in India. In tropical moist deciduous forests, the most significant negative changes were discovered.

The acquired results are consistent with the majority of studies, but some locations show different results. This can be explained by the possibility that trends have shifted over unaccounted years.

5. Conclusions

The most popular trend analysis on hydro-climatological variables is the linear and non-parametric Mann–Kendall trend test. The question of the existence of monotonic linearity have prompted to apply other methods like the Bayesian Ensemble Algorithm to the recent researches. This study has used the Bayesian method to examine the variations in groundwater levels over 385 wells across Kerala state during the JAAN months for 21 years from 1996 to 2016. The variations were obtained using an inference in the form of the percentage area under the probability curve for the trend changes to be positive, zero, and negative. The results showed variations in groundwater levels in 19, 26, 32, and 18 well sites. From the viewpoint of climatological variables, such as temperature and rainfall, as well as surface characteristics, such as soil moisture and normalized differential vegetation index, the groundwater levels had to be studied in detail. The matching trend percentages in terms of the relationship with the variables are identified. These matching trend percentages account to about 63.2% by Soil Moisture during January, 59.4% by Soil Moisture in April, 76.9% by Temperature in August, and 66.7% equally by Temperature and Normalized Difference Vegetation Index in November. It is observed that there was no reliance on a specific variable or a particular month in the trend analysis of these dynamic layers in connection to changes in groundwater level. This prompted the creation of the SMVITERA index, which is based on these dynamic factors. The developed index has a higher average proportion of roughly 56.7% matching, according to analysis of the matching percentages of each layer individually and collectively. The individual average matching percentages were observed to be 44.25% for Rainfall, 53.23% for Temperature, 51.38% for Soil Moisture, and 50.45% for Normalized Difference Vegetation Index. The static properties of the non-matching well locations were investigated. The site has been deemed overexploited if the lithological and geomorphologic properties continue to influence the amount of water while the trend result is positive. Identification and maintenance of these places are required in a predicted-to-be water-scarce Earth.

Author Contributions: Conceptualization, methodology, formal analysis, and writing—original draft preparation is done by K.A., writing—review and editing, supervision is done by A.N. All authors have read and agreed to the published version of the manuscript.

Funding: This research received no external funding.

Institutional Review Board Statement: Not applicable.

Informed Consent Statement: Not applicable.

Data Availability Statement: The study is on freely accessible datasets and the sources are mentioned in the document.

Acknowledgments: The authors wish to acknowledge Copernicus Global Land Services, Climate Prediction Center (CPC), Indian Meteorological Department (IMD) and Central Ground Water Board (CGWB) for the freely accessible datasets. The authors would like to thank the anonymous reviewers for their valuable time and comments that helped to improve the manuscript.

Conflicts of Interest: The authors declare no conflict of interest.

References

1. Varma, A. *Groundwater Resource and Governance in Kerala Groundwater Resource and Governance in Kerala: Status, Issues and Prospects*; Forum Policy Dialogue Water Conflicts: Pune, India, 2017.

2.	Pullare, N.; Ground, C.; Board, W. Changes in Ground Water Utilization in Kerala—Causes & Consequences. In Proceedings of the 4th National Ground Water Congress, Kerala, India, January 2012. [CrossRef]

3.	Ziolkowska, J.R.; Reyes, R. Groundwater Level Changes Due to Extreme Weather-an Evaluation Tool for Sustainable Water Management. *Water* **2017**, *9*, 117. [CrossRef]

4.	Hua, Z.; Cheng, W.; Yi, S.; Jiang, Q. Geostatistical Analysis of Spatial and Temporal Variations of Groundwater Depth in Shule River. In Proceedings of the 2009 WASE International Conference on Information Engineering (ICIE 2009), Taiyuan, China, 10–11 July 2009; Volume 2, pp. 453–457.

5.	Bhanja, S.N.; Mukherjee, A. In Situ and Satellite-Based Estimates of Usable Groundwater Storage across India: Implications for Drinking Water Supply and Food Security. *Adv. Water Resour.* **2019**, *126*, 15–23. [CrossRef]

6.	Bhanja, S.N.; Mukherjee, A.; Saha, D.; Velicogna, I.; Famiglietti, J.S. Validation of GRACE Based Groundwater Storage Anomaly Using In-Situ Groundwater Level Measurements in India. *J. Hydrol.* **2016**, *543*, 729–738. [CrossRef]

7.	Machiwal, D.; Jha, M.K.; Mal, B.C. Assessment of Groundwater Potential in a Semi-Arid Region of India Using Remote Sensing, GIS and MCDM Techniques. *Water Resour. Manag.* **2011**, *25*, 1359–1386. [CrossRef]

8.	Mengistu, H.A.; Demlie, M.B.; Abiye, T.A. Review: Groundwater Resource Potential and Status of Groundwater Resource Development in Ethiopia. *Hydrogeol. J.* **2019**, *27*, 1051–1065. [CrossRef]

9.	Srivastava, S.; Singh, J.; Shirsath, P.B. Sustainability of Groundwater Resources at the Subnational Level in the Context of Sustainable Development Goals. *Agric. Econ. Res. Rev.* **2018**, *31*, 79. [CrossRef]

10.	Koundouri, P. Potential for Groundwater Management: Gisser-Sanchez Effect Reconsidered. *Water Resour. Res.* **2004**, *40*, W06S16. [CrossRef]

11.	Thakur, G.S.; Thomas, T. Analysis of Groundwater Levels for Detection of Trend in Sagar District, Madhya Pradesh. *J. Geol. Soc. India* **2011**, *77*, 303–308. [CrossRef]

12.	Knapp, K.C.; Weinberg, M.; Howitt, R.; Posnikoff, J.F. Water Transfers, Agriculture, and Groundwater Management: A Dynamic Economic Analysis. *J. Environ. Manag.* **2003**, *67*, 291–301. [CrossRef] [PubMed]

13.	Ali, R.; Rashid Abubaker, S.; Othman Ali, R. Spatio-Temporal Pattern in the Changes in Availability and Sustainability of Water Resources in Afghanistan View Project Water Resources Problem in Huai River Basin View Project Trend Analysis Using Mann-Kendall, Sen's Slope Estimator Test and Innovative. *Int. J. Eng. Technol.* **2019**, *8*, 110–119. [CrossRef]

14.	Xing, L.; Huang, L.; Chi, G.; Yang, L.; Li, C.; Hou, X. A Dynamic Study of a Karst Spring Based on Wavelet Analysis and the Mann-Kendall Trend Test. *Water* **2018**, *10*, 698. [CrossRef]

15.	Hamidov, A.; Khamidov, M.; Ishchanov, J. Impact of Climate Change on Groundwater Management in the Northwestern Part of Uzbekistan. *Agronomy* **2020**, *10*, 1173. [CrossRef]

16.	Dinpashoh, Y. Trend Analysis of Groundwater Level, Using Mann-Kendall Non Parametric Method (Case Study: Tabriz Plain). *J. Water Soil Sci.* **2019**, *23*, 335–348. [CrossRef]

17.	Meggiorin, M.; Passadore, G.; Bertoldo, S.; Sottani, A.; Rinaldo, A. Assessing the Long-Term Sustainability of the Groundwater Resources in the Bacchiglione Basin (Veneto, Italy) with the Mann–Kendall Test: Suggestions for Higher Reliability. *Acque Sotter. Ital. J. Groundw.* **2021**, *10*, 35–48. [CrossRef]

18.	Wilopo, W.; Putra, D.P.E.; Hendrayana, H. Impacts of Precipitation, Land Use Change and Urban Wastewater on Groundwater Level Fluctuation in the Yogyakarta-Sleman Groundwater Basin, Indonesia. *Environ. Monit. Assess.* **2021**, *193*, 76. [CrossRef] [PubMed]

19.	Ndlovu, M.S.; Demlie, M. Statistical Analysis of Groundwater Level Variability across KwaZulu-Natal Province, South Africa. *Environ. Earth Sci.* **2018**, *77*, 739. [CrossRef]

20.	Valois, R.; MacDonell, S.; Núñez Cobo, J.H.; Maureira-Cortés, H. Groundwater Level Trends and Recharge Event Characterization Using Historical Observed Data in Semi-Arid Chile. *Hydrol. Sci. J.* **2020**, *65*, 597–609. [CrossRef]

21.	Goyal, S.K.; Chaudhary, B.S.; Singh, O.; Sethi, G.K.; Thakur, P.K. Variability Analysis of Groundwater Levels—AGIS-Based Case Study. *J. Indian Soc. Remote Sens.* **2010**, *38*, 355–364. [CrossRef]

22.	Anand, B.; Karunanidhi, D.; Subramani, T.; Srinivasamoorthy, K.; Suresh, M. Long-Term Trend Detection and Spatiotemporal Analysis of Groundwater Levels Using GIS Techniques in Lower Bhavani River Basin, Tamil Nadu, India. *Environ. Dev. Sustain.* **2020**, *22*, 2779–2800. [CrossRef]

23.	Panda, D.K.; Mishra, A.; Kumar, A.; Mishra, D.K.; Kumar, A. Quantification of Trends in Groundwater Levels of Gujarat in Western India. *Hydrol. Sci. J.* **2012**, *57*, 1325–1336. [CrossRef]

24.	Krishan, G.; Rao, M.S.; Loyal, R.S.; Lohani, A.K.; Tuli, N.K.; Takshi, K.S.; Kumar, C.P.; Semwal, P.; Kumar, S. Groundwater Level Analyses of Punjab, India: A Quantitative Approach. *Octa J. Environ. Res.* **2014**, *2*, 221–226.

25.	Sishodia, R.P.; Shukla, S.; Graham, W.D.; Wani, S.P.; Garg, K.K. Bi-Decadal Groundwater Level Trends in a Semi-Arid South Indian Region: Declines, Causes and Management. *J. Hydrol. Reg. Stud.* **2016**, *8*, 43–58. [CrossRef]

26.	Tirkey, A.S.; Pandey, A.C.; Nathawat, M.S. Groundwater Level and Rainfall Variability Trend Analysis Using GIS in Parts of Jharkhand State (India) for Sustainable Management of Water Resources. *Int. Res. J. Environ. Sci.* **2012**, *1*, 24–31.

27.	Dinpashoh, Y.; Mirabbasi, R.; Jhajharia, D.; Abianeh, H.Z.; Mostafaeipour, A. Effect of Short-Term and Long-Term Persistence on Identification of Temporal Trends. *J. Hydrol. Eng.* **2014**, *19*, 617–625. [CrossRef]

28.	Kumar, S.; Merwade, V.; Kam, J.; Thurner, K. Streamflow Trends in Indiana: Effects of Long Term Persistence, Precipitation and Subsurface Drains. *J. Hydrol.* **2009**, *374*, 171–183. [CrossRef]

29. Drápela, K.; Drápelová, I.; Drápela, D.I.K. Application of Mann-Kendall Test and the Sen's Slope Estimates for Trend Detection in Deposition Data from Bílý Kříž (Beskydy Mts., the Czech Republic) 1997–2010. *Beskydy* **2011**, *4*, 133–146.
30. Patakamuri, S.K.; Muthiah, K.; Sridhar, V. Long-Term Homogeneity, Trend, and Change-Point Analysis of Rainfall in the Arid District of Ananthapuramu, Andhra Pradesh State, India. *Water* **2020**, *12*, 211. [CrossRef]
31. Satish Kumar, K.; Venkata Rathnam, E. Analysis and Prediction of Groundwater Level Trends Using Four Variations of Mann Kendall Tests and ARIMA Modelling. *J. Geol. Soc. India* **2019**, *94*, 281–289. [CrossRef]
32. Swain, S.; Sahoo, S.; Taloor, A.K.; Mishra, S.K.; Pandey, A. Exploring Recent Groundwater Level Changes Using Innovative Trend Analysis (ITA) Technique over Three Districts of Jharkhand, India. *Groundw. Sustain. Dev.* **2022**, *18*, 100783. [CrossRef]
33. Zakwan, M. Trend Analysis of Groundwater Level Using Innovative Trend Analysis. In *Groundwater Resources Development and Planning in the Semi-Arid Region*; Springer: Cham, Switzerland, 2021; pp. 389–405.
34. Abbas, M.; Arshad, M.; Shahid, M.A. Charectarization of Groundwater Level Zones Using Innovative Trend & Regression Analysis: Case Study at Rechna Doab-Pakistan. Available online: https://www.researchsquare.com/article/rs-2140740/v1 (accessed on 5 November 2022).
35. Li, J.; Li, Z.L.; Wu, H.; You, N. Trend, Seasonality, and Abrupt Change Detection Method for Land Surface Temperature Time-Series Analysis: Evaluation and Improvement. *Remote Sens. Environ.* **2022**, *280*, 113222. [CrossRef]
36. Yang, X.; Tian, S.; You, W.; Jiang, Z. Reconstruction of Continuous GRACE/GRACE-FO Terrestrial Water Storage Anomalies Based on Time Series Decomposition. *J. Hydrol.* **2021**, *603*, 127018. [CrossRef]
37. Xu, X.; Yang, J.; Ma, C.; Qu, X.; Chen, J.; Cheng, L. Segmented Modeling Method of Dam Displacement Based on BEAST Time Series Decomposition. *Measurement* **2022**, *202*, 111811. [CrossRef]
38. Tingwei, C.; Tingxuan, H.; Bing, M.; Fei, G.; Yanfang, X.; Rongjie, L.; Yi, M.; Jie, Z.; Tingwei, C.; Tingxuan, H.; et al. Spatiotemporal Pattern of Aerosol Types over the Bohai and Yellow Seas Observed by CALIOP. *Infrared Laser Eng.* **2021**, *50*, 20211030–20211031. [CrossRef]
39. Duke, N.C.; Mackenzie, J.R.; Canning, A.D.; Hutley, L.B.; Bourke, A.J.; Kovacs, J.M.; Cormier, R.; Staben, G.; Lymburner, L.; Ai, E. ENSO-Driven Extreme Oscillations in Mean Sea Level Destabilise Critical Shoreline Mangroves—An Emerging Threat. *PLoS Clim.* **2022**, *1*, e0000037. [CrossRef]
40. Zannat, F.; Islam, A.R.M.T.; Rahman, M.A. Spatiotemporal Variability of Rainfall Linked to Ground Water Level under Changing Climate in Northwestern Region, Bangladesh. *Eur. J. Geosci.* **2019**, *1*, 35–56. [CrossRef]
41. Shaji, E. Groundwater Quality Management in Kerala. *Int. Inte-Res. J.* **2013**, *3*, 63–68.
42. Tabari, H.; Nikbakht, J.; Shifteh Some'e, B. Investigation of Groundwater Level Fluctuations in the North of Iran. *Environ. Earth Sci.* **2011**, *66*, 231–243. [CrossRef]
43. Oleszczuk, R.; Jadczyszyn, J.; Gnatowski, T.; Brandyk, A. Variation of Moisture and Soil Water Retention in a Lowland Area of Central Poland—Solec Site Case Study. *Atmosphere* **2022**, *13*, 1372. [CrossRef]
44. Naga Rajesh, A.; Abinaya, S.; Purna Durga, G.; Lakshmi Kumar, T.V. Long-Term Relationships of MODIS NDVI with Rainfall, Land Surface Temperature, Surface Soil Moisture and Groundwater Storage over Monsoon Core Region of India. *Arid L. Res. Manag.* **2022**, 1–20. [CrossRef]
45. Halder, S.; Roy, M.B.; Roy, P.K. Analysis of Groundwater Level Trend and Groundwater Drought Using Standard Groundwater Level Index: A Case Study of an Eastern River Basin of West Bengal, India. *SN Appl. Sci.* **2020**, *2*, 507. [CrossRef]
46. Sahoo, S.; Swain, S.; Goswami, A.; Sharma, R.; Pateriya, B. Assessment of Trends and Multi-Decadal Changes in Groundwater Level in Parts of the Malwa Region, Punjab, India. *Groundw. Sustain. Dev.* **2021**, *14*, 100644. [CrossRef]
47. Babre, A.; Kalvāns, A.; Avotniece, Z.; Retiķe, I.; Bikše, J.; Jemeljanova, K.P.M.; Zelenkevičs, A.; Dēliņa, A. The Use of Predefined Drought Indices for the Assessment of Groundwater Drought Episodes in the Baltic States over the Period 1989–2018. *J. Hydrol. Reg. Stud.* **2022**, *40*, 101049. [CrossRef]
48. Guo, M.; Yue, W.; Wang, T.; Zheng, N.; Wu, L. Assessing the Use of Standardized Groundwater Index for Quantifying Groundwater Drought over the Conterminous US. *J. Hydrol.* **2021**, *598*, 126227. [CrossRef]
49. Khaira, A.; Dwivedi, R.K. A State of the Art Review of Analytical Hierarchy Process. *Mater. Today Proc.* **2018**, *5*, 4029–4035. [CrossRef]
50. Singh, R.P.; Nachtnebel, H.P. Analytical Hierarchy Process (AHP) Application for Reinforcement of Hydropower Strategy in Nepal. *Renew. Sustain. Energy Rev.* **2016**, *55*, 43–58. [CrossRef]
51. Goepel, K.D. Comparison of Judgment Scales of the Analytical Hierarchy Process—A New Approach. *Int. J. Inf. Technol. Decis. Mak.* **2019**, *18*, 445–463. [CrossRef]
52. Azizkhani, M.; Vakili, A.; Noorollahi, Y.; Naseri, F. Potential Survey of Photovoltaic Power Plants Using Analytical Hierarchy Process (AHP) Method in Iran. *Renew. Sustain. Energy Rev.* **2017**, *75*, 1198–1206. [CrossRef]
53. Thanki, S.; Govindan, K.; Thakkar, J. An Investigation on Lean-Green Implementation Practices in Indian SMEs Using Analytical Hierarchy Process (AHP) Approach. *J. Clean. Prod.* **2016**, *135*, 284–298. [CrossRef]
54. Joseph, E.J.; Anitha, A.B.; Jayakumar, P.; Sushanth, C.M.; Jayakumar, K.V. *Climate Change and Sustainable Water Resources Management in Kerala*; Centre of Water Resources Development and Management: Kozhikode, India, 2011.
55. Huang, J.; Van Den Dool, H.M.; Georgakakos, K.P. Analysis of Model-Calculated Soil Moisture over the United States (1931-1993) and Applications to Long-Range Temperature Forecasts. *J. Clim.* **1996**, *9*, 1350–1362. [CrossRef]

56. Fan, Y.; Van Den Dool, H. Climate Prediction Center Global Monthly Soil Moisture Data Set at 0.5° Resolution for 1948 to Present. *J. Geophys. Res.* **2004**, *109*, 10102. [CrossRef]
57. Sajjad, M.M.; Wang, J.; Abbas, H.; Ullah, I.; Khan, R.; Ali, F. Impact of Climate and Land-Use Change on Groundwater Resources, Study of Faisalabad District, Pakistan. *Atmosphere* **2022**, *13*, 1097. [CrossRef]
58. Baret, F.; Weiss, M.; Lacaze, R.; Camacho, F.; Makhmara, H.; Pacholcyzk, P.; Smets, B. GEOV1: LAI and FAPAR Essential Climate Variables and FCOVER Global Time Series Capitalizing over Existing Products. Part1: Principles of Development and Production. *Remote Sens. Environ.* **2013**, *137*, 299–309. [CrossRef]
59. León-Tavares, J.; Roujean, J.-L.; Smets, B.; Wolters, E.; Toté, C.; Swinnen, E. Correction of Directional Effects in VEGETATION NDVI Time-Series. *Remote Sens.* **2021**, *13*, 1130. [CrossRef]
60. Toté, C.; Swinnen, E.; Sterckx, S.; Clarijs, D.; Quang, C.; Maes, R. Evaluation of the SPOT/VEGETATION Collection 3 Reprocessed Dataset: Surface Reflectances and NDVI. *Remote Sens. Environ.* **2017**, *201*, 219–233. [CrossRef]
61. Toté, C.; Swinnen, E.; Sterckx, S.; Adriaensen, S.; Benhadj, I.; Iordache, M.-D.; Bertels, L.; Kirches, G.; Stelzer, K.; Dierckx, W.; et al. Evaluation of PROBA-V Collection 1: Refined Radiometry, Geometry, and Cloud Screening. *Remote Sens.* **2018**, *10*, 1375. [CrossRef]
62. Pai, D.; Rajeevan, M.; Sreejith, O.; Mukhopadhyay, B.; Satbha, N. Development of a New High Spatial Resolution ($0.25° \times 0.25°$) Long Period (1901-2010) Daily Gridded Rainfall Data Set over India and Its Comparison with Existing Data Sets over the Region. *Mausam* **2021**, *65*, 1–18. [CrossRef]
63. Srivastava, A.K.; Rajeevan, M.; Kshirsagar, S.R. Development of a High Resolution Daily Gridded Temperature Data Set (1969-2005) for the Indian Region. *Atmos. Sci. Lett.* **2009**, *10*, 249–254. [CrossRef]
64. Saaty, R.W. The Analytic Hierarchy Process-What It Is and How It Is Used. *Math. Model.* **1987**, *9*, 161–176. [CrossRef]
65. Saaty, T.L. How to Make a Decision: The Analytic Hierarchy Process. *Eur. J. Oper. Res.* **1990**, *48*, 9–26. [CrossRef]
66. David, F.N.; Kendall, M.G. Rank Correlation Methods. *J. R. Stat. Soc. Ser. A* **1956**, *119*, 90. [CrossRef]
67. Mann, H.B. Nonparametric Tests Against Trend. *Econometrica* **1945**, *13*, 245. [CrossRef]
68. Sen, P.K. Estimates of the Regression Coefficient Based on Kendall's Tau. *J. Am. Stat. Assoc.* **1968**, *63*, 1379–1389. [CrossRef]
69. Theil, H. A Rank-Invariant Method of Linear and Polynomial Regression Analysis, 1–2; Proceedings of the Koninklijke Nederlandse Akademie Wetenschappen, Series A Mathematical Sciences. 1950, 53, pp. 386–392, 521–525. Available online: https://www.scirp.org/(S(351jmbntvnsjt1aadkposzje))/reference/ReferencesPapers.aspx?ReferenceID=1245706 (accessed on 5 November 2022).
70. Theil, H. A Rank-Invariant Method of Linear and Polynomial Regression Analysis, 3; Proceedings of the Koninklijke Nederlandse Akademie Wetenschappen, Series A Mathematical Sciences. 1950, 53, pp. 1397–1412. Available online: https://www.scirp.org/(S(351jmbntvnsjt1aadkposzje))/reference/ReferencesPapers.aspx?ReferenceID=1245706 (accessed on 5 November 2022).
71. Zhao, K.; Wulder, M.A.; Hu, T.; Bright, R.; Wu, Q.; Qin, H.; Li, Y.; Toman, E.; Mallick, B.; Zhang, X.; et al. Detecting Change-Point, Trend, and Seasonality in Satellite Time Series Data to Track Abrupt Changes and Nonlinear Dynamics: A Bayesian Ensemble Algorithm. *Remote Sens. Environ.* **2019**, *232*, 111181. [CrossRef]
72. White, J.H.R.; Walsh, J.E.; Thoman, R.L. Using Bayesian Statistics to Detect Trends in Alaskan Precipitation. *Int. J. Climatol.* **2021**, *41*, 2045–2059. [CrossRef]
73. Verbesselt, J.; Hyndman, R.; Zeileis, A.; Culvenor, D. Phenological Change Detection While Accounting for Abrupt and Gradual Trends in Satellite Image Time Series. *Remote Sens. Environ.* **2010**, *114*, 2970–2980. [CrossRef]
74. Jiang, B.; Liang, S.; Wang, J.; Xiao, Z. Modeling MODIS LAI Time Series Using Three Statistical Methods. *Remote Sens. Environ.* **2010**, *114*, 1432–1444. [CrossRef]
75. Zhao, K.; Valle, D.; Popescu, S.; Zhang, X.; Mallick, B. Hyperspectral Remote Sensing of Plant Biochemistry Using Bayesian Model Averaging with Variable and Band Selection. *Remote Sens. Environ.* **2013**, *132*, 102–119. [CrossRef]
76. Sajeena, S.; Kurien, E.K. Hydrogeological Characteristics and Groundwater Scenario of Kadalundi River Basin, Malappuram District, Kerala. *Trends Biosci.* **2017**, *10*, 2193–2200.
77. Jayasankar, P.; Babu, M.N.S. An Assessment of Ground Water Potential for State of Kerala, India: A Case Study. *AE Int. J. Sci. Technol.* **2017**, *5*.
78. Jagadeesh, P.; Anupama, C. Statistical and Trend Analyses of Rainfall: A Case Study of Bharathapuzha River Basin, Kerala, India. *ISH J. Hydraul. Eng.* **2014**, *20*, 119–132. [CrossRef]
79. Jaman Basheer Ahamed, M.; Ravichandran, C.; Ebraheem, A.M.A. Analysis of Trend and Magnitude Using Mann-Kendall and Sen's Slope Test in 115 Years Annual Rainfall Data of South India. *Adv. Appl. Math. Sci.* **2022**, *21*, 3419–3429.
80. Sai, K.V.; Joseph, A. Trend Analysis of Rainfall of Pattambi Region, Kerala. *Int. J. Curr. Microbiol. Appl. Sci.* **2018**, *7*, 3274–3281. [CrossRef]
81. Brema, J. John Anie Rainfall Trend Analysis by Mann-Kendall Test for Vamanapuram River Basin, Kerala. *Int. J. Civ. Eng. Technol.* **2018**, *9*, 1549–1556.
82. Anjali, K.; Roshni, T. Linking Satellite-Based Forest Cover Change with Rainfall and Land Surface Temperature in Kerala, India. *Environ. Dev. Sustain.* **2022**, *24*, 11282–11300. [CrossRef]
83. George, J. Long-Term Changes in Climatic Variables over the Bharathapuzha River Basin, Kerala, India. *Theor. Appl. Climatol.* **2020**, *142*, 269–286. [CrossRef]

84. Varughese, A.; Hajilal, M.S.; George, B. Analysis of Historical Climate Change Trends in Bharathapuzha River Basin, Kerala, India. *Nat. Environ. Pollut. Technol.* **2017**, *16*, 237.
85. Subash, N.; Sikka, A.K. Trend Analysis of Rainfall and Temperature and Its Relationship over India. *Theor. Appl. Climatol.* **2014**, *117*, 449–462. [CrossRef]
86. Kabbilawsh, P.; Sathish Kumar, D.; Chithra, N.R. Trend Analysis and SARIMA Forecasting of Mean Maximum and Mean Minimum Monthly Temperature for the State of Kerala, India. *Acta Geophys.* **2020**, *68*, 1161–1174. [CrossRef]
87. Bhimala, K.R.; Rakesh, V.; Prasad, K.R.; Mohapatra, G.N. Identification of Vegetation Responses to Soil Moisture, Rainfall, and LULC over Different Meteorological Subdivisions in India Using Remote Sensing Data. *Theor. Appl. Climatol.* **2020**, *142*, 987–1001. [CrossRef]
88. Parida, B.R.; Pandey, A.C.; Patel, N.R. Greening and Browning Trends of Vegetation in India and Their Responses to Climatic and Non-Climatic Drivers. *Climate* **2020**, *8*, 92. [CrossRef]
89. Chakraborty, A.; Seshasai, M.V.R.; Reddy, C.S.; Dadhwal, V.K. Persistent Negative Changes in Seasonal Greenness over Different Forest Types of India Using MODIS Time Series NDVI Data (2001–2014). *Ecol. Indic.* **2018**, *85*, 887–903. [CrossRef]

Article

Reclassifying the Spring Maize Drought Index on the Loess Plateau under a Changing Climate

Shujie Yuan [1], Nan Jiang [2], Jinsong Wang [3,*], Liang Xue [1] and Lin Han [1]

[1] School of Atmospheric Sciences, Chengdu University of Information Technology, Chengdu 610225, China; ysj@cuit.edu.cn (S.Y.); 3210101053@stu.cuit.edu.cn (L.X.); hanlin@cuit.edu.cn (L.H.)
[2] Inner Mongolia Meteorological Service Center, Hohhot 010051, China; nmgqxzx@163.com
[3] Key Laboratory of Arid Climate Change and Disaster Reduction of Gansu Province, Institute of Arid Meteorology, China Meteorological Administration, Lanzhou 730020, China
* Correspondence: wangjs@iamcma.cn

Abstract: Drought is the main meteorological disaster that affects the yield and quality of spring maize on the Loess Plateau. This study used data of the spring maize growth period, relative soil humidity, and yield at 18 agricultural meteorological observation stations on the Loess Plateau from 1997 to 2013 to determine the drought category based on the yield reduction rate. Through the drought index according to the conformity rate of category standard and individual case verification, a refined suitability drought index of spring maize on the Loess Plateau was constructed, and the spatial distribution characteristics of drought in different growth stages of spring maize were analyzed. The results showed the following: (1) The number of days in the whole growth period of spring maize in all regions of the Loess Plateau has been extended. The average sowing date of spring maize in the northwest region of the Loess Plateau was 9 April, and that in the east and central regions was 26 April. In terms of spatial distribution, each growth period was gradually delayed from west to east. (2) The correlation between relative soil humidity and yield of spring maize at the jointing stage and heading stage was the best, followed by the milky stage and mature stage, and the relative soil humidity at the sowing stage and emergence stage had little effect on the yield. (3) According to the national drought category standard "Drought Category of Spring Maize in the North", based on the data of yield reduction rate, the drought index of spring maize on the Loess Plateau was refined by region and growth stage. The drought category index values of spring maize in different growth stages and regions changed according to the revised drought category standard, with 71.4% of the sites in the sowing seedling stage and 85.7% of the sites in the seedling jointing stage, and the revised drought category was more severe than the national drought category standard, while at 57.1% of the sites in the jointing and tasseling stages and 71.4% in the tasseling and milking stages, the revised drought category was less severe than the national drought category standard. (4) Based on the revised refined drought index for spring maize on the Loess Plateau, the spatial distribution of drought occurrence frequency across different growth stages of spring maize on the Loess Plateau was analyzed. The frequency of drought occurrence during the seeding and emergence stages was 25–75%. With the change in growth stages, the high-value area of drought occurrence frequency gradually moved northward, and the overall frequency of drought occurrence decreased. For the milky mature stage, the frequency of drought occurrence in a few regions was around 42%, and the drought frequency in most regions was between 8% and 33%.

Keywords: Loess Plateau; spring maize; relative soil moisture; drought index

Citation: Yuan, S.; Jiang, N.; Wang, J.; Xue, L.; Han, L. Reclassifying the Spring Maize Drought Index on the Loess Plateau under a Changing Climate. *Atmosphere* **2023**, *14*, 1481. https://doi.org/10.3390/atmos14101481

Academic Editor: Haibo Liu

Received: 16 June 2023
Revised: 13 August 2023
Accepted: 23 August 2023
Published: 25 September 2023

1. Introduction

The Loess Plateau, a climate-change-sensitive area and drought-prone zone in China, has been severely affected by drought, which has impacted the development of local agricultural production [1]. Maize is the main grain crop on the Loess Plateau, which has important significance for ensuring food security on the Loess Plateau [2]. Previous research results have shown that drought can lead to maize plant dwarfing, hindered growth and development, deteriorated ear traits, and ultimately, a significant decrease in economic yield and biomass [3]. Since the 1990s, the annual precipitation on the Loess Plateau has shown a significant decrease, with droughts becoming increasingly frequent [4], the frequency and degree of droughts increasing and deepening [5–7], and the impact of drought on maize intensifying [8,9]. Therefore, timely and objective monitoring of the spatiotemporal distribution characteristics of maize drought on the Loess Plateau, analyzing the impact of climate change on it, has important practical significance for improving maize's disaster prevention and reduction capabilities.

Objectively and effectively monitoring and estimating maize drought and accurately reflecting the degree, scope, and duration of maize drought require appropriate and accurate maize drought indicators [10–14]. Relative soil humidity can directly reflect the available water status of maize and has a good correlation with yield [15], which is a key factor affecting the growth, development, and yield of maize. In addition, during the analyzed period, the relative soil humidity observation stations of the China Meteorological Administration Agricultural Meteorological Observatory were widely distributed, and the time series of the observation data was long. Therefore, this study selected relative soil humidity as a monitoring indicator, which could better reflect the degree of maize drought on the Loess Plateau [16].

Although the "China Meteorological Administration's Northern Spring Maize Drought Rating" (QX/T 259-2015) [17] stipulates the standard of relative soil humidity drought rating in different growth periods of maize (Table 1), the standard does not take regional differences into account, and the same maize drought index values were used in different regions. The application of this standard to study the spatiotemporal distribution characteristics of spring maize drought on the Loess Plateau was not sufficiently precise. In order to more accurately monitor and evaluate the impact of climate change on spring maize drought on the Loess Plateau and analyze the spatiotemporal distribution characteristics of spring maize drought on the Loess Plateau, this study used 18 representative spring maize agricultural meteorological observation stations on the Loess Plateau from 1997 to 2013 to observe the growth period and yield of spring maize, manually observing the moisture content (relative soil humidity) data of maize every ten days and determining the drought level based on the yield reduction rate. We revised Table 1 "Drought Levels of Spring Maize in the North (QX/T 259-2015)" by region and growth period to construct a new refined drought index for the relative soil humidity of spring maize on the Loess Plateau and, based on this, study the impact of climate change on spring maize drought on the Loess Plateau.

Table 1. Relative soil humidity drought category standard for maize at different growth stages.

Drought Categories	Relative Soil Moisture at Different Developmental Stages (%)				
	Sowing–Seedling	Seedling–Jointing	Jointing–Tasseling	Tasseling–Milking	Milking–Maturing
Drought-free	R > 65	R > 60	R > 70	R > 75	R > 65
Light drought	55 < R ≤ 65	50 < R ≤ 60	60 < R ≤ 70	65 < R ≤ 75	55 < R ≤ 65
Moderate drought	45 < R ≤ 55	40 < R ≤ 50	50 < R ≤ 60	55 < R ≤ 65	45 < R ≤ 55
Severe drought	35 < R ≤ 45	30 < R ≤ 40	40 < R ≤ 50	45 < R ≤ 55	35 < R ≤ 45
Extreme drought	R ≤ 35	R ≤ 30	R ≤ 40	R ≤ 45	R ≤ 35

2. Materials and Methods

2.1. Study Area

Considering the main planting areas of spring maize, the Loess Plateau in this study spanned Shaanxi Province, Shanxi Province, Ningxia, the middle east of Gansu Province, and the east of Qinghai Province. Due to its wide coverage and obvious differences in crop growth periods in different regions, the study area was divided into three regions based on China's agricultural natural division, referred to as the northwest Loess Plateau (Zone I), the central region of the Loess Plateau (Zone II), and the eastern region of the Loess Plateau (Zone III) (Figure 1).

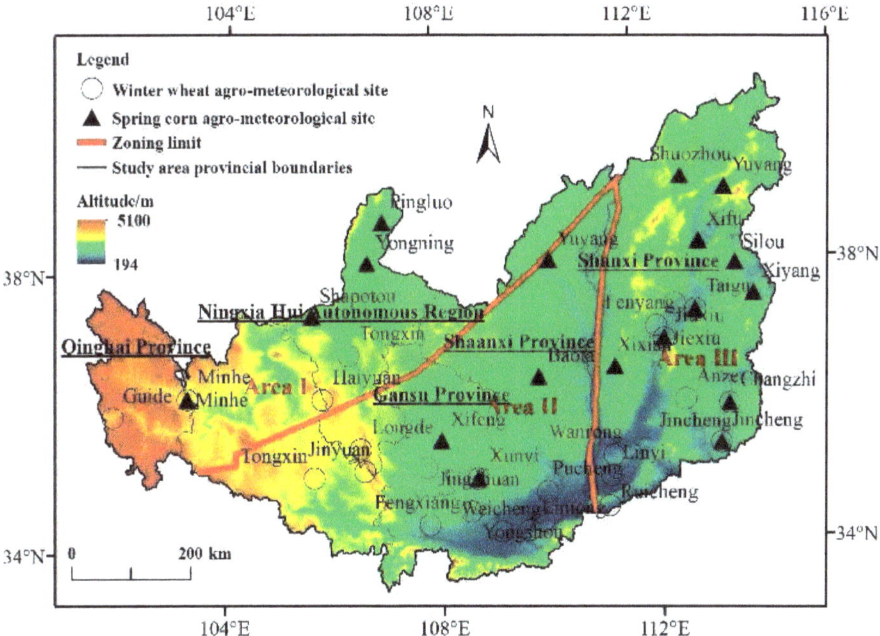

Figure 1. Regional division of the Loess Plateau and distribution map of agrometeorological stations.

2.2. Data

(1) Relative soil humidity data were extracted from the China Meteorological Data Network (http://data.cma.cn/, accessed on 12 August 2023). In this study, 18 spring maize agrometeorological observation stations (Minhe, Pingluo, Yongning, Shapotou, Xifeng, Xunyi, Yuyang, Baota, Shuozhou, Fanshi, Xinfu, Shilou, Xiyang, Taigu, Jiexiu, Xixian, Changzhi, and Jincheng) from 1997 to 2013 were selected, and relative soil humidity data with depths of 10, 20, and 50 cm were collected. The soil moisture data were observed by an agrometeorological observation station according to the "Specifications for Agrometeorological Observation" of the China Meteorological Administration, which satisfied the requirements of this study.

(2) Data of the growth period (seeding, seedling emergence, jointing, tasseling, milking, and ripening) of 18 spring maize agrometeorological observation stations from 1997 to 2013 were sourced from the China Meteorological Data Network.

(3) Production data were collected from the "Decadal Data Set of China's Crop Yield Data" and collated by the China Meteorological Administration, then extracted from the China Meteorological Data Network, including the actual production data of 16 spring maize agricultural meteorological observation stations (ibid., Shapotou and Xunyi) from 1997 to 2013. There were 272 spring maize data samples in total (sample

number = 17 years $\times$ 16 sites); however, after excluding some missing data, the final number of valid samples involved in the analysis was 190.

(4) The agrometeorological drought disaster data were obtained from the statistical results of some documents and the "China Agrometeorological Disaster Data Set" collected and sorted by the China Meteorological Administration and obtained by the China Meteorological Data Network. In this study, a drought disaster record of the data set at each agrometeorological observation station on the Loess Plateau region was defined as a drought event. Drought disaster data for the stations and years not recorded in the data set were supplemented by consulting relevant documents.

2.3. Research Methods

2.3.1. Relative Humidity of Soil (R)

The relative humidity of the soil (R) was calculated according to Equation (1).

$$R = \left(\frac{W_g}{f_c} \right) \times 100\% \tag{1}$$

where R is the relative humidity (%) of the soil, W_g is the soil weight water content (g/g), and f_c is the soil field water capacity (g/g).

2.3.2. Correlation Analysis

In this study, the Pearson correlation coefficient between the relative soil humidity and the production of spring maize in each ten-day growth period was calculated to reveal the influence of relative soil humidity on production in different development periods. The correlation was calculated according to Equation (2).

$$\gamma(a,b) = \frac{Cov(a,b)}{\sqrt{Var[a]\,Var[b]}} \tag{2}$$

where $\gamma(a,b)$ is the correlation coefficient of a and b, $Cov(a,b)$ is the covariance of a and b, $Var[a]$ is the variance of a, $Var[b]$ is the variance of b, and the value range of correlation coefficient γ is $[-1, 1]$; if $\gamma > 0$, it means that there is a positive correlation between the two variables.

2.3.3. Yield Reduction Rate

The difference between the annual actual output and the trend output was the negative value of the percentage of the trend output (%). Many methods can be used to calculate the production reduction rate. If the data skip a long time, the trend production method could be used. Generally, this method is more accurate. If the span of the data is short (less than 15 years) and the trend production cannot be calculated, the average method could be used.

(1) Trend production method [18]

Trend production was calculated according to Equation (3).

$$W = -\frac{Y - Y_t}{Y_t} \times 100\% \, (Y < Y_t) \tag{3}$$

where W is the production reduction rate, Y is the actual production, and Y_t is the trend mean.

(2) Mean method [19]

Trend production was calculated according to Equation (4).

$$W = -\frac{Y - \overline{Y}}{Y_t} \times 100\% \left(Y < \overline{Y} \right) \tag{4}$$

where W is the production reduction trend, Y is the actual production, and Y is the average production.

Light drought refers to a reduction of less than 10%, moderate drought refers to a 10% $\leq W < 20\%$ reduction, severe drought refers to a 20% $\leq W < 30\%$ reduction, and extreme drought refers to a $\geq 30\%$ reduction.

2.3.4. Frequency of Drought

The drought frequency was calculated according to Equation (5). If the total number of years of crop planting was N at a certain site, i, and the number of years of drought in a certain growth period was n, then the frequency of drought occurrence, F_i, could be expressed as:

$$F_i = \frac{n}{N} \times 100\% \tag{5}$$

2.3.5. Comparison with National Standards

In the verification process, the new drought classification results were compared with the national standards. The comparing results were divided into three categories, namely, compliance (where the drought level of the two treatments was consistent), basic compliance (where the difference between the two treatments was one level), and non-conformance (where the difference between two treatments was greater than two levels). Using the statistics of the three categories, the percentage of the number of time periods that agreed over the total number of time periods was the compliance rate.

3. The Impact of Climate Change on the Growth Period of Spring Maize in the Loess Plateau

3.1. Temporal Changes in Key Growth Periods of Spring Maize on the Loess Plateau

The changes in the growth period of spring maize have altered the length of its nutritional and reproductive growth periods, resulting in changes in the water and heat conditions experienced by spring maize during the important stage of yield formation. This study used data of the spring maize growth period observed by 18 agricultural meteorological stations on the Loess Plateau from 1997 to 2013, and average values of the spring maize growth period in three regions of the Loess Plateau from 1997 to 2013 were calculated using the partitioning method depicted in Figure 1.

As shown in Figure 2, the multi-year average sowing date of spring maize in Zone Ir was 9 April. The earliest was 2 April and the latest was 28 April. The average maturity date was 15 September, the earliest was 7 September, and the latest was 28 September. The number of days in the whole growth period (seeding–maturity period) was the shortest in 2013: 144 days. The longest period, in 2010, was 173 days. The number of days in the whole growth period tended to be longer. There were different degrees of delay from sowing to maturity. It can be seen in Figure 2b that the multi-year average sowing date of spring maize in Zone II of the Loess Plateau was 26 April, the earliest was 18 April, and the latest was 5 May. The average maturity date was 10 September, the earliest was 2 September, and the latest was 23 September. The number of days in the whole reproductive period was the shortest in 2002, 127 days. The longest year in 2009 was 147 days. The number of days in the whole growth period showed a trend of lengthening, and the inter-annual changes in the growth period in Zone II and Zone I was similar, with a trend of postponing from the sowing period to the mature period of spring maize. Figure 2c shows that the annual average sowing date of spring maize in Zone III of the Loess Plateau was 26 April, the earliest was 20 April, and the latest was 2 May. The average maturity date was September 18, the earliest was September 11, and the latest was 28 September. The number of days in the whole reproductive period was the shortest in 2001, 139 days, while the longest was 2013: 156 days. The sowing date, tasseling date, and ripening date exhibited a tendency to delay, and the emergence, jointing, and milk ripening dates were becoming earlier, although the trend was not significant. Figure 2d shows that the average sowing date of spring maize in the whole region was 24 April, the earliest was 17 April, while the latest was 28 April. The multi-year average maturity date was 16 September, the earliest was 11 September, and the latest was 23 September. The shortest maturity in the whole reproductive period was 142 days, and the longest was 150 days.

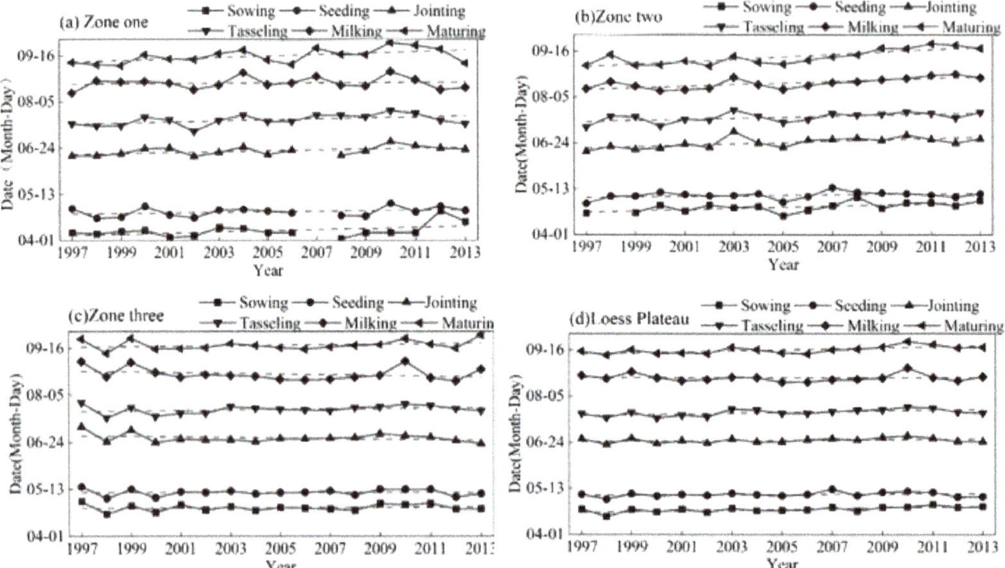

Figure 2. Temporal variation in spring maize growth periods on the Loess Plateau.

3.2. Spatial Distribution of Key Growth Periods of Spring Maize on the Loess Plateau

Figure 3 shows the spatial distribution of the average value (1997–2013) of the Julian date of sowing, emergence, jointing, tasseling, maturity, and the whole growth periods of spring maize on the Loess Plateau.

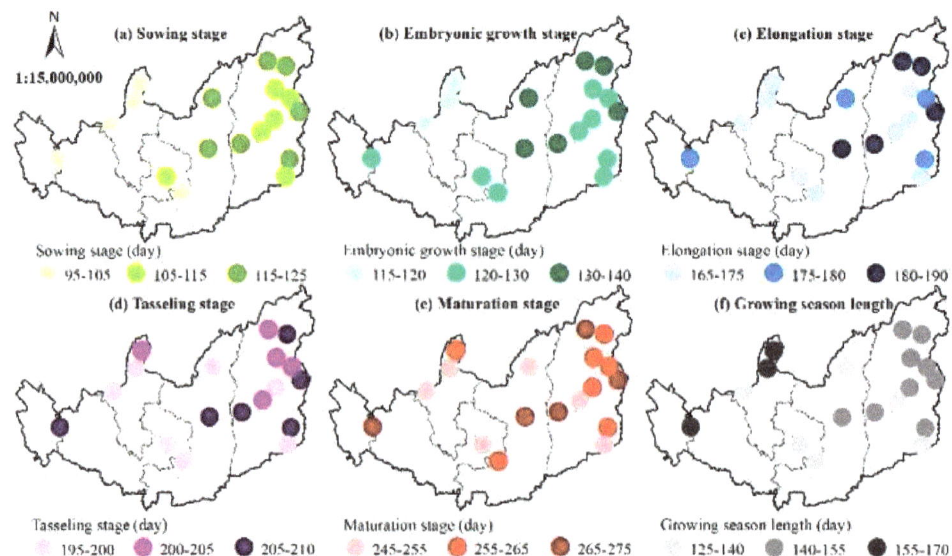

Figure 3. Spatial distribution of daily sequence averages of critical fertility stages of spring maize on the Loess Plateau.

Figure 3 shows that the sowing period of spring maize was gradually postponed from west to east, with Ningxia and Qinghai concentrated in early April, Gansu and central Shaanxi mainly in mid-to-late April, and Shanxi and northern Shaanxi mainly in late April to early May. The emergence period also showed a delay in the east–west direction, i.e., the emergence period gradually became later from west to east, and the emergence date was mainly from late April to early May. The jointing period of spring maize was postponed from west to east in mid-June to early July. The jointing period of spring maize in Shanxi and Shaanxi was generally later, mainly distributed in late June to early July. The tasseling stage and jointing stage demonstrated the same pattern, gradually becoming later from west to east, concentrated in mid-to-late July. The maturity period was mainly from early to late September. The number of days during the entire growth period was the shortest in the central region of the Loess Plateau (less than 140 days), the longest in the western region (155–170 days), and the longest in the eastern region.

4. Effects of Relative Soil Humidity on Spring Maize Production

The whole growth period of maize can be divided into the vegetative growth stage (seeding stage–jointing stage), transitional stage (jointing stage–tasseling stage), and reproductive growth stage (tasseling stage–mature stage). The results showed that the water demand for maize growth at the seedling stage is low. If drought occurs at the vegetative growth stage, the conditions will be better at the later stage, and the crops will exhibit characteristics of recovering or compensating for these effects, which would have a minor impact on the growth of spring maize. If drought occurs in the reproductive growth stage of maize, after the jointing stage, especially in the early stage of reproductive growth, even if the water conditions improve in the later stage, it will have a considerable impact on the growth of maize: the harvest could even fail [19,20]. We determine the key growth period of spring maize by analyzing the relationship between relative soil humidity and yield of spring maize in different periods on the Loess Plateau.

Due to the long duration of each growth period of maize, in order to more accurately determine the key period for the effects of relative soil humidity on the growth and development of maize, this study analyzed the correlation between the relative soil humidity and yield in the whole growth period of maize. Figure 4 shows the correlation coefficient between relative soil humidity and the production of spring maize in different growth stages on the Loess Plateau. Figure 4 shows that the relationship between relative soil humidity and production was the best in May, late June, early July, and September. The relative soil humidity at each depth passed the 0.05 significance test, and the relative soil humidity at each depth in early July and early September passed the 0.01 significance test. In addition, the relative humidity of 50 cm soil passed the 0.05 significance test in mid- and late July, and the relationship between the relative humidity of 10 cm and 20 cm soil and production passed the 0.01 significance test in late August. From the growth period perspective, the relationship between relative soil humidity and production at each depth from jointing stage to mature stage passed the 0.05 significance test, and the soil depth at jointing stage and heading stage passed the 0.01 significance test. The results showed that the relative humidity of soil at the jointing and heading stages of spring maize had the best correlation with yield, which was superior to the relative humidity at other growth stages, followed by the milky and mature stages, and the relative humidity of soil at the sowing and seedling stages had little impact on the yield. According to previous research results [20,21], tasseling–milking is the reproductive growth stage of maize and the key period of maize growth and development, which was similar to the results of this study.

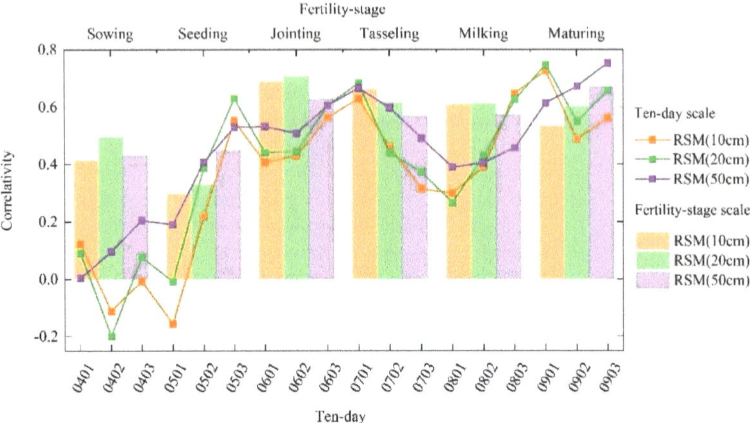

Figure 4. Correlation coefficients between relative soil humidity and yield at different fertility stages of spring maize on the Loess Plateau.

5. Construction of Refined Drought Indicators for Spring Maize at Different Growth Stages on the Loess Plateau

5.1. Revision of Refined Drought Indicator of Spring Maize at Different Growth Stages

Table 1 details the "Drought Grade of Northern Spring Maize (QX/T259-2015) " [22] issued by the China Meteorological Administration, which presents the standards of relative soil humidity drought grades in different growth periods of maize; the standard of maize in all growth periods was the same in all regions. The standard of relative soil humidity drought grade in different growth periods of maize was correlated to the region and local, in order to make the standard of drought grade of spring maize on the Loess Plateau more accurate. Refined revisions needed to be made based on drought and disaster data from various regions and maize yield reduction rates. This study used the actual observation data of relative soil humidity, maize yield, and drought disasters from 16 spring maize agricultural meteorological observation stations on the Loess Plateau from 1997 to 2013 to analyze the impact of relative soil humidity on maize growth, development, and yield year by year. The standard compliance rate of relative soil humidity on drought disaster record data was tested, and the drought grade index of relative soil humidity at different growth stages of maize at this station was revised. The results are shown in Table 2. The relative soil humidity data of spring maize used in the study showed that the average relative humidity was 0–20 cm during the sowing emergence period, and the average relative soil humidity was 0–50 cm during other growth periods [21].

Table 2. Drought rating criteria for relative soil humidity at different fertility stages of maize.

Station	Drought Categories	Relative Soil Humidity at Different Stages (%)				
		Sowing–Seedling	Seedling–Jointing	Jointing–Tasseling	Tasseling–Milking	Milking–Maturing
	Drought-free	R > 70	R > 80	R > 70	R > 75	R > 70
	Light drought	$60 < R \leq 70$	$70 < R \leq 80$	$60 < R \leq 70$	$65 < R \leq 75$	$60 < R \leq 70$
Minhe	Moderate drought	$40 < R \leq 60$	$60 < R \leq 70$	$50 < R \leq 60$	$55 < R \leq 65$	$50 < R \leq 60$
	Severe drought	$30 < R \leq 40$	$50 < R \leq 60$	$40 < R \leq 50$	$45 < R \leq 55$	$40 < R \leq 50$
	Extreme drought	$R \leq 30$	$R \leq 50$	$R \leq 40$	$R \leq 45$	$R \leq 40$
	Drought-free	R > 70	R > 75	R > 60	R > 70	R > 75
	Light drought	$60 < R \leq 70$	$65 < R \leq 75$	$50 < R \leq 60$	$60 < R \leq 70$	$60 < R \leq 75$
Pingluo	Moderate drought	$40 < R \leq 60$	$55 < R \leq 65$	$40 < R \leq 50$	$40 < R \leq 60$	$40 < R \leq 60$
	Severe drought	$30 < R \leq 40$	$45 < R \leq 55$	$35 < R \leq 40$	$30 < R \leq 40$	$30 < R \leq 40$
	Extreme drought	$R \leq 30$	$R \leq 45$	$R \leq 35$	$R \leq 30$	$R \leq 30$

Table 2. *Cont.*

Station	Drought Categories	Relative Soil Humidity at Different Stages (%)				
		Sowing–Seedling	Seedling–Jointing	Jointing–Tasseling	Tasseling–Milking	Milking–Maturing
Suozhou	Drought-free	R > 60	R > 60	R > 50	R > 50	R > 60
	Light drought	53 < R ≤ 60	50 < R ≤ 60	40 < R ≤ 50	40 < R ≤ 50	55 < R ≤ 60
	Moderate drought	45 < R ≤ 53	40 < R ≤ 50	35 < R ≤ 40	35 < R ≤ 40	40 < R ≤ 55
	Severe drought	40 < R ≤ 45	30 < R ≤ 40	30 < R ≤ 35	30 < R ≤ 35	35 < R ≤ 40
	Extreme drought	R ≤ 40	R ≤ 30	R ≤ 30	R ≤ 30	R ≤ 35
Baota	Drought-free	R > 65	R > 65	R > 60	R > 70	R > 65
	Light drought	60 < R ≤ 65	60 < R ≤ 65	50 < R ≤ 60	60 < R ≤ 70	55 < R ≤ 65
	Moderate drought	40 < R ≤ 60	50 < R ≤ 60	40 < R ≤ 50	40 < R ≤ 60	45 < R ≤ 55
	Severe drought	35 < R ≤ 40	45 < R ≤ 50	35 < R ≤ 40	30 < R ≤ 40	35 < R ≤ 45
	Extreme drought	R ≤ 35	R ≤ 45	R ≤ 35	R ≤ 30	R ≤ 35
Fanshi	Drought-free	R > 62	R > 65	R > 59	R > 58	R > 68
	Light drought	60 < R ≤ 62	62 < R ≤ 65	57 < R ≤ 59	55 < R ≤ 58	65 < R ≤ 68
	Moderate drought	55 < R ≤ 60	60 < R ≤ 62	50 < R ≤ 57	50 < R ≤ 55	60 < R ≤ 65
	Severe drought	40 < R ≤ 55	45 < R ≤ 60	45 < R ≤ 50	45 < R ≤ 50	50 < R ≤ 60
	Extreme drought	R ≤ 40	R ≤ 45	R ≤ 45	R ≤ 45	R ≤ 50
	Drought-free	R > 63	R > 65	R > 70	R > 75	R > 75
	Light drought	55 < R ≤ 63	60 < R ≤ 65	68 < R ≤ 70	65 < R ≤ 75	70 < R ≤ 75
Xiyang	Moderate drought	45 < R ≤ 55	55 < R ≤ 50	55 < R ≤ 68	55 < R ≤ 65	60 < R ≤ 70
	Severe drought	40 < R ≤ 45	50 < R ≤ 55	45 < R ≤ 55	45 < R ≤ 55	50 < R ≤ 60
	Extreme drought	R ≤ 40	R ≤ 50	R ≤ 45	R ≤ 45	R ≤ 50
Xifeng	Drought-free	R > 65	R > 65	R > 60	R > 55	R > 60
	Light drought	60 < R ≤ 65	55 < R ≤ 65	55 < R ≤ 60	50 < R ≤ 55	56 < R ≤ 60
	Moderate drought	55 < R ≤ 60	50 < R ≤ 55	50 < R ≤ 55	45 < R ≤ 50	53 < R ≤ 56
	Severe drought	45 < R ≤ 55	45 < R ≤ 50	45 < R ≤ 50	40 < R ≤ 45	50 < R ≤ 53
	Extreme drought	R ≤ 45	R ≤ 45	R ≤ 45	R ≤ 40	R ≤ 50

Due to the complex of reasons for the annual yield reduction, the maize yield reduction rate was calculated year by year for each station, and the relative soil humidity during the maize growth period was also calculated. According to the annual drought disaster records of each agricultural meteorological observation station on the Loess Plateau recorded in the Ten-day Data Set of China's Agrometeorological Disasters collected and collated by the China Meteorological Administration (a drought disaster record is defined as a drought event, and the drought disaster data of stations and years not recorded in the data set are supplemented by consulting the relevant literature), it was determined whether drought occurred in a certain year. The reduction in maize production that year was identified as a result of drought. Although there may be some errors in reductions in maize production caused by drought in dry years, as well as other factors, it could generally be considered in agricultural meteorological research that this reduction in maize production was caused by drought.

Due to the change in relative soil humidity observation method, the time period of relative soil humidity recorded by the agrometeorological observation stations in the China Meteorological Data Network was restricted within the years 1997–2013. It was reasonable and effective to determine the drought index of spring maize refinement suitability on the Loess Plateau with a time span of 17 years. The change in relative soil humidity was slow, and the ten-day data were reasonable and effective.

According to the relative soil humidity and production data of 18 spring maize agrometeorological observation stations on the Loess Plateau from 1997 to 2013, the production reduction rate of maize was calculated year by year. Although the yield of maize in 12 stations decreased, the relative soil humidity in each growth period of maize did not meet the drought category standards of relative soil humidity in different growth periods of maize specified in the Drought Category of Spring Maize in the North (QX/T 259-2015). It could be considered that the maize production reductions were not caused by drought [23]. There were seven stations (Minhe, Pingluo, Shuozhou, Baota, Fanshi, Xiyang, and Xifeng) at which the maize yield reduction rate corresponded to the drought standards of relative soil humidity. The Drought Category of Spring Maize in North China (QX/T

259-2015) was revised by combining the yield reduction rate and the drought disaster data on the Loess Plateau.

According to the statistics of relative soil humidity and production data of 18 spring maize agrometeorological observation stations on the Loess Plateau from 1997 to 2013, the production reduction rate of maize was calculated year by year [24]. Although the production of maize in 12 stations decreased, the relative soil humidity in each growth period of maize did not meet the drought category standard of relative soil humidity in different growth periods of maize specified in the Drought Category of Spring Maize in the North (QX/T 259-2015). It could be considered that the maize production reductions were not caused by drought. There were seven stations (Minhe, Pingluo, Shuozhou, Baota, Fanshi, Xiyang, and Xifeng) at which the maize yield reduction rate corresponded to the drought standard of relative soil humidity. The Drought Category of Spring Maize in North China (QX/T 259-2015) was subsequently revised by combining the production reduction rate and the drought disaster data on the Loess Plateau.

The indexes of drought-free and light drought in the seed–seedling stage and milk–maturing stage after revision in Pingluo increased by 5–10% compared with the original national standard value, and the indexes of moderate drought, severe drought, and extreme drought decreased by 5% compared with the original national standard value. The drought indexes of emergence and jointing increased by 5% compared with the original national standard values. The drought indexes of jointing–tasseling and tasseling–milking decreased by 5–10% compared with the original national standard values. Compared with the original national standard value, the indicators of drought-free, light drought, and moderate drought in Shuozhou decreased by 2–5%, and the indicators of severe drought and extreme drought increased by 5%. The drought indexes of jointing–tasseling and tasseling–milking decreased by 10–25% compared with the original national standard values. The drought indexes of the seedling–jointing stage and milk–maturing stage were the same as the national standard. The drought indexes in the periods of sowing–emergence, jointing–tasseling, and tasseling–milking in Baota decreased by 5–15% compared with the original national standard. The drought index in the period from seedling emergence to jointing increased by 5–15% compared with the original national standard value. The drought index of the milk–ripening period was the same as the national standard. The drought indexes of the seed–seedling stage, jointing–tasseling stage, and tasseling–milking stage in Fanshi decreased by 3–17% compared with the original national standard values. The drought indexes of the seedling–jointing stage and the milk–maturing stage increased by 3–15% compared with the original national standard values. Compared with the original national standard values, the indexes of drought-free and light drought in the seed–seedling stage and milk–maturing stage in Xiyang decreased by 2–15%, and the indexes of moderate drought, severe drought, and extreme drought increased by 5–15%. The drought indexes of the seedling–jointing stage and jointing–tasseling stage increased by 3–20% compared with the original national standard value. The drought index in the period from tasseling to milking was the same as the national standard. The drought indexes of the Xifeng seed–seedling stage and the seed–jointing stage increased by 10–25% compared with the original national standard value. The drought indexes of the jointing–tasseling stage, tasseling–milking stage, and milking–ripening stage decreased by 3–20% compared with the original national standard values.

In summary: (1) the drought standard value of relative soil humidity had increased by 3–25% at 71.4% of the stations in the seed–seedling stage and at 85.7% of the stations in the seed–jointing stage, which indicates that the national standard index of this growth period underestimates the drought category and the disaster degree of drought; (2) the drought standard value of relative soil humidity had decreased by 3–25% at 27.1% of the sites in the jointing–tasseling stage and at 71.4% sites in the tasseling–milking stage, indicating that the national standard index of this growth period overestimates the disaster degree of drought.

5.2. Reclassification of the Drought Index of Spring Maize on the Loess Plateau

Using the yield reduction rate and drought disaster data, we validated the revised relative soil humidity drought category standard for spring maize at different growth stages at seven stations (Minhe, Pingluo, Xifeng, Baota, Shuozhou, Fanshi, and Xiyang). The validation period of this study was 1997–2013, a total of 17 years, and each year was divided into five fertility stages, a total of 85 periods. During the validation process, the compliance was divided into three types, namely, compliance (where the drought level of the two is the same), basic compliance (where the difference between the two is one level), and non-conformance (where the difference between the two is two levels or more). Through the statistics of the three cases, the percentage of time periods satisfying the three cases in the total number of time periods was obtained; then, the changes in percentages before and after category correction were compared (Figure 5).

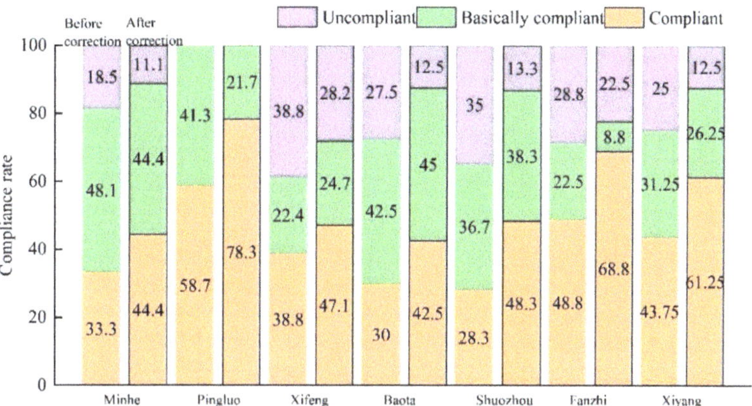

Figure 5. Compliance rate of drought-level criteria for soil moisture at different growth periods of maize on the Loess Plateau (%).

Figure 5 shows that the compliance degree of the relative soil humidity suitability index after correction had been significantly improved, and the compliance rate of each station category had increased by 8.2–20%, with Pingluo and Fanshi exhibiting the most obvious improvement in the compliance rate. Non-conformance rates of each station decreased by varying degrees, among which the rate of non-compliance at Shuozhou decreased the most, reaching 21.7%, and the rate of non-compliance at Fanshi decreased the least, by only 6.3%. After correction of the index category, the proportion of actual drought at the seven stations in line with the calculated results was significantly increased, while the proportion of non-conformity was significantly decreased. This showed that the revised category was more suitable for the study areas and could be used as the drought grading for studying the different growth stages of spring maize on the Loess Plateau.

5.3. Individual Validation of the Drought Index

Taking Shuozhou, Shanxi Province, as an example, the drought disaster data showed that moderate drought occurred in spring corn in July 2005, light drought occurred in spring maize in August 2008, and severe drought occurred in spring maize in 2010. The drought was judged using the revised relative soil humidity drought category standard, as shown in Table 3, and the drought degree of the corresponding growth period in the drought month was reduced, which was more in line with the actual drought situation.

Table 3. Drought disaster data and relative soil humidity drought level determination in Shuozhou (before/after revision).

Actual Drought Conditions	Jointing–Tasseling	Tasseling–Milking
July 2005, Moderate Drought	Extreme Drought/Severe Drought	
August 2008, Light Drought		Extreme Drought/Light Drought
July 2010, Severe Drought	Extreme Drought/Severe Drought	

6. Spatial Distribution of Spring Maize Drought Frequency on the Loess Plateau

According to the revised drought category standard of relative soil humidity of spring maize at different growth stages in different regions of the Loess Plateau, the occurrence frequencies of drought and above-drought events in 18 stations from 1997 to 2013 were calculated, and spatial distribution maps of the frequency of drought occurrence in different growth stages of spring maize on the Loess Plateau was obtained (Figure 6).

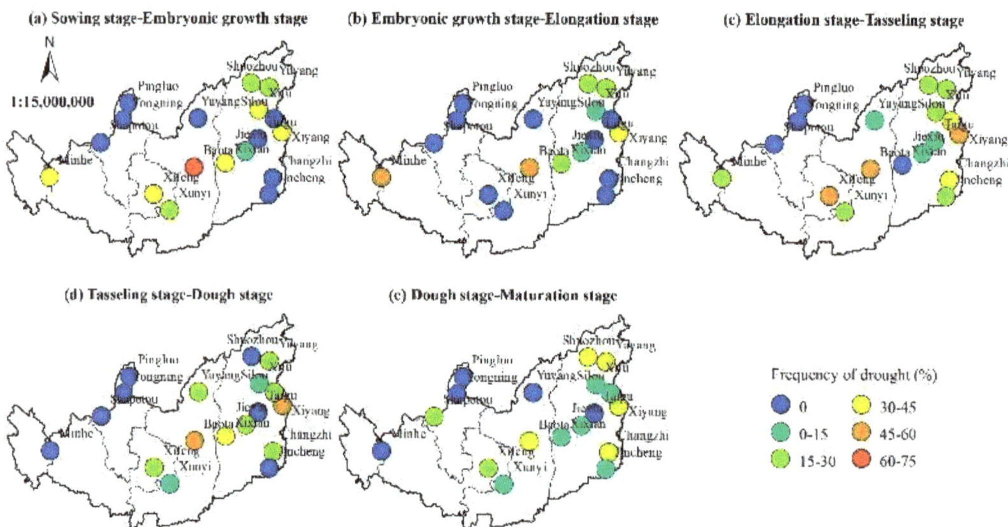

Figure 6. Spatial distribution of drought frequency in different growth stages of spring maize in the Loess Plateau.

Figure 6 shows that drought in the emergence period mainly occurred in central and southern Shaanxi, northern Shanxi, central and eastern Gansu, and some regions of Qinghai Province, with a frequency of 25–75%. The maximum drought frequency was 75%, which occurred in pagodas. Figure 6b indicates that drought in the emergence and jointing stage mainly occurred in central Shaanxi, northeastern Shanxi, central and eastern Gansu, and some regions of Qinghai Province, with the drought frequency reaching 10–53%. The maximum drought frequency occurred in Xifeng: approximately 53%.

Figure 6c shows that the high-frequency drought area in the jointing–tasseling period was concentrated in the eastern part of the Loess Plateau, with a frequency of 20–41%, and drought occurred in other areas at a frequency below 20%, characterized by higher frequency in the east and lower frequency in the west. Figure 6d shows that drought in the period from tasseling to milking mainly occurred in central Shaanxi, eastern Shanxi, and most of Gansu, with a frequency of 6–41%. The maximum drought frequency occurred in Xi County: approximately 41%. Figure 6e shows that the high drought frequency in the milky and mature period occurred in the northeast of the Loess Plateau; the drought frequency in Xiyang and Changzhi was about 42%; and the drought frequency in other regions was 8–33%, occupying a large area of the Loess Plateau. The overall characteristics of high drought frequency in the northeast and low drought frequency in the central and western regions were presented. With the change in growth period, the area of high drought frequency gradually moved northward, the overall drought frequency decreased, and the degree of drought decreased. Ningxia Province did not exhibit obvious drought in each growth period; the overall occurrence frequency of each growth period was high in the east and low in the west.

7. Conclusions

We used the relative soil humidity and yield data at stations on the Loess Plateau for spring maize growth periods from 1997 to 2013 to refine the drought categories based on the yield reduction rate. The findings are as follows:

(1)   The multi-year average sowing date of spring maize on the Loess Plateau in Zone I was 9 April, whereas that in Zone II and Zone III was 26 April. The growth period of spring maize was mainly delayed—the delaying trend of the tasseling and maturity period was significant—and the number of days the whole growth period of spring maize lasted was prolonged. In terms of spatial distribution characteristics, it was shown that the east–west direction of each growth period was delayed, and gradually become later from west to east.

(2)   The relative soil humidity in May and September had a significant impact on the yield of spring maize. The soil depths in the jointing and heading stages of spring maize before and after July passed the 0.01 significance test; the relative soil humidity in the jointing and heading stages of spring maize exhibited the best correlation with the yield, which was superior to the relative humidity in other growth stages, followed by the milky and mature stages; and the relative soil humidity in the sowing and seedling stages had a minor impact on the yield.

(3)   Based on the yield reduction rate of spring maize on the Loess Plateau, the original national drought standard was revised, and the drought categories of the relative soil humidity of spring maize were derived. The results showed that the relative soil humidity index of spring maize in each growth period increased or decreased by different degrees at each station, and the category at each station fluctuated within 25% compared with the national standard. Through the verification of the compliance rate of the drought index category standard and individual cases, it was shown that the relative soil humidity drought category classification standard of each station determined in this study can better indicate the severity of agricultural drought on the Loess Plateau. After correction, the compliance rate of the relative soil humidity suitability index was significantly improved, and the compliance rate of each station category was increased by 8.2–20%, with the most obvious changes in the compliance rate at Pingluo and Fanshi. Non-conformance rates of each station decreased by varying degrees, and the rate of non-conformance of the category at Shuozhou decreased the most, reaching 21.7%. The revised category was more suitable for the study area, and could be used for grading spring maize drought in different growth stages on the Loess Plateau.

(4)   The drought frequency on the Loess Plateau was determined using the revised standard of relative soil humidity drought categories of spring maize. The results showed that moderate drought, severe drought, and extreme drought in the sowing and seedling stage of spring maize on the Loess Plateau from 1997 to 2013 mainly occurred in central and southern Shaanxi, northern Shanxi, central and eastern Gansu, and parts of Qinghai. The drought frequency was 25–75%, and the maximum drought frequency occurred in the pagoda of Gansu, which was 75%. With the change in the growth period of spring maize, the area with high drought frequency gradually moved northward, the drought frequency decreased overall, and the degree of

drought decreased. The drought frequency in most regions during the milk–ripening period was less than 20%; the drought frequency in Xiyang and Changzhi in Shanxi Province was about 42%.

8. Discussion

Liu Qingqing [21] used the standardized precipitation evapotranspiration index (SPEI) and soil moisture deficit index (SMDI) to analyze the spatial–temporal evolution of agricultural drought and meteorological drought during the spring/summer maize growth period in Northeast, Northwest, and North China. The research results indicate that there was a certain degree of similarity in the spatiotemporal variation patterns of agricultural drought and meteorological drought during the maize growth period in different regions, but there were certain differences. Therefore, this study addressed the response law of yield-related factors of maize on the Loess Plateau to drought, analyzed the impact of meteorological and agricultural drought on maize growth and yield during the maize growth period, determined the key months and time scales of drought's impact on maize growth and yield, and selected the drought indicators that were most suitable for reflecting maize yield response and corrected them, which has practical application and promotion significance.

Yan C. et al. (2023) calculated the standardized precipitation evapotranspiration index (SPEI) of maize growth periods at different scales based on measured meteorological data from 32 meteorological stations in Shaanxi Province from 1971 to 2020. They analyzed the spatiotemporal characteristics of drought in different maize growth periods and the impact of drought on the meteorological yield of maize [25]. Research has shown that the meteorological yield of maize is closely related to dry and wet conditions during the flowering period; the maize yield in most regions of Shaanxi Province was considerably affected by the dry and wet conditions throughout the entire growth period. Zhang X. F. (2022) utilized the daily climate data of 156 meteorological stations in dry farming areas of northern China from 1961 to 2018 and spring maize growth period data from 1991 to 2013, to divide the dryland area in northern China into four sub-areas according to their aridity index. Based on the crop water deficit index (CWDI), the temporal and spatial characteristics of spring maize drought in the dryland area of northern China were revealed by analyzing water supply and demand, inter-annual variation in CWDI, drought station ratio, and frequency. The results showed that in terms of spatial distribution, drought grade and frequency showed an obvious east–west distribution [26]. This demonstrated that exploring the temporal and frequency domain correlation between maize meteorological yield and drought indicators in different regions, and finally quantifying the relationship between dry and wet conditions and yield through regression analysis, could effectively provide a reference basis for ensuring disaster prevention and reducing agricultural production in the research area, as well as optimizing irrigation decision-making.

Author Contributions: Formal analysis and Writing—original draft, S.Y.; Conceptualization, Supervision, and Writing—review and editing, N.J. and L.X.; Formal analysis and Writing—review and editing, L.H. and J.W. The results of previous studies were similar to those of this article. All authors have read and agreed to the published version of the manuscript.

Funding: "Research on Quantitative Assessment of Climate Impact of Northwest Sichuan Ecological Demonstration Area on Surrounding Regions" (2022 NSFSC0208), a project funded by the Sichuan Natural Science Foundation; "Study on the Applicability of Refined Drought Indicators for Maize on the Loess Plateau", project of Lanzhou Institute of Drought Meteorology, China Meteorological Administration (IAM202004).

Institutional Review Board Statement: Not applicable.

Informed Consent Statement: Not applicable.

Data Availability Statement: Not applicable.

Conflicts of Interest: The authors declare that they have no known competing financial interests or personal relationships that could have appeared to influence the work reported in this paper.

References

1. Hou, Q.Q.; Pei, T.T.; Chen, Y.; Ji, Z.X.; Xie, B.P. Variations of drought and its trend in the Loess Plateau from 1986 to 2019. *Chin. J. Appl. Ecol.* **2021**, *32*, 649–660.
2. Dong, C.Y.; Liu, Z.J.; Yang, X.L. Effects of different category drought on grain yield of spring maize in Northern China. *Trans. Chin. Soc. Agric. Eng.* **2015**, *31*, 157–164.
3. Yang, W.J. *Research on Corn Drought Monitoring Based on Landsat8 Growth Time Series Remote Sensing Information*; Shihezi University: Shihezi, China, 2017.
4. Wang, J.R.; Sun, C.J.; Zheng, Z.J.; Li, X.M. Drought characteristics of the Loess Plateau in the past 60 years and its relationship with changes in atmospheric circulation. *Acta Ecol. Sin.* **2021**, *41*, 5340–5351.
5. Zhang, Q.; Yao, Y.B.; Li, Y.H.; Huang, J.P.; Ma, Z.G.; Wang, Z.L.; Wang, S.P.; Wang, Y.; Zhang, Y. Progress and prospect on the study of causes and variation regularity of droughts in China. *Acta Meteorol. Sin.* **2020**, *78*, 500–521. [CrossRef]
6. Cheng, W.J. Causes and Prevention Measures of Drought Disaster in the Loess Plateau. *J. Smart Agric.* **2022**, *3*, 10–12.
7. Li, M.; Deng, Y.Y.; Ge, C.H.; Wang, G.W.; Chai, X.R. Characteristics of meteorological drought across the Loess Plateau and their linkages with large-scale climatic factors during 1961–2017. *Ecol. Environ. Sci.* **2020**, *29*, 2231–2239.
8. Li, J.; Jiang, B.; Hu, W. Characteristics of deep soil desiccation on rainfed G rain Croplands in different rainfall areas of the Loess Plateau of China. *J. Nat. Resour.* **2009**, *24*, 2124–2134.
9. Zhang, J.Q. Risk assessment of drought disaster in the maize-growing region of Songliao Plain, China. *Agric. Ecosyst. Environ.* **2004**, *102*, 133–153. [CrossRef]
10. Wang, P.X.; Gong, J.Y.; Li, X.W. Advances in drought monitoring by using remotely sensed normalized difference vegetation index and land surface temperature products. *Adv. Earth Sci.* **2003**, *18*, 527–533.
11. Ghulam, A.; Qin, Q.; Zhan, Z. Designing of the perpendicular drought index. *Environ. Geol.* **2007**, *52*, 1045–1052. [CrossRef]
12. Shi, X.X. Variations of Soil Moisture During the Growth Phase of Main Crops in Semi—Arid Region in Gansu. *J. Arid Meteorol.* **2011**, *29*, 461–465.
13. Yao, X.Y.; Pu, J.Y.; Yao, R.H. Assessment of maize water adequacy in Loess Plateau of Gansu Province. *Acta Ecol. Sin.* **2010**, *30*, 6242–6248.
14. Yin, H.X.; Zhang, B.; Zhang, J.X.; Zhang, T.F.; Li, X.Y.; Jin, Z.B. Spatial-Temporal Characteristics of Drought and Spring Maize in Eastern Gansu. *Resour. Sci.* **2012**, *34*, 2347–2355.
15. An, X.L.; Wu, J.J.; Zhou, H.K.; Li, X.H.; Liu, L.Z.; Yang, J.H. Assessing the relative soil moisture for agricultural drought monitoring in Northeast China. *Geogr. Res.* **2017**, *36*, 837–849.
16. Martínez-Fernández, J.; González-Zamora, A.; Sánchez, N.; Gumuzzio, A. A soil water-based index as a suitable agricultural drought indicator. *J. Hydrol.* **2015**, *522*, 265–273. [CrossRef]
17. *QX/T259-2015*; Meteorological Industry Standard of the People's Republic of China, Drought Category of Spring Maize in North China. China Standards Publishing House: Beijing, China, 2015.
18. Zhang, J.; Mu, Q.; Huang, J. Assessing the remotely sensed Drought Severity Index for agricultural drought monitoring and impact analysis in North China. *Ecol. Indic.* **2016**, *63*, 296. [CrossRef]
19. Bai, X.L.; Sun, S.X.; Yang, G.H.; Liu, M.; Zhang, Z.P.; Qi, H. Effect of Water Stress on Maize Yield During Different Growing Stages. *J. Maize Sci.* **2009**, *17*, 60–63.
20. Hou, Q.; Li, J.J.; Wang, H.M. Dynamic Indexes of Water-saving irrigation based on Maize growth characteristics. *J. Irrig. Drain.* **2015**, *34*, 34.
21. Liu, Q.Z.; Li, Y. *Evaluation of the Impact of Spatial and Temporal Evolution of Drought on Maize Growth and Yield*; College of Water Resources and Architectural Engineering: Xian, China, 2022.
22. *GB_ T 32136-2015*; Meteorological Industry Standard of the People's Republic of China, Agricultural Drought Rating. China Standards Publishing House: Beijing, China, 2015.
23. He, Y.T.; Li, W.H.; Lang, H.O. Study on the characteristics of precipitation resources and the afforestation suitability in the Loess Plateau. *Arid Zone Res.* **2009**, *26*, 406–412. [CrossRef]
24. Tang, Q.; Xu, Y.; Liu, Y. Spatial Difference Analysis of Land Use Dynamic Change in the Loess Plateau Region. *Resour. Environ. Arid. Areas* **2010**, *24*, 15–21.
25. Yan, C.; Zhang, X.; Sun, Y.; Wei, Y.T.; Hu, X.M.; Luo, S.Q. Analysis of drought characteristics and its effect on maize yield in Shaanxi Province based on SPEI. *Water Sav. Irrig.* **2023**, *1*, 10–18.
26. Zhang, X.F.; Zhang, B.; Ma, S.Q.; Huang, H.; Chen, J.; Zhou, J. Analysis on Temporal and Spatial Variation of Spring Maize Drought in Dry Farming Area of Northern China Based on Crop Water Deficit Index. *Chin. J. Agrometeorol.* **2022**, *43*, 749–760.

Article

Estimating Precipitation Using LSTM-Based Raindrop Spectrum in Guizhou

Fuzeng Wang [1,2], Yaxi Cao [1], Qiusong Wang [1], Tong Zhang [2] and Debin Su [1,*]

[1] College of Electronic Engineering, Chengdu University of Information Technology, Chengdu 610225, China; wangfz@cuit.edu.cn (F.W.)

[2] Key Laboratory of Land Surface Process and Climate Change in Cold and Arid Regions, Chinese Academy of Sciences, Lanzhou 730000, China; zhangt@lzb.ac.cn

* Correspondence: sudebin@cuit.edu.cn

Abstract: The change in raindrop spectrum characteristics is an important factor affecting the accuracy of estimations of precipitation. The in-depth study of raindrop spectrum characteristics is of great interest for understanding precipitation process and improving quantitative radar precipitation estimation. In this paper, the raindrop size distributions at Longli (57913), Puding (57808) and Luodian (57916) stations in Guizhou were analyzed from the perspective of precipitation microphysical characteristics. The results showed that the raindrop size distribution was different among different regions. The correlation coefficients of the mass-weighted average diameter for the rain intensities at these three stations were 46.89%, 49.51%, and 47.03%, respectively, which were slightly lower than the normal correlation coefficients of the average volume diameter for the rain intensities: 67.80%, 71.28%, and 71.46%, respectively. Based on the data from the Guiyang weather radar, raindrop spectrometer, and automatic rain gauge, the dynamic Z-I relationship method and the LSTM neural network method were used to estimate precipitation. The correlation coefficients of the dynamic Z-I relationship method and the LSTM neural network method at the three stations studied were 0.8432, 0.7763, and 0.8658 and 0.8745, 0.9125, and 0.8676, respectively. Regarding the process of stratiform cloud precipitation, the correlation coefficients of the dynamic Z-I relationship method and LSTM neural network method at the three stations were 0.6933, 0.0902, and 0.1409 and 0.7114, 0.4984, and 0.4902, respectively. In the estimation of cumulative precipitation for 45 days from 1 July to 16 August 2020, the relative errors of the neural network estimation at the three stations were −4.25%, −11.35%, and −8.68% and the relative errors of the dynamic Z-I relationship estimation were −2.68%, −7.41%, and −21.23%, respectively. The final relative error of the neural network was slightly worse than that of the dynamic Z-I relationship in the cumulative precipitation estimations of Longli station and Puding station, but the overall correlation coefficients of the LSTM neural network method were better than those of the dynamic Z-I relationship method.

Keywords: raindrop spectrum; radar; dynamic Z-I; LSTM neural network; precipitation estimation

Citation: Wang, F.; Cao, Y.; Wang, Q.; Zhang, T.; Su, D. Estimating Precipitation Using LSTM-Based Raindrop Spectrum in Guizhou. *Atmosphere* **2023**, *14*, 1031. https://doi.org/10.3390/atmos14061031

Academic Editors: Haibo Liu and Tomeu Rigo

Received: 20 April 2023
Revised: 25 May 2023
Accepted: 10 June 2023
Published: 15 June 2023

1. Introduction

Raindrop spectrum analysis is an important part of cloud precipitation physics. Through the analysis of the raindrop spectrum, the accuracy of radar quantitative estimations of precipitation can be further improved. The accuracy of precipitation is of great significance for predicting and preventing flood disasters; for forecasting hydrological water resource systems, agricultural crop production, and ecological environmental changes; and for providing decision-making support for the work of agricultural and water conservancy and other relevant departments, as well as for the formulation of corresponding policy measures. As a device for directly detecting the ground raindrop spectrum, many scholars have studied the distribution characteristics of the raindrop spectrum. Huang et al. [1] classified the raindrop spectrum data of Anxi in Fujian Province from 2017 to

2020 based on the precipitation season and type. The results showed that the raindrop spectrum's spectral width and number concentrations in summer are the highest in the precipitation season. The raindrop spectrum of stratiform clouds is narrower than that of convective clouds in their precipitation types. The average particle size is smaller and the number and concentration are lower. Chakravarty et al. [2] statistically analyzed the particle distribution characteristics before, during, and after the monsoon in Pune, India. The results showed that the strong updraft affected convective precipitation after the monsoon period, which reduced the small particle size and increased the average particle diameter. In order to study the influences of underlying surface, precipitation type, and other factors on the characteristics of the raindrop spectrum, Zhao et al. [3] analyzed the parts of the raindrop spectrum in the summer of 2017–2018 in mountainous and plain areas of Beijing. The results showed that the particle size of the peak concentration of the raindrop spectrum in mountainous regions is larger, and convective cloud precipitation has a larger mass-weighted average diameter and spectral width than stratiform cloud precipitation. Zhang et al. [4] analyzed the raindrop size distribution and microphysical characteristics of different precipitation types in summer at the southern foot of the Qilian Mountains. It was concluded that the number, concentration, average diameter, and maximum diameter of convective cloud precipitation are higher than those of stratiform cloud precipitation. The GAMMA distribution was closer to the actual raindrop size distribution. Chen [5] used the machine learning method to establish a convolutional neural network model with which to identify the type of precipitation based on the base data of the ground-based raindrop spectrometer. The model was distinguished by extracting the characteristics of particle size, particle velocity, and number density distribution in the raindrop spectrum distribution image. Compared with the traditional measurement method, the accuracy rate was improved by 11.87%.

With the deepening of the research on the raindrop spectrum, many scholars have applied raindrop spectrometer data to radar quantitative precipitation estimation in order to improve the accuracy of precipitation estimation. Jing et al. [6] used the observation data of large-area heavy precipitation processes, collected with the ground raindrop spectrometer and S-band weather radar in Guangzhou, to analyze and correct the deviation between the aerial reflectance observed by radar and the ground reflectance obtained with the raindrop spectrometer. The results showed that the relative error of precipitation estimated with the raindrop spectrometer combined with the weather radar method was 25% higher than that of the traditional Z-I relationship method. Liu et al. [7] used a ground raindrop spectrometer combined with dual-polarization weather radar data for precipitation inversion research. In the analysis of inversion results, it was found that the inversion effect was related to the mass-weighted average diameter of the precipitation process and the size range of precipitation intensity. Zhang et al. [8] used the raindrop spectrometer network method and the traditional radar quantitative estimation method to estimate precipitation. The results showed that the two methods have a better estimation effect on stratiform cloud precipitation, and the raindrop spectrometer network method is more accurate. Zhou et al. [9] analyzed the characteristics of the raindrop spectrum in the stratocumulus mixed cloud precipitation process using the raindrop spectrometer's observation data. The results showed that the precipitation intensity mainly depends on the maximum raindrop diameter. It is also positively correlated with the raindrop concentration, but has little relationship with the average diameter. The fitting parameters of Γ distribution provide the exact change trend with time, and the fluctuation of each appropriate parameter during the heavy precipitation period is slow. The values remain at the same level.

Meteorological data have the characteristics of solid periodicity, space–time, uncertainty, and high attribute correlation, the variability of which make it challenging to analyze and process meteorological data by conventional methods (Peng Yuzhong et al. [10]). In recent years, with the continuous development of machine learning technology, researchers are increasingly applying it to natural science. Machine learning is an artificial intelligence algorithm different from traditional prediction methods. Because the machine learning model has strong expression and fitting ability, it is especially suitable for solving nonlinear problems. Kang et al. [11] used nine input variables in order to construct a short-term long memory (LSTM) model. They refined the selected meteorological variables, according to the relative importance of the input variables, to reconstruct the LSTM model. The LSTM model of the final selected input variable was used to predict precipitation, and the performance was compared with other classical statistical and machine learning algorithms. The results show that LSTM is suitable for precipitation prediction. Shen et al. [12] used long-term and short-term memory networks in order to predict precipitation. The results showed that the prediction capability of the LSTM network is better than those of the stepwise regression, BP neural network, and model output methods. The LSTM network has a specific prediction ability for the overall precipitation situation. Tang et al. [13] used the LSTM method to predict short-term rainfall, and collected debris flow data and daily cumulative rainfall data. The statistical classification method was used to delineate the precipitation warning threshold of debris flow, and the predicted value and point were compared. The results showed that the threshold warning accuracy of rainfall-induced debris flow exceeds 90%, and the error of LSTM-predicted rainfall results is less than 1.5 mm. Liu et al. [14] used a long short-term memory neural network (LSTM) to predict monthly precipitation on the Qinghai–Tibet Plateau. The results showed that the spatial range of R2 ≥ 0.6 in the LSTM model was much larger than that in the traditional model, and the RMSE value of LSTM prediction was lower than that of other models under different prediction lengths. Chen et al. [15] used the LSTM neural network in order to construct a monthly precipitation prediction model. They compared the precipitation prediction accuracy based on LSTM and random forest (RF) models under different covariates. The results show that the LSTM model has higher accuracy when using climatic factors (C) and historical precipitation data (H) as input variables than under single-input C and H variables. Compared with the RF model, the LSTM-based precipitation prediction model has higher prediction performance. Yashon O. Ouma et al. [16] applied the LSTM neural network and wavelet neural network (WNN) in the spatiotemporal prediction of precipitation and runoff time series trends in hydrological basins. The results showed that LSTM performed better than WNN. Yang et al. [17] predicted precipitation with reasonable accuracy by establishing a coupling model with the ensemble empirical mode decomposition and long short-term memory neural network (LSTM).

Firstly, this paper analyzes the characteristics of Longli (57913), Puding (57808), and Luodian (57916) stations in Guizhou, and finds the diameter parameter with a strong correlation with precipitation intensity to pave the way for precipitation estimation. Thereafter, based on the multi-source data from the Guiyang weather radar, raindrop spectrometer, and automatic rain gauge, the precipitation is estimated using the LSTM neural network method and the dynamic Z-I method.

2. Data and Quality Control

2.1. Data Sources

2.1.1. Weather Radar Data

The weather radar data comes from Guiyang CINRAD/CD new generation Doppler weather radar in Guizhou Province, China. The station number is Z9851, located at (106.7264° E, 26.5903° N), with an altitude of 1255.7 m, a reflectivity factor distance resolution of 250 m, and a maximum detection distance of 250 km. The radar volume scanning mode of this department is VCP21; that is, one body scan can scan 9 elevation angles in 6 min, and the elevation angles are 0.5°, 1.5°, 2.4°, 3.4°, 4.3°, 6.0°, 9.9°, 14.6°, and 19.5°.

2.1.2. Raindrop Spectrometer and Rain Gauge Data

The raindrop spectrometers used in this paper are all OTT-Parsivel laser raindrop spectrometers, which detect the particle size and particle velocity by measuring the width and time of the laser beam with a wavelength of 650 mm, through the plane of 30 mm wide and 180 mm long. The rain gauge data used are from the tipping bucket rain gauge of the automatic meteorological observation station in Guizhou Province, and the detection accuracy is 0.1 mm. In order to ensure the spatial position coincidence of the raindrop spectrometer and rain gauge, the data from the two instruments used in this paper are from the same meteorological station. The detection time resolutions of the raindrop spectrometer and of the rain gauge are set to 1 min, which ensures the time matching of the two data. Table 1 lists the basic site information.

Table 1. Basic information of raindrop spectrum stations.

Station Number	Station Name	Geographical Coordinates	Distance from Radar
57913	Longli	106.98° E, 26.45° N	30.21 km
57808	Puding	105.74° E, 26.31° N	99.95 km
57916	Luodian	106.76° E, 25.43° N	129.03 km

2.2. Data Quality Control

2.2.1. Data Quality Control of Raindrop Spectrometer

Firstly, the gross error is eliminated. The research shows that raindrop particles more than 6 mm in nature are extremely rare, so this part is destroyed. Subsequently, in order to ensure the stability of the droplet spectrum measurement, the data in which the number of detected raindrops is less than 10 in a single sample are removed as noise. Finally, according to the local D-V relationship, the errors of particle overlap and raindrop spectrometer rain cover splashing particles are eliminated.

ATLA D et al. [18] (1973) found that the final velocity formula of precipitation particles under ideal windless conditions is as follows:

$$\begin{cases} v = 0, x < 0.03 \text{ mm} \\ v = 4.323 \times (x - 0.03) & , 0.03 \text{ mm} < x \leq 0.6 \text{ mm} \\ v = 9.65 - 10.3 \times e^{-0.6x}, x > 0.6 \text{ mm} \end{cases} \tag{1}$$

In the formula, x is the particle diameter, whose unit is mm; v represents the final velocity of the particle, whose unit is m/s.

By counting different types of precipitation samples, the D-V curve of local precipitation particles is fitted, representing the particle size and particle velocity characteristics of regional precipitation. Referring to the data method proposed by Kruger et al. [19] to eliminate the excessive dispersion of velocity sampling, the threshold is set to eliminate the particles in the interval with significant deviation from the reference curve, and the precipitation particles can be retained.

2.2.2. Weather Radar Data Quality Control

Weather radar will detect both meteorological and non-meteorological echoes in the detection process. For some isolated echoes, or thin line-shaped echoes caused by birds, insects, aircraft, etc., speckle filtering can be used to eliminate them. The formula is as follows:

$$P_X = N / N_{\text{total}} \tag{2}$$

In the formula, N_{total} is the total number of distance libraries contained in the 5×5 grid centered on the X-th distance library, N is the number of libraries with radar reflectivity values in N_{total}, and P_x is the percentage of radar reflectivity in the grid. When P_x is less than a certain threshold (default 75% in this paper), the X-distance library calibrated in the radar base data is regarded as the non-meteorological echo being eliminated.

However, it is difficult to eliminate large-scale ground clutter in some areas close to the radar using speckle filtering alone. For this part of the clutter, texture filtering is needed to eliminate it. Firstly, the horizontal texture feature parameter T_{dBZ} is calculated. The horizontal texture feature parameter T_{dBZ} can describe the smoothness of adjacent distance and radial data in the same elevation layer. The formula is as follows:

$$T_{dBZ} = \sum_{j=1}^{N_A} \sum_{i=1}^{N_B} (Z_{i,j} - Z_{i,j+1})^2 / (N_A \times N_B) \tag{3}$$

In the formula, $Z_{i,j}$ is the radar reflectivity intensity, whose unit is dBZ; i and j are the distance library serial number and the azimuth serial number in the reflectivity library, respectively. N_A and N_R are the numbers of calculation libraries centered on the calibration distance library in distance and azimuth, respectively.

The vertical texture feature parameter V_{dBZ} can describe the continuity of the echo in the adjacent elevation layers of the same distance library. The V_{dBZ} of the calibrated distance library can be calculated using the following formula:

$$V_{dBZ} = (Z_i - Z_{i+1}) / (H_{i+1} - H_i) \tag{4}$$

In the formula, Z_i and Z_{i+1} are the radar reflectivity intensity of the upper elevation angle of the calibration range library and the calibration range library, respectively, and their unit is dBZ. H_i and H_{i+1} correspond to the elevation angles of the upper layer of the calibration distance bank and the calibration distance bank, respectively.

The horizontal texture features can distinguish stratiform cloud precipitation echo and convective cloud echo, but cannot distinguish precipitation echo and non-precipitation echo. The vertical texture features can distinguish non-precipitation echo and stratiform cloud echo, but cannot distinguish non-precipitation echo and convective cloud echo. Therefore, it is necessary to combine the two. The non-precipitation echo formula is as follows:

$$\begin{cases} V_{dBZ} > 20 dB/km \\ T_{dBZ} > 10 dB^2 \end{cases} \tag{5}$$

It can be seen from Figure 1 that the ground clutter in the area near the radar is effectively suppressed, and some sporadic echoes at the edge of the precipitation echo are also eliminated. After quality control, a complete precipitation echo area can still be obtained.

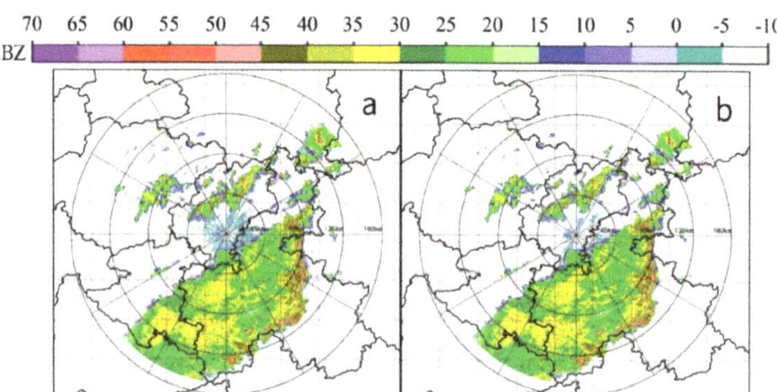

Figure 1. Quality control effect diagram of weather radar base data (**a**) before quality control and (**b**) after quality control.

3. Calculation and Characteristic Analysis of Raindrop Spectrum Parameters

3.1. Calculation of Raindrop Spectrum Parameters

3.1.1. Number Density

The particle number density $N(D)$ represents the total number of raindrop particles per unit volume, and its unit is $m^{-3}m^{-1}$. The calculation formula is as follows:

$$N(D) = \sum_{i=1}^{32} \sum_{j=1}^{32} \frac{n_{ij}}{A \times \Delta T \times V_j} \tag{6}$$

In Equation (6), n_{ij} represents the number of particles in the i-th particle diameter and the j-th particle velocity; A is the sampling base area of raindrop spectrometer 5400 mm^2; ΔT is the sampling time of 60 s; V_j is the sampling particle velocity in m/s.

3.1.2. Precipitation Intensity

The precipitation intensity is the amount of precipitation per unit of time, and the unit is mm/h. The calculation formula was proposed by Pruppacher and Klett [20] as follows:

$$I = \frac{6\pi}{10^4} \sum_{i=1}^{32} D_i^3 V(D_i) N(D_i) \tag{7}$$

In Equation (7), D_i represents the diameter of the sampled particle, and $N(D_i)$ is the number of particles at the current particle diameter and particle velocity.

3.1.3. Mean Diameter

The average diameter is the sum of the diameters of all raindrops divided by the total number of raindrops. The formula is as follows:

$$D_l = \frac{\sum\limits_{i=1}^{32} N(D_i) D_i}{\sum\limits_{i=1}^{32} N(D_i)} \tag{8}$$

3.1.4. Mass-Weighted Average Diameter

The mass-weighted average diameter is the average diameter of the weighted mass of all particles in the unit volume relative to the total mass of the particles. The unit is mm. The calculation formula is as follows:

$$D_m = \frac{\sum\limits_{i=1}^{32} N(D_i) D_i^4}{\sum\limits_{i=1}^{32} N(D_i) D_i^3} \tag{9}$$

3.1.5. VMD

The average volume diameter represents the diameter of the equivalent raindrop whose volume is equal to the average raindrop volume. Its unit is mm. The calculation formula is as follows:

$$D_v = \left[\frac{\sum\limits_{i=1}^{32} N(D_i) D_i^3}{\sum\limits_{i=1}^{32} N(D_i)} \right]^{\frac{1}{3}} \tag{10}$$

3.1.6. M-P Distribution and GAMMA Distribution

Marshall and Palmer [21] first proposed the M-P distribution. They found that the exponential function can be used to describe the distribution of raindrop size distribution, and established the following formula:

$$N(D) = N_0 \times \exp(-\lambda D) \tag{11}$$

In Equation (11), the unit of the particle number density parameter N_0 is $\mathrm{mm}^{-1}\mathrm{m}^{-3}$, and the unit of the particle size parameter λ is mm^{-1}.

It can be seen from previous studies that the M-P distribution formula is simple and easy to calculate. Still, when describing the small-particle-size and large-particle-size regions, there is an inevitable error between the actual observation data. To illustrate the raindrop size distribution more accurately, Ulbrich [22] proposed a method based on the M-P distribution to regard the raindrop size distribution as the GAMMA distribution. This method can correct the distribution pattern in the regions with small particle size or large particle size. At this time, the raindrop spectrum description function changes from the original two-parameter exponential function to the three-parameter function form. The formula is as follows:

$$N(D) = N_0 \times D^{\mu} \times \exp(-\lambda D) \tag{12}$$

The shape factor u in Equation (12) is a dimensionless parameter. When $\mu > 0$, the function curve bends upward. When $\mu < 0$, the function curve bends downward. When $\mu = 0$, the formula becomes an M-P distribution.

3.2. Analysis of Raindrop Spectrum Characteristics

In Figure 2, the horizontal axis is the particle diameter, the vertical axis represents the final velocity of the particle, and different particle number densities are distinguished by color. It can be seen from the average distribution of particles from March to August 2020 in the figure that the corresponding particle size terminal velocities and particle distribution patterns of the three stations are similar, and they are concentrated in the small-particle-size area. The difference is that the particle concentration from the Longli (57913) station, in the range of 1–1.5 mm, is more dispersed than those of the other two stations, and the spectrum width of the Luodian (57916) station is wider than those of the other two stations.

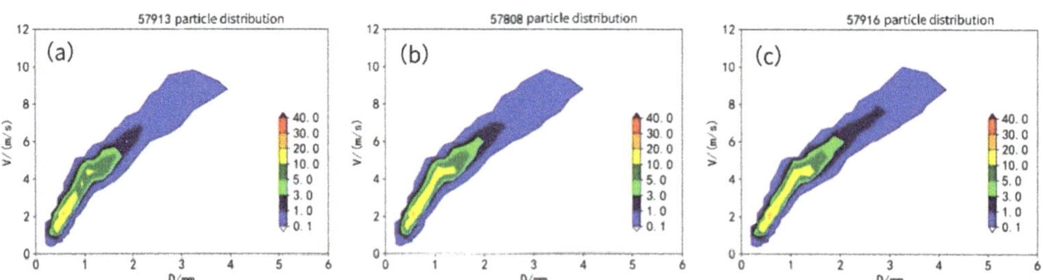

Figure 2. Particle distribution map: (**a**) Longli (57913) station, (**b**) Puding (57808) station, and (**c**) Luodian (57916) station.

Figure 3 shows the particle number density distribution corresponding to each particle size in the average raindrop spectrum. The horizontal axis represents the particle size, and the vertical axis represents the particle number density. The fitting coefficients of the three-site M-P distribution and the GAMMA distribution are shown in Table 2.

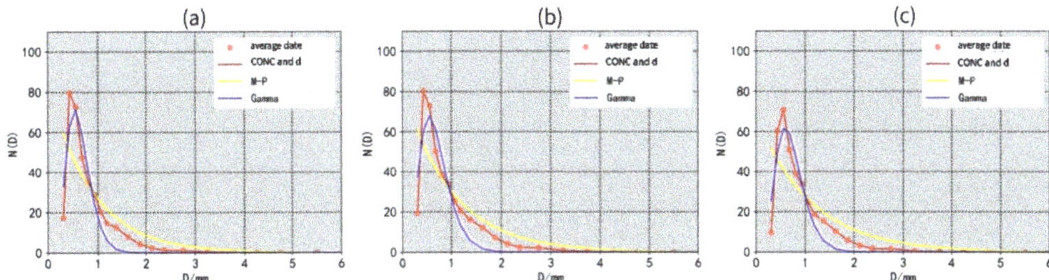

Figure 3. Average raindrop size distribution, M-P distribution, and GAMMA distribution fitting: (**a**) Longli (57913) station, (**b**) Puding (57808) station, and (**c**) Luodian (57916) station.

Table 2. Mean raindrop spectrum M-P distribution fitting parameters.

Station Name Station Number	N_0	λ	Correlation Coefficient
Longli (57913)	85.5219	−1.1507	83.52%
Puding (57808)	84.5301	−1.0243	85.25%
Luodian (57916)	69.1213	−0.9158	81.69%

The correlation coefficient parameters in Tables 2 and 3 show that the GAMMA distribution's fitting effect is significantly more potent than that of the M-P distribution. The three-parameter GAMMA distribution is more flexible than the two-parameter M-P distribution, which can depict more curve details. At the same time, the natural raindrop spectrum of the three stations is unimodal, resulting in a significant error between the M-P distribution and the actual value in the small-particle-size area. When the GAMMA distribution fits the unimodal particle spectrum, the steeper the curve on both sides of the peak, the larger the μ and λ values needed. Still, this would cause a great N_0 value to reduce the error between the peak value of the fitting curve and the actual value. It would also lead to underestimation in the large-particle-size ($D > 1$ mm) area, where the slope of the actual value curve is relatively moderate, and the M-P distribution in the large particle-size-area performs better.

Table 3. GAMMA distribution fitting parameters of average raindrop size distribution.

Station Name Station Number	N_0	μ	λ	Correlation Coefficient
Longli(57913)	1,177,999.13	6.0209	11.1092	95.65%
Puding(57808)	154,920.62	5.1694	8.6301	95.84%
Luodian(57916)	62,872.31	4.3027	7.7396	94.82%

The average diameter, mass-weighted average diameter, and average volume diameter are used as parameters that can reflect the characteristics of the particle spectrum. Finding the parameters with the highest consistency with rainfall intensity lays the foundation for estimating precipitation later. The correlation coefficient between each diameter and rainfall intensity is shown in Table 4.

The data used in Table 4 are the precipitation data of three stations, taken from March to August 2020. To ensure the quality of the sample, the samples with rainfall intensity of less than 1 mm/h are eliminated. The data in the table show that the correlation coefficient between the average diameter and rainfall intensity is the lowest, and the correlation between the average volume diameter and rainfall intensity is the highest. Figure 4 selects a heavy precipitation process in order to find the area where the mass-weighted average diameter and the average volume diameter differ.

Table 4. Correlation coefficients between diameter and rainfall intensity.

Station Name Station Number	Correlation Coefficient between Average Diameter and Rainfall Intensity	Correlation Coefficient between Mass Weighted Average Diameter and Rainfall Intensity	Correlation Coefficient between Average Volume Diameter and Rainfall Intensity
Longli (57913)	25.04%	46.89%	67.80%
Puding (57808)	28.45%	49.51%	71.28%
Luodian (57916)	18.13%	47.03%	71.46%

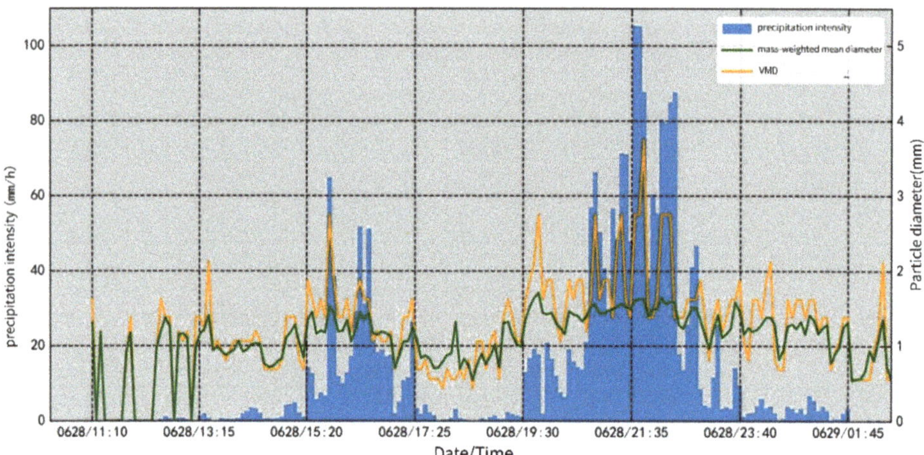

Figure 4. Changes in mass-weighted average diameter and average volume diameter with precipitation intensity.

Figure 4 shows a heavy precipitation process on 28 June. The horizontal axis represents the precipitation process time, the left axis represents the precipitation intensity, and the right axis represents the particle diameter of the mass-weighted average diameter and the average volume diameter. The precipitation intensity value of this process shows a bimodal distribution. It can be seen that, when the precipitation intensity is not high early, late, and between the two peaks of precipitation, the curves of the mass-weighted average diameter and average volume diameter almost coincide, which are in good agreement with the precipitation intensity value. However, the performance of the average volume diameter in the heavy precipitationarea is better than that of the mass-weighted average diameter. The correlation coefficients between the mass-weighted average diameter and average volume diameter, and the precipitation intensity were 54.84% and 69.15%, respectively.

4. Precipitation Estimation Method and Result Analysis

4.1. Precipitation Estimation Method

4.1.1. Dynamic Z-I Relationships

The dynamic Z-I relationship method is based on the $Z = aI^b$ index relationship, and the fast real-time updated radar-automatic rainfall station data adjust the a and b coefficients in the relationship. Different Z-I relationships can be used to estimate precipitation with the change in the precipitation process. This method uses the weather radar detection data and the automatic rainfall station data from the previous hour to calculate the optimal parameters of the current Z-I relationship. It uses this relationship to estimate the recent precipitation. Firstly, the value range of the a and b coefficients is determined. Wang et al. [23] pointed out in their study that the value range of a is 16–1200, and that of b is 1–2.87. In order to save calculation time, when calculating the optimal coefficient, 60 a values are taken, from 16 to 1200 with an interval of 20, and 38 b values are taken, from 1 to 2.87 with an interval of 0.05, to calculate 60×38 groups of Z = amIbn (m = 1, 2, ..., 59, 60; n = 1, 2, ..., 37, 38), and simultaneously calculate 60×38 discriminant functions, CTF2 (Zhang Peichang et al. [24]), as follows.

$$CTF2 = \min\left\{ \sum\left((H_i - G_i)^2 + (H_i - G_i) \right) \right\} \tag{13}$$

In Equation (13), Hi is the inversion precipitation of the Z-I relationship, G_i is the precipitation measured at the automatic rainfall station, and i is the sequence of the automatic rainfall station. The Z-I relationship with the lowest *CTF2* value is the optimal one to apply to radar echo estimation precipitation.

4.1.2. Neural Network Method

Considering the temporal characteristics of precipitation, this paper uses the LSTM neural network to build and train the model. In terms of data, Guiyang weather radar data, raindrop spectrometer, and the automatic rain gauge data from Longli (57913), Puding (57808), and Luodian (57916) automatic weather stations were used. The data overview is shown in Table 5.

Table 5. Data Overview.

Data	State	Start Time	Deadline	Sample Size
Raindrop spectrometer, weather radar,	train	15 April 2019 0:00 3 March 2020 0:04	17 July 2019 23:58 30 June 2020 23:57	49,297
automatic rain gauge	inspection	1 July 2020 0:03	16 August 2020 06:05	12,126

The construction of the data set is significant for using the neural network to estimate precipitation. The closer the relationship between the input eigenvalues and the calibration values to be output, the more accurate the final training results. The calibration value to be estimated in this paper is precipitation. Therefore, the reflectivity data in the weather radar, as well as the particle diameter, particle falling speed, and particle number density parameters in the inversion formula of the raindrop spectrum precipitation intensity were selected as the input eigenvalues of the neural network. Considering the distance between the raindrop spectrum station and the radar station, and drawing on experience in the process of model debugging, select 1–3 lamination scan data as feature value input, as shown in Table 6.

Table 6. Estimated neural network input overview.

Eigenvalue	Estimated Value
Radar reflectivity intensity (1–3 layers) Inversion of particle number density with raindrop spectrometer Average particle velocity inversion with raindrop spectrometer Raindrop spectrometer inversion average volume diameter	A rain gauge measures precipitation

When establishing the neural network input data set, the time sliding construction is used to enrich the data set, and the input diagram of the estimated data set is shown in Figure 5.

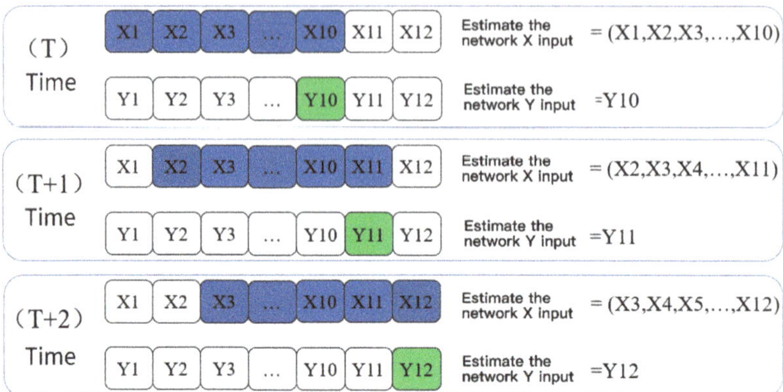

Figure 5. Estimate network and benchmark network input diagram.

As shown in Figure 5, the X input of the estimation network keeps the ten sample sizes unchanged and slides on the time axis in turn. The Y input maintains the time corresponding to the last sample of the X input—that is, using the data from the previous hour to estimate the current time's precipitation.

The basic structure of the neural network is shown in Figure 6.

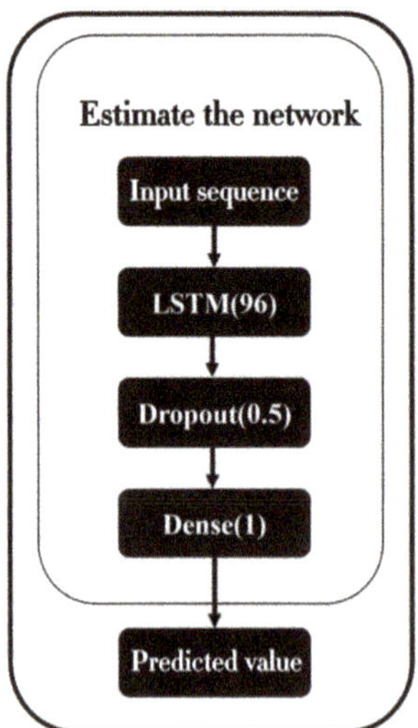

Figure 6. The basic structure diagram of the estimated neural network.

The data are first entered into the LSTM layer for temporal feature extraction, and then the dropout layer is used to increase the network's generalization ability. Finally, the predicted value is output in the BP layer of a single neuron.

In this paper, according to the method of dividing convective clouds and stratiform clouds proposed by Tokay et al. [25], two representative precipitation processes and 45-day cumulative precipitation were selected to estimate rainfall using the dynamic Z-I method and neural network method, respectively. The data were used to compare the effects of the Longli (57913), Puding (57808), and Luodian (57916) automatic weather stations. The station information is shown in the table.

In terms of the estimation accuracy evaluation index, correlation coefficient (*Cor*), mean relative error (*MRE*), mean absolute error (*MAE*), and root mean square error (*RMSE*) are used. The calculation formula is as follows:

$$Cor = \frac{\sum_{i=1}^{n} \left(R_{est}(i) - \overline{R_{est}}\right)\left(R_{real}(i) - \overline{R_{real}}\right)}{\sqrt{\sum_{i=1}^{n} \left(R_{est}(i) - \overline{R_{est}}\right)^2 \left(R_{real}(i) - \overline{R_{real}}\right)^2}} \tag{14}$$

$$MRE = \frac{1}{n}\sum_{i=1}^{n} \frac{|R_{est}(i) - R_{real}(i)|}{R_{real}(i)} \times 100\% \tag{15}$$

$$MAE = \frac{1}{n}\sum_{i=1}^{n} |R_{est}(i) - R_{real}(i)| \tag{16}$$

$$RMSE = \sqrt{\frac{1}{n}\sum_{i=1}^{n}(R_{est}(i) - R_{real}(i))^2} \tag{17}$$

The correlation coefficient can compare the linear correlation between the two sequences. The higher the correlation is, the closer the correlation coefficient is to 1. The average absolute error is the average value of the fundamental error, which can reflect the positive and negative offset parts of the error. The average relative error is the average value of the absolute error percentage to the actual value, representing the measurement results' reliability. The root means the square error is the square root of the mean fair sum of the deviation between the observed value and the actual value, and is used to measure the divergence between the observed value and the actual value. It is more sensitive to singular values.

4.2. Analysis of Precipitation Estimation Results

Firstly, a convective cloud precipitation process occurring on 18 July 2020 was selected for analysis. The weather development process is as follows.

As shown in Figure 7, at 5:57 on 18 July 2020, an unmistakable echo appeared near the junction of Bijie, Zunyi, and Luzhou in Sichuan Province, in the northwest direction from Guiyang. The maximum combined reflectivity reached 53.2 dBZ, and the echo continued to develop and move southeastward. At about 8:00 BST, a strong echo band appeared over the junction of southwest Anshun and southwest Guizhou in the southwest of Guiyang, and quickly moved to the northeast. From Figure 7d, it can be seen that the echo gradually evolved into a northeast–southwest squall line echo. At this time, the lower atmosphere in southern Guizhou showed a strong wind field convergence phenomenon. At 14:00, with the continuous development of the weather phenomenon, the squall line the squall line gradually gradually weakens and tends to die, but the rear part of the squall line still maintains a high reflectivity intensity. The echo area covered most of southern Guizhou, and the two independent echoes were connected and began to merge, affecting the region. At 18:00, the echo over the Qiannan area dissipated, but the precipitation cloud was continuously transported from the solid southeast wind to the Qiannan area. Within a few hours, the echo area slowly moved southward, the precipitation echo was evenly distributed, and the intensity remained unchanged with short-term strong wind. Until 5 o'clock on 19 July 2020, the precipitation area continued to dissipate.

The convective cloud precipitation estimated with the neural network method and dynamic Z-I relationship method is shown in Figure 8.

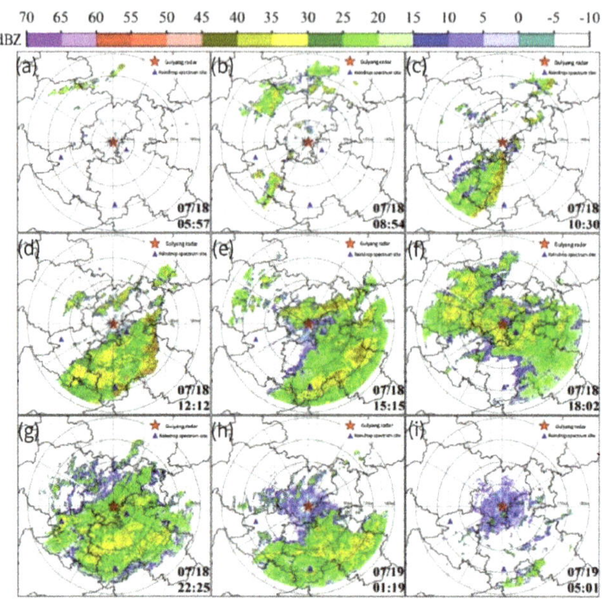

Figure 7. Radar echo of precipitation process on 18 July 2020: (**a**) 07/18 05:57, (**b**) 07/18 08:54, (**c**) 07/18 10:30, (**d**) 07/18 12:12, (**e**) 07/18 15:15, (**f**) 07/18 18:02, (**g**) 07/18 22:25, (**h**) 07/19 01:19, and (**i**) 07/19 05:01.

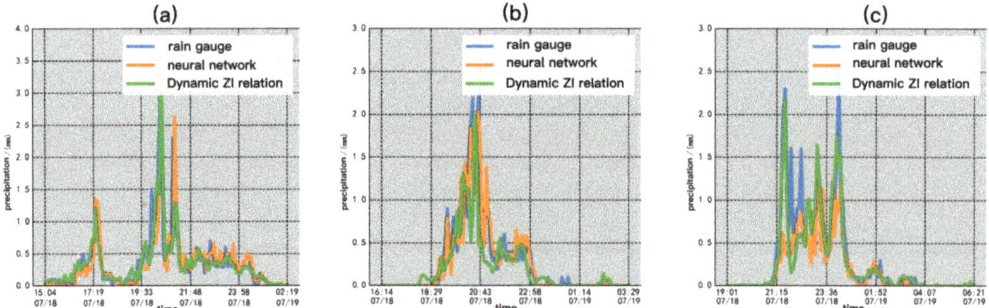

Figure 8. Rainfall estimation of convective cloud precipitation process: (**a**) Longli (57913) station, (**b**) Puding (57808) station, and (**c**) Luodian (57916) station.

The real-time estimation of the rainfall evaluation index is shown in Table 7.

Table 7. Evaluation index of rainfall estimation in convective cloud precipitation process.

Site	Estimating Method	Real-Time Correlation Coefficient	MRE	MAE	RMSE
Longli (57913)	Dynamic Z-I	0.8432	0.5046	0.1462	0.2745
	Neural network	0.8745	0.4646	0.1228	0.2454
Puding (57808)	Dynamic Z-I	0.7763	0.8039	0.1324	0.2962
	Neural network	0.9125	0.7628	0.0935	0.1884
Luodian (57916)	Dynamic Z-I	0.8658	0.7799	0.1357	0.3379
	Neural network	0.8676	0.7986	0.1372	0.3412

From the fitting curve in Figure 8 and the data in Table 7, it can be seen that the neural network method outperforms the dynamic Z-I relationship method at all three stations in terms of real-time correlation coefficients. For MRE, MAE, and RMSE, the error of the neural network method in the Longli (57913) and Puding (57808) stations is smaller than that of the dynamic Z-I relationship method, and slightly higher than that of the dynamic Z-I relationship method at Luodian (57916) station. Dynamic Z-I relationship: The Z-I relationship is corrected using the first ten body sweep data and raingauge data. When the automatic rain gauge does not detect precipitation information, the Z-I relationship is set $Z = 16 \times I^{1.0}$. Only a tiny intensity of radar reflectivity can invert a certain precipitation intensity. The advantage of this setting is that it can make up for the residual error from the tipping bucket of the automatic rain gauge to a certain extent. Still, it is limited to the quality of the radar data and the automatic rain gauge data. If the radar detection echo is shorter before the precipitation data, it will prevent the precipitation from being overestimated. The neural network method is less effective at estimating the precipitation start period, and there is a specific lag phenomenon. This is because the neural network needs to learn from rich samples in order to establish the model, including precipitation samples, non-precipitation samples, and some samples with mismatched precipitation information. Therefore, for the neural network to determine the precipitation start time, the input sample needs to contain a sufficient amount of precipitation samples, so it will affect the determination of the precipitation start time. When the precipitation intensity in the latter part of the precipitation is large and the precipitation is continuous, the actual precipitation can be accurately reflected regardless of the neural network or the dynamic Z-I relationship. The dynamic Z-I relationship and the neural network method can accurately describe the convective cloud precipitation process, independent of the distance from the station to the radar.

Secondly, a continuous stratiform cloud precipitation process on 19 July 2020 was selected for analysis. The weather process is as follows.

As shown in Figure 9, at 12:00 on 19 July, scattered echoes appeared and developed slowly over the junction of southern Guizhou, Guangxi, and Yunnan. During this period, the wind force in Guizhou was low, and the wind field was relatively stable. At 17:00, the echo development gradually covered the target site. At this time, under the influence of slight southeast wind, the echo development accelerated but remained stable, and the space did not change much. The echo gradually attenuated and dissipated by 0:00 on 20 July. Then, a slight southerly wind carried a large area of echoes and continued to affect the southern part of Guizhou until the end of the weather process at 13:00 on 20 July. From the radar echo map, the weather phenomenon has low precipitation intensity and a simple process. The echo is not seen in the second and third layers of the radar volume scan, it indicating that the development of precipitation clouds is thin, and the wind direction and wind speed are relatively stable during the entire duration of precipitation, and there is no apparent convergence phenomenon.

The convective cloud precipitation estimated with the neural network method and dynamic Z-I relationship method is shown in Figure 10.

Table 8 shows the real-time rainfall estimation evaluation indicators.

Table 8. Rainfall estimation and evaluation index of the stratiform cloud precipitation process.

Site	Estimating Method	Real-Time Correlation Coefficient	MRE	MAE	RMSE
Longli (57913)	Dynamic Z-I	0.6933	0.8068	0.0267	0.0479
	Neural network	0.7114	0.8008	0.0401	0.1047
Puding (57808)	Dynamic Z-I	0.0902	0.9794	0.0704	0.1721
	Neural network	0.4984	0.8771	0.0564	0.1416
Luodian (57916)	Dynamic Z-I	0.1409	0.9396	0.0880	0.1445
	Neural network	0.4902	0.8409	0.1183	0.3211

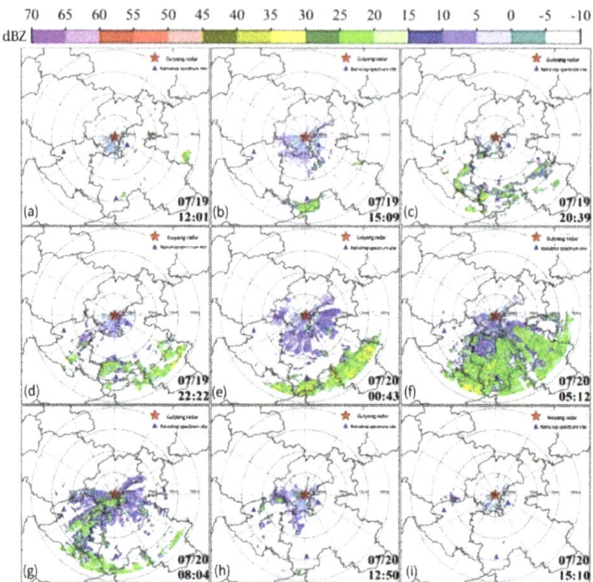

Figure 9. Radar echo of precipitation process on 19 July 2020: (**a**) 07/19 12:01, (**b**) 07/19 15:09, (**c**) 07/19 20:39, (**d**) 07/19 22:22, (**e**) 07/20 00:43, (**f**) 07/20 05:12, (**g**) 07/20 08:04, (**h**) 07/20 12:50, (**i**) 07/19 15:10.

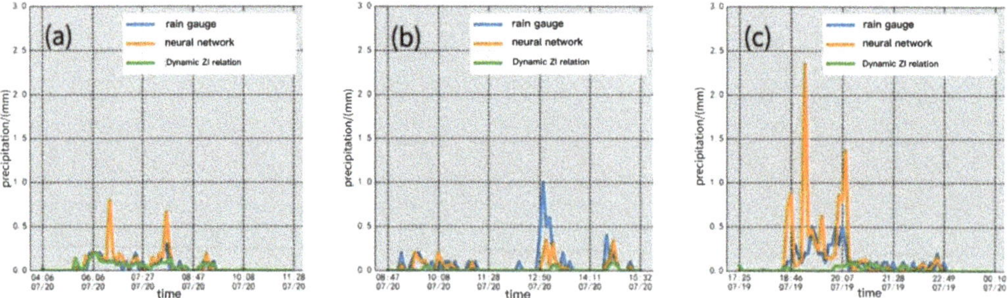

Figure 10. Estimated precipitation of stratiform cloud precipitation process: (**a**) Longli (57913) station, (**b**) Puding (57808) station, and (**c**) Luodian (57916) station.

It can be seen from the data shown in Table 8 that, except for the similar results obtained using the two methods at Longli (57913) station, which is close to the radar station, the neural network method is far better than the dynamic Z-I method in the precipitation estimation of the other two stations. Still, the overall estimation effect is far less than the convective cloud process estimation. Figure 10b,c show that the dynamic Z-I relationship method does not invert precipitation on the time series estimated at the Puding (57808) and Luodian (57916) stations. From the weather background, it can be seen that the precipitation process is weak, and the clouds are thin. Examining the radar base data, it is found that the two stations far from the radar station can detect very little echo information, which directly affects the estimation effect of the dynamic Z-I relationship. The neural network method uses multi-source data from radar and a raindrop spectrometer for estimation. Compared with the single-source data, the dependence is reduced, and the estimation effect is improved compared with the dynamic Z-I relationship method. However, due to the lack of precipitation information detected during the stratiform cloud precipitation process and the frequent interruption of precipitation, the estimation effect still needs to be significantly different from that of convective cloud precipitation.

Finally, the neural network and the dynamic Z-I methods were used to estimate the total precipitation of the three stations for 45 days from 1 July 2020, to 16 August 2020, respectively. Calculating cumulative rainfall in an area can reflect the applicability of the estimation method and the universality of different types of precipitation. The cumulative precipitation of rain gauges at the Longli (57913), Puding (57808), and Luodian (57916) stations during the selected period were 340.0 mm, 312.0 mm, and 264.6 mm, respectively. The results are shown in Figure 11.

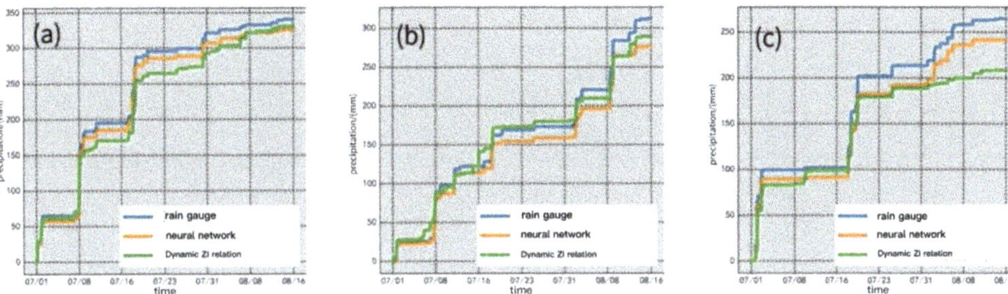

Figure 11. Comparison of estimated cumulative rainfall: (**a**) Longli (57913) station, (**b**) Puding (57808) station, and (**c**) Luodian (57916) station.

Table 9 lists the estimated cumulative rainfall evaluation indexes of the three stations.

Table 9. Estimated cumulative rainfall evaluation index.

Site	Estimating Method	Correlation Coefficient	MRE	MAE	RMSE	Relative Error
Longli (57913)	Dynamic Z-I	0.9951	0.0816	19.4211	22.0374	−2.68%
	Neural network	0.9998	0.0526	10.3156	10.5963	−4.25%
Puding (57808)	Dynamic Z-I	0.9938	0.0986	9.3672	11.4746	−7.41%
	Neural network	0.9992	0.0911	14.3672	16.8846	−11.35%
Luodian (57916)	Dynamic Z-I	0.9862	0.1542	26.6712	32.5039	−21.23%
	Neural network	0.9996	0.1122	16.4154	17.3465	−8.68%

From the precipitation measured with the rain gauge, the rainfall estimated using the neural network, and the precipitation estimated using the dynamic Z-I relationship shown in Figure 11, along with the comprehensive evaluation index, it can be seen that the rainfall estimates made using the dynamic Z-I relationship method and the neural network method have a firm consistency with the measured rainfall. Still, the dynamic Z-I relationship is reflected in some weak precipitation processes, but the dynamic Z-I relationship did not reflect satisfactory effects in some weak precipitation processes, such as Puding (57808); the dynamic Z-I relationship of multiple inefficient precipitation processes after August in Luodian (57916) station also shows apparent underestimation, and a weak precipitation process on July 10 was overestimated. By examining the base data and radar echo map on 10 July, it is found that the radar detected a strong echo lasting about one hour before the automatic rainfall station detected the precipitation. Because the dynamic Z-I relationship was $Z = 16 \times I^{10}$ when the rainfall data of the rainfall station is 0, the strong echo in the period of 0 times the rainfall station data caused the weak precipitation process to be overestimated. The relative errors of Longli (57913), Puding (57808) and Luodian (57916) stations using dynamic Z-I relationships to estimate cumulative precipitation were −2.68%, −7.41% and −21.23%, respectively, and the relative errors estimated by neural networks were −4.25%, −11.35% and −8.68%. With regards to the cumulative precipitation of the

Longli and Puding stations, the final relative error of the neural network is slightly worse than that of the dynamic Z-I relationship. Still, the neural network is better than the dynamic Z-I relationship in terms of the correlation coefficient index. Combined with Figure 11a, it can be found that the dynamic Z-I relationship shows a certain underestimation in several heavy precipitation processes. Still, it shows apparent overestimation in several weak precipitation processes in August, and finally offers a small cumulative error and low correlation coefficient. From Figure 11c, it can be obtained that dynamic Z-I is affected by the quality of radar detection data, and the estimation effect of stratomorph cloud precipitation is poor at a distance from the radar station, which is directly reflected in the cumulative precipitation error of -21.23% at Luodian (57916) station.

5. Conclusions

In this paper, the characteristics of the raindrop spectra of Longli (57913), Puding (57808), and Luodian (57916) stations in Guizhou Province were analyzed to find the diameter parameters with a strong correlation with rainfall intensity. Thereafter, the multi-source data from weather radar, raindrop spectrometer, and surface rain gauge in Guiyang, Guizhou Province, were used to estimate precipitation. The conclusions are as follows:

(1) After analyzing the raindrop spectral distribution at Longli, Puding, and Luodian stations, it was found that the peak value and spectral width of particle distribution at Luodian (57916) station were different from those of the other two stations, indicating that the raindrop spectral distribution was further between different regions. The correlation coefficients of the three sites, as fitted using the M-P distribution, were 83.52%, 85.25%, and 81.69%, respectively. The correlation coefficients provided using the GAMMA distribution were 95.65%, 95.84%, and 94.82%, indicating that the fitting effect of the GAMMA distribution was better. The correlation coefficients between the mass-weighted average diameter of the three stations and the rainfall intensity were 46.89%, 49.51%, and 47.03%, respectively, slightly lower than the 67.80%, 71.28%, and 71.46% of the mass-weighted average diameter. Using a heavy rainfall example, it was found that the two diameters had the same fitting effect on the rainfall intensity at low rainfall intensity. Still, the average volume diameter was more sensitive at heavy rainfall intensity.

(2) Based on the three-source data from the Guiyang weather radar, Parsivel raindrop spectrometer, and automatic rain gauge, the neural network and dynamic Z-I relationship methods were used to estimate the precipitation. In the process of convective cloud precipitation, the correlation coefficients of the neural network method and the dynamic Z-I relationship method at the three stations were 0.8745, 0.9125, and 0.8676 and 0.8432, 0.7763, and 0.8658, respectively. This shows that the dynamic Z-I relationship and the LSTM neural network method are both consistent with the measured rainfall of the automatic rain gauge.

(3) In stratiform cloud precipitation, the correlation coefficients of the neural network and dynamic Z-I relationship methods at the three stations were 0.7114, 0.4984, and 0.4902 and 0.6933, 0.0902, and 0.1409, respectively. The dynamic Z-I relationship method is greatly affected by weather radar detection data. At stations where the echo development is thin and far away from the weather radar, the weather radar cannot stably provide high-quality detection data. Currently, the neural network method using multi-source data has more advantages.

(4) Based on the three-source data from the Guiyang weather radar, Parsivel raindrop spectrometer, and automatic rain gauge, the neural network method and the dynamic Z-I relationship method were used to estimate the total precipitation of the three stations for 45 days. In the comparison of cumulative estimated rainfall, it was found that there is a strong consistency between the dynamic Z-I relationship method and the neural network method to estimate precipitation and the measured precipitation. The relative errors of the three sites using neural network estimation were -4.25%, -11.35%, and -8.68%, respectively. The relative errors of the cumulative precipitation estimated using the dynamic Z-I relationship were -2.68%, -7.41%, and -21.23%, respectively. The final comparable mistake of the neural network in the cumulative precipitation of the Longli

and Puding stations was slightly worse than that of the dynamic Z-I relationship. Still, the correlation coefficient between the precipitation estimated using the neural network and the real value remained higher than that of the dynamic Z-I relationship.

Author Contributions: Conceptualization, F.W., Y.C., Q.W., T.Z. and D.S.; methodology, T.Z.; writing—original draft preparation, F.W., Y.C. and Q.W.; writing—review and editing, F.W.; supervision, D.S.; funding acquisition, F.W. All authors have read and agreed to the published version of the manuscript.

Funding: This work was supported by the Sichuan Provincial Natural Science Foundation (2022NS-FSC0208) and the Open Fund Project of the Key Laboratory of Land Surface Processes and Climate Change in Cold and Arid Regions, Chinese Academy of Sciences (LPCC2020009).

Institutional Review Board Statement: Not applicable for studies not involving humans or animals.

Informed Consent Statement: Not applicable for studies not involving humans.

Data Availability Statement: The data used to support the findings of this study are available from the corresponding author upon request.

Conflicts of Interest: The authors declare no conflict of interest.

References

1. Huang, Z.W.; Peng, S.Y.; Zhang, H.R.; Zheng, J.F.; Zeng, Z.M.; Wang, Y.Y. Characteristics of raindrop size distribution in Anxi, Fujian. *J. Appl. Meteorol. Sci.* **2022**, *33*, 205–217.
2. Chakravarty, K.; Raj, P.E. Raindrop size distributions and their association with characteristics of clouds and precipitation during monsoon and post-monsoon periods over a tropical Indian station. *Atmos. Res.* **2013**, *124*, 181–189. [CrossRef]
3. Zhao, C.C.; Zhang, L.J.; Liang, H.H.; Li, L.; Liu, Y.L. Analysis of summer raindrop spectrum characteristics in Beijing mountainous and plain areas. *Meteorol. Mon.* **2021**, *47*, 830–842.
4. Zhang, Y.X.; Han, H.B.; Guo, S.Y.; Tian, J.B.; Tang, W.T. Characteristics of raindrop size distribution and Z-R relationship of different precipitation clouds in summer at the southern foot of Qilian Mountains. *Arid Zone Res.* **2021**, *38*, 1048–1057.
5. Chen, S. *Research on Precipitation Phenomenon Recognition Technology Based on Raindrop Spectrum and Machine Learning*; Nanjing University of Information Science and Technology: Nanjing, China, 2021.
6. Jing, G.F.; Luo, L.; Guo, J.; Cui, X.L. Application of raindrop spectrometer in weather radar precipitation estimation. *Foreign Electron. Meas. Technol.* **2020**, *39*, 11–18. [CrossRef]
7. Liu, S.N.; Wang, G.L. The influence of DSD parameters on precipitation estimation by dual-frequency radar. *Plateau Meteorol.* **2020**, *39*, 570–580.
8. Zhang, Y.; Liu, L.P.; He, J.X.; Wen, H. Application of raindrop spectrometer network data in radar quantitative precipitation estimation. *Torrential Rain Disasters* **2016**, *35*, 173–181.
9. Zhou, L.M.; Zhang, H.S.; Wang, J.; Wang, Q.; Chen, X.L. Characteristics of raindrop spectrum of a typical stratocumulus mixed cloud precipitation process. *Meteorol. Sci. Technol.* **2010**, *22*, 73–77.
10. Peng, Y.Z.; Wang, Q.; Yuan, C.A.; Lin, K.P. Application of data mining technology in meteorological forecast research. *J. Drought Meteorol.* **2015**, *33*, 19–27.
11. Kang, J.L.; Wang, H.M.; Yuan, F.F.; Wang, Z.P.; Huang, J.; Qiu, T. Prediction of Precipitation Basedon Recurrent Neural Networks in Jingdezhen, Jiangxi Province, China. *Atmosphere* **2020**, *11*, 246. [CrossRef]
12. Shen, H.J.; Luo, Y.; Zhao, Z.C.; Wang, H.J. Research on summer precipitation prediction in China based on LSTM network. *Clim. Chang. Res.* **2020**, *16*, 263–275.
13. Tang, W.; Ma, S.C.; Chen, R. The early warning method of rainfall-induced debris flow in northwestern Sichuan based on LSTM. *J. Guilin Univ. Technol.* **2020**, *40*, 719–725.
14. Liu, X.; Zhao, N.; Guo, J.Y.; Guo, B. Prediction of monthly precipitation over the Tibetan Plateau based on LSTM neural network. *J. Earth Inf. Sci.* **2020**, *22*, 1617–1629.
15. Chen, H.; Qi, Z.F. Monthly precipitation prediction in eastern Sichuan based on LSTM network. *Water Resouces Power* **2021**, *39*, 14–17+13.
16. Yashon, O.O.; Rodrick, C.; Alice, N.W. Rainfall and Runoff Time-Series Trend Analysis Using LSTM Recurrent Neural Network and Wavelet Neural Network with Satellite-Based Meteorological Data: Case Study of Nzoia Hydrologic Basin. *Complex Intell. Syst.* **2021**, *8*, 213–236.
17. Yang, Q.; Qin, L.; Gao, P.; Zhang, R.B. Based on the EEMD-LSTM model, the annual precipitation prediction of the economic belt on the northern slope of the Tianshan Mountains. *Arid Zone Res.* **2021**, *38*, 1235–1243.
18. Atlas, D.; Srivastava, R.C.; Sekhon, R.S. Doppler characteristics of precipitation at vertical incidence. *Rev. Geophys. Space Phys.* **1973**, *11*, 1–35. [CrossRef]
19. Kruger, A.; Krajewski, W.F. Two-dimensional video disdrometer: A description. *J. Atmos. Ocean. Technol.* **2002**, *19*, 602–617. [CrossRef]

20. Pruppacher, H.R.; Klett, J.D.; Wang, P.K. *Microphysics of Clouds and Precipitation*; Springer: Berlin/Heidelberg, Germany, 1998.
21. Marshall, J.S.; Palmer, W.M. The distribution of raindrops with size. *J. Atmos. Sci.* **1948**, *5*, 165–166. [CrossRef]
22. Ulbrich, C.W. Natural variations in the analytical form of the raindrop size distribution. *J. Clim. Appl. Meteorol.* **1983**, *22*, 1764–1775. [CrossRef]
23. Wang, Y.; Feng, Y.R.; Cai, J.H.; Hu, S. Z-I relationship estimation method of radar quantitative precipitation dynamic classification. *J. Trop. Meteorol.* **2011**, *27*, 601–608.
24. Zhang, P.C.; Du, B.Y.; Dai, T.P. *Radar Meteorology*; Meteorological Press: Beijing, China, 2001; pp. 181–187.
25. Tokay, A.; Petersen, W.A.; Gatlin, P.; Matthew, W. Comparison of Raindrop Size Distribution Measurements by Collocated Disdrometers. *J. Atmos. Ocean. Technol.* **2013**, *30*, 1672–1690. [CrossRef]

Article

Relationship between South China Sea Summer Monsoon and Western North Pacific Tropical Cyclones Linkages with the Interaction of Indo-Pacific Pattern

Shengyuan Liu [1,2,3], Jianjun Xu [2,3,*], Shifei Tu [2,3,*], Meiying Zheng [1,2,3] and Zhiqiang Chen [1,2,3]

1   College of Ocean and Meteorology, Guangdong Ocean University, Zhanjiang 524088, China
2   China Meteorological Administration-Guangdong Ocean University (CMA-GDOU) Joint Laboratory for Marine Meteorology, Guangdong Ocean University, Zhanjiang 524088, China
3   South China Sea Institute of Marine Meteorology, Guangdong Ocean University, Zhanjiang 524088, China
*   Correspondence: jxu@gdou.edu.cn (J.X.); tusf@gdou.edu.cn (S.T.)

Abstract: The South China Sea (SCS) summer monsoon (SCSSM) and Western North Pacific tropical cyclones (TCs) are both tropical systems that interact with each other on multiple scales. This study examines the differences in TCs activity characteristics between anomalous strong and weak SCSSM years, and explores the possible mechanisms behind these differences through the coupling relationship between tropical atmospheric circulation and oceanic surface conditions. Results show that the destructiveness of TCs over the Western North Pacific is stronger during weak SCSSM years than in strong years, whereas the opposite occurs for TCs over the SCS. The interaction between the tropical Indo-Pacific ocean and atmosphere plays a key role in the relationship between SCSSM intensity and TCs activity. In strong (weak) SCSSM years, the sea surface temperature anomaly (SSTA) in the tropical Pacific Ocean tends to correspond to a La Niña-like (El Niño-like) distribution, whereas the tropical Indian Ocean shows an Indian Ocean dipole-negative (positive) phase distribution. Moreover, Walker circulations in both the Indian and Pacific Oceans are coupled during these years, which creates a seesaw-like relationship in the conditions for TCs formation between the SCS and the Western Pacific Ocean. During weak SCSSM years, the formation and activity of TCs over the SCS are suppressed due to the weakened water vapor transport caused by abnormal easterly winds from the eastern Indian Ocean to the southern SCS. Meanwhile, the higher SSTA in the Western Pacific Ocean enhances the TCs activity. In strong SCSSM years, the enhanced monsoon drives a stronger monsoon trough, improving the convective environment over the SCS, whereas in contrast, the Western Pacific Ocean is covered by colder water, resulting in poorer conditions for TCs genesis.

Keywords: cyclones power dissipation index (PDI); monsoon trough; El Niño southern oscillation (ENSO); Indian Ocean dipole (IOD); Walker circulation

Citation: Liu, S.; Xu, J.; Tu, S.; Zheng, M.; Chen, Z. Relationship between South China Sea Summer Monsoon and Western North Pacific Tropical Cyclones Linkages with the Interaction of Indo-Pacific Pattern. *Atmosphere* **2023**, *14*, 645. https://doi.org/10.3390/atmos14040645

Academic Editor: Haibo Liu

Received: 10 February 2023
Revised: 24 March 2023
Accepted: 27 March 2023
Published: 29 March 2023

1. Introduction

The South China Sea (SCS) is located at the center of the Asian–Australian monsoon system, connecting the East Asian, South Asian, and Western North Pacific monsoon regions. The South China Sea summer monsoon (SCSSM) is an important component of the Asian monsoon [1,2] and profoundly affects the climates of East and Southeast Asia [3]. The Western North Pacific (WNP) is the most active area for tropical cyclones (TCs) in the world, with approximately 30 TCs generated each year [4,5], of which approximately ten TCs are generated in, or active in the SCS [6]. The SCSSM and WNP TCs are parts of the tropical system, and the main weather systems causing significant rainfall processes along the coast of South China in the pre- and post-flood periods, respectively. They bring heavy precipitation to the coastal areas of South China and are profoundly related to their economic activities and agricultural production. In years of active monsoons, strong convection is frequent along the coast, often with strong and destructive weather processes, causing catastrophic impacts on society, the environment, and ecology. Understanding how these two tropical systems interact can improve the climate prediction of monsoons and TCs, which is important for disaster prevention and mitigation in coastal areas.

Previous studies have shown that approximately 70–80% of TCs are generated in monsoon troughs or are directly related to monsoon circulation [7,8], and the presence of thermodynamic and dynamic conditions in monsoon troughs that are favorable for tropical cyclone generation reveals an important relationship between monsoons and TCs. Numerous studies have shown that on interannual and seasonal scales, significant correlations exist between the frequency of TC occurrence in the WNP and the East Asian summer monsoon (EASM) [9–11], the Western North Pacific monsoon (WNPM) [12,13], and the South Asian monsoon (SAM) [14]. These monsoon systems influence the environment of TCs generation through monsoon troughs, subtropical high-level jet streams, and subtropical high or teleconnection wave trains, which in turn affect the frequency of TCs genesis [15,16]. On the interdecadal scale, TCs and monsoon troughs are also related [17–19]. Huangfu et al. [20] found that the weakening and westward extension of the Western Pacific monsoon trough after the late 1990s could explain the interdecadal decrease in the number of TCs generated over the WNP.

Recently, the interaction between the SCSSM and TCs over the SCS has received increasing attention. Huangfu et al. [21] and Hu et al. [22] found that the onset date and the withdrawal date of SCSSM are both correlated with the number of TCs over the WNP. Wang et al. [23] found that seasonal-scale monsoon characteristics largely determine the distribution of TCs over the SCS. Chen et al. [15] found that the onset of the SCSSM coincided with the onset of the typhoon season in the WNP and generalized the circulation pattern of the SCSSM before the season. Wang and Chen [24] found a significant relationship between SCSSM onset and landfalling TCs frequency in China. These suggest that interactions exist between SCSSM and TCs. However, one more question remains to be answered: how do TCs activities differ in different SCSSM intensity backgrounds, and by what mechanisms are they connected?

In this study, we reveal the different activity characteristics of TCs over WNP between the strong and weak SCSSM years. The remainder of this paper is organized as follows. Section 2 introduces the data and the methods used in this study. The different statistical characteristics and spatial patterns of the TCs in the strong and weak SCSSM anomalous backgrounds are illustrated in Section 3. Section 4 explains the mechanisms of the coupling relationship through tropical ocean–atmosphere interactions. At the end of the paper, a conclusion and a discussion are given in Section 5.

2. Data and Methods

2.1. Data

The atmospheric reanalysis data used in this study are from the ERA5 monthly averaged data of the European Centre for Medium-Range Weather Forecasts (ECMWF), including the U-component of wind, V-component of wind, vertical velocity, relative vorticity, total column water vapor (TCWV), with a 0.25° latitude × 0.25° longitude spatial resolution [25]. Sea surface temperature (SST) data were obtained from the Met Office Hadley Centre Sea Ice and Sea Surface Temperature dataset, HadISST1, with a 1° latitude × 1° longitude horizontal resolution [26]. To calculate the intensity of the SCSSM, a Monthly Outgoing Longwave Radiation (OLR) dataset from the National Oceanic and Atmospheric Administration (NOAA), with a 2.5° latitude × 2.5° longitude horizontal resolution [27–29], was utilized. All of the above monthly data cover the period 1979–2020 (42 years).

The best-track data used in this study were obtained from the Joint Typhoon Warning Center (JTWC), archived in the International Best Track Archive for Climate Stewardship (IBTrACS) v04r00 [30]. Our focus was on TCs that occurred during the TC season, which runs from May to December [23], and had intensities of at least 25 knots (kts), excluding non-TC systems. Hereafter, we refer to these TCs as WNPTC, which are generated over the WNP basin (105°–180° E, 0°–50° N).

For the purposes of this study, we also independently analyzed TCs that were active over the SCS region (5°–20° N, 105°–120° E) during their lifetimes, which we refer to as SCSTC. These were a subset of the WNPTC that we studied (Figure 1).

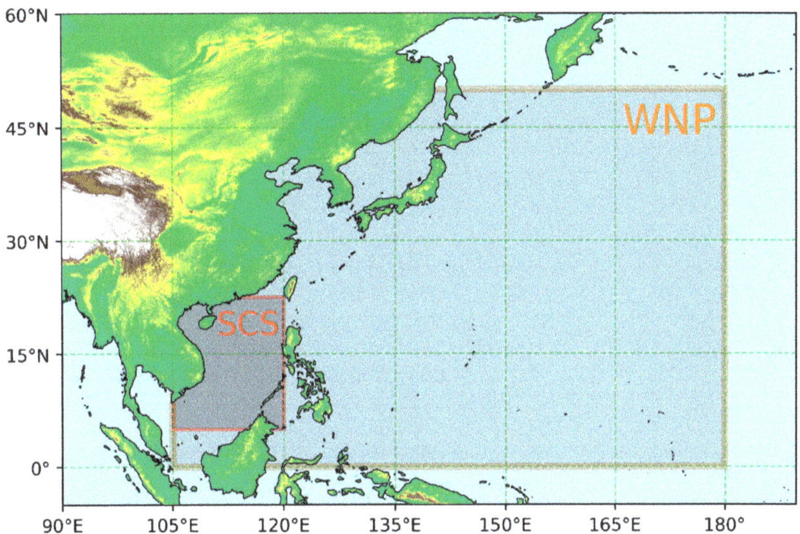

Figure 1. Diagram of the Western North Pacific (WNP) basin (orange box, 0°–50° N, 105°–180° E) and the South China Sea (SCS) region (red box, 5°–20° N, 105°–120° E).

2.2. Methods

2.2.1. Indices for TCs Activity

In this study, the total cyclone power dissipation index (PDI) is utilized to examine the destructive potential of TCs [31]. For each TC season, the PDI is calculated as follows:

$$\text{PDI} = \sum_{i=1}^{N} \int_{0}^{\tau_i} V_{max}^3 dt \tag{1}$$

where V_{max} is the maximum sustained surface wind speed every 6 h [32], τ_i indicates the lifetime of each TC, and N is the TCs count in every TC season. The larger the PDI, the stronger the combined TC activity and the greater the destructiveness.

To examine the spatial pattern of the PDI, this study also calculates the regional PDI, which is defined in the same way as the PDI, except that the research area is gridded into $2° \times 2°$ grid points, and the PDI within each grid point is calculated as the Regional PDI. The Regional PDI depends mainly on the intensity and track density of TCs entering the grid point [33].

To compare the maximum wind speeds during the lifetimes of TCs within different TC seasons, this work used the average Lifetime Maximum Intensity (LMI) [34] as follows:

$$\text{LMI} = \frac{1}{N} \sum_{i=1}^{N} (V_{smax})_i \tag{2}$$

where V_{smax} is the maximum sustained surface wind speed during the lifetime of each TC [34]. In addition, the research also counts the frequency of TCs genesis, average lifetime, and average generation location as indicators of TCs activity.

2.2.2. Criteria for SCSSM Intensity and Anomalous Years

Current studies of climate variability in the South China Sea lack a consensus method for defining summer monsoon intensity. Most of the existing methods are calculated from either dynamic (e.g., zonal wind difference, vorticity, or divergence) or thermodynamic elements (e.g., precipitation, equivalent potential temperature, or OLR) [35–39]. In this study, the SCSSM index refers to the definition of Wu and Liang [40]. By combining dynamic and thermodynamic parameters, the southwest wind, and the OLR, the intensity of SCSSM (I_{SM}) is characterized as follows:

$$I_{SM} \equiv I_{Vsw} + I_{OLR} = \frac{Vsw - \overline{Vsw}}{\delta_{Vsw}} - \frac{OLR - \overline{OLR}}{\delta_{OLR}} \tag{3}$$

where I_{Vsw} is the standardized southwest wind index and I_{OLR} is the standardized OLR index. Specifically, $Vsw = \frac{1}{\sqrt{2}}(u + v)$ is a projection of the monthly average 850 hPa winds in the southwest direction over the SCS (Figure 1); OLR is the average OLR value over the month. $\overline{Vsw}$ and $\overline{OLR}$ are multi-year means over the month, and δ_{Vsw} and δ_{OLR} are the standard deviations of Vsw and OLR, respectively. Physically, the intensity of the SCSSM can be indicated by the magnitude of the 850 hPa southwest wind and the intensity of convection. The stronger (weaker) the southwest wind, and the smaller (larger) the OLR value, indicating stronger (weaker) convection and stronger (weaker) SCSSM.

Based on the multi-year average monthly I_{SM} seasonal circulation curve (Figure 2), the I_{SM} from May to October is typically greater than zero. Therefore, to obtain the annual I_{SM}, we calculated the average value of the monthly I_{SM} from May to October for each year. This time period coincides with the months of onset and withdrawal of SCSSM identified in previous studies [2,41]. The resulting annual I_{SM} was normalized and detrended, and used as a year-by-year time series of the SCSSM intensity index (Figure 3). In this study, a standard deviation of one is used as the criterion to identify years with anomalously strong or weak SCSSM activity. Based on this criterion, eight years were identified as anomalously strong (1984, 1985, 1999, 2000, 2009, 2012, 2013, and 2016), whereas eight years were identified as anomalously weak (1982, 1983, 1987, 1993, 1997, 2004, 2015, and 2019) during the period 1979–2020. The remaining years were categorized as neutral.

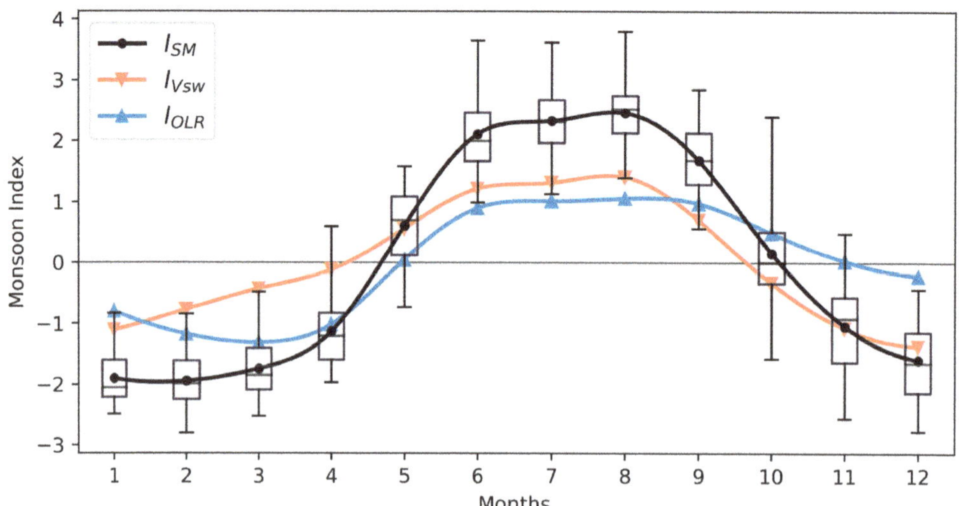

Figure 2. Seasonal circulation curves of the SCSSM intensity index (I_{SM}, black curve), southwest wind index (I_{Vsw}, orange curve), and the OLR index (I_{OLR}, blue curve) for each month of the multi-year average. The boxes and whiskers are the monthly I_{SM} of multi-year ranges.

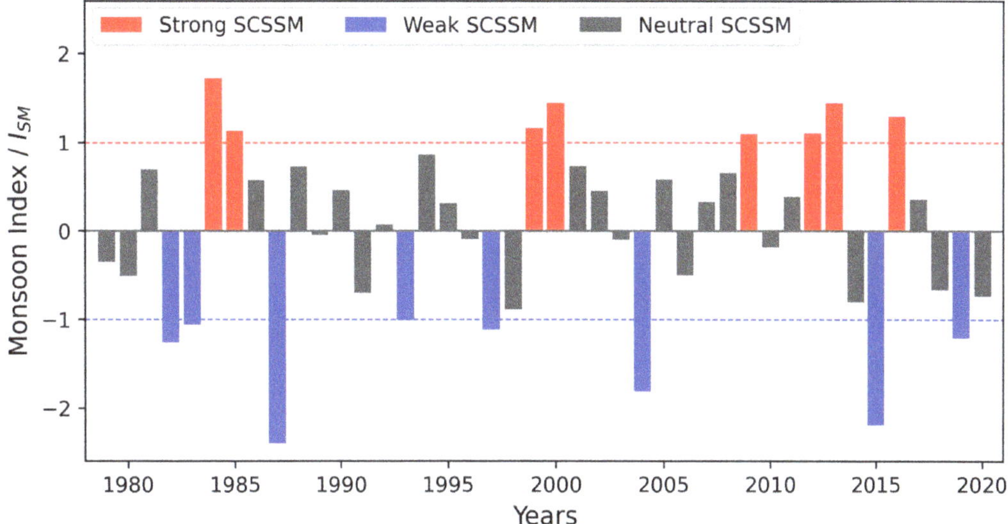

Figure 3. Time series of SCSSM intensity index (I_{SM}) from 1979–2020. The red columns represent years with SCSSM activity exceeding one standard deviation, indicating anomalously strong years. The blue columns represent years with SCSSM activity below negative one standard deviation, indicating anomalously weak years.

In addition, to filter out high-frequency variations in the spatial distribution, a spatial 9-point filter was applied [33].

3. Characteristics of TCs activity in SCSSM Anomalous Years

To investigate the differences in TCs activity during SCSSM anomalous years, we conducted composite analyses of the annual frequency, intensity, lifetime, destructive potential, and spatial patterns of TCs activity for the anomalously strong and weak years, respectively.

3.1. General Indicators

The statistical characteristics of the general indicators of WNPTC and SCSTC activity during the strong and weak SCSSM years, respectively, are shown in Table 1. The frequency of WNPTC during the TC season is not significantly different between the strong and weak years, which both deviated from the neutral years within 5%. However, the frequency of SCSTC in the weak years is only 8.75, which is 2.21 (~20.2%) less than the neutral years, and 2.75 (~23.9%) less than the strong years. The LMI is used to examine the intensity characteristics of the TCs. In both WNPTC and SCSTC, the LMI in the weak years is significantly higher than that in strong years, and the differences between them are 6.07 m/s (16.0%) and 5.16 m/s (14.6%) over WNP and SCS, respectively. During the SCSSM strong years, in both WNPTC and SCSTC, lifetimes are significantly shorter than neutral years, by 0.8 days (12.6%) and 0.4 days (7.3%), respectively. The weak years displayed significantly longer lifetimes, by 0.94 days (14.8%) and 1.2 days (21.8%), respectively. As expected, the differences in TCs lifetimes between the strong and weak years are significant at approximately 31% in both the WNP and SCS.

To characterize the destructive potential of TCs, we also examined the cyclone power dissipation index (PDI) in the two types of anomalous SCSSM years. For WNP, the PDI is lower in the strong years and higher in the weak years, with a significant difference of 38.4% between these two types. For SCS, we calculated the PDI of SCSTC in the entire WNP and in only the SCS, respectively. The results of both areas influenced by SCSSM are consistent, being stronger in the strong years and weaker in the weak years, although the differences are not significant.

The above results indicate that there are notable differences in TCs activity between anomalously strong and weak SCSSM years. TCs occur more frequently during strong SCSSM years, but have weaker maximum intensities and shorter lifetimes compared to weak SCSSM years. These findings hold true for both WNPTC and SCSTC, with the most significant differences observed in lifetimes. However, the destructive potentials of WNPTC and SCSTC exhibit opposite patterns in strong and weak SCSSM years. Specifically, the PDI is lower for WNPTC but higher for SCSTC during strong years, indicating that TCs activity is more concentrated in the SCS, whereas the opposite is true during weak years. Spatial pattern analysis is used to further explore the underlying reasons for these differences.

Table 1. Statistical characteristics of indicators for TC activity in SCSSM anomalous years.

Type	Index	Neutral SCSSM	Strong SCSSM	Weak SCSSM	Difference Percent (%)	*p*-Values
WNPTC	Count (N)	27.50	28.88	26.50	8.2	0.16
	LMI (m/s)	39.22	37.92	43.99	−16.0 *	0.02
	Lifetime (d)	6.37	5.57	7.31	−31.3 *	<0.01
	PDI (10^{11} m^3/s^2)	7.48	6.71	9.28	−38.4 *	0.05
SCSTC	Count (N)	10.96	11.50	8.75	23.9 *	0.04
	LMI (m/s)	35.31	35.30	40.46	−14.6 *	0.08
	Lifetime (d)	5.51	5.11	6.71	−31.4 *	0.02
	PDI in WNP (10^{11} m^3/s^2)	1.91	1.99	1.76	11.2	0.67
	PDI in SCS (10^{11} m^3/s^2)	0.71	0.75	0.61	19.3	0.32

Note: An asterisk (*) indicates statistical significance at the 90% confidence level.

3.2. Spatial Patterns of TCs Activity

Firstly, we analyze the locations of TCs genesis and the range of their activity. As shown in Figure 4, during SCSSM strong years, TCs generally generate close to the East Asian continental coast, with the average location of cyclone genesis at 16.2° N, 135.8° E, and a higher frequency of TCs genesis over the SCS region, which confirms the findings in Section 3.1. On the other hand, during weak years, the latitudinal spread of TCs genesis widens, with the average genesis location shifting closer to the tropical Central Pacific at 13.4° N, 142.6° E, whereas fewer and weaker TCs generate around the East Asian coast. The average TCs genesis location shows a significant difference between the strong and weak years. Additionally, it is noticeable that the low latitude region, located east of 150° E and south of 15° N (green box), only generates 2.5 TCs per year during strong years, compared to 7.4 TCs per year in weak years, which represents a nearly threefold difference, indicating an unfavorable environment for TCs genesis in strong years, but a favorable one in weak years. Possible mechanisms will be described in Section 4.1.

To reveal spatial differences in TCs activity between the two types of SCSSM anomalous years, we calculated the TCs track density, average intensity (represented by the grid point average TCs maximum wind speed), and regional PDI for strong and weak years, respectively. The spatial distributions of the differences between them are shown in Figure 5. From the distribution of TCs track density differences (Figure 5a), it is evident that the TCs track densities along the Asian continental coast in strong SCSSM years are higher than those in weak years, especially in the northern SCS off the coast of Southern China, whereas those over the broad sea area east of the Philippines are significantly lower. The amount of TCs activity close to the continent responds to the SCSSM just opposite the broad sea area east of the Philippines. A similar pattern exists in the different distributions of TCs average intensity (Figure 5b), with TCs in the northern SCS being significantly stronger in the strong years, while being weaker over most areas of the WNP and southern Japan.

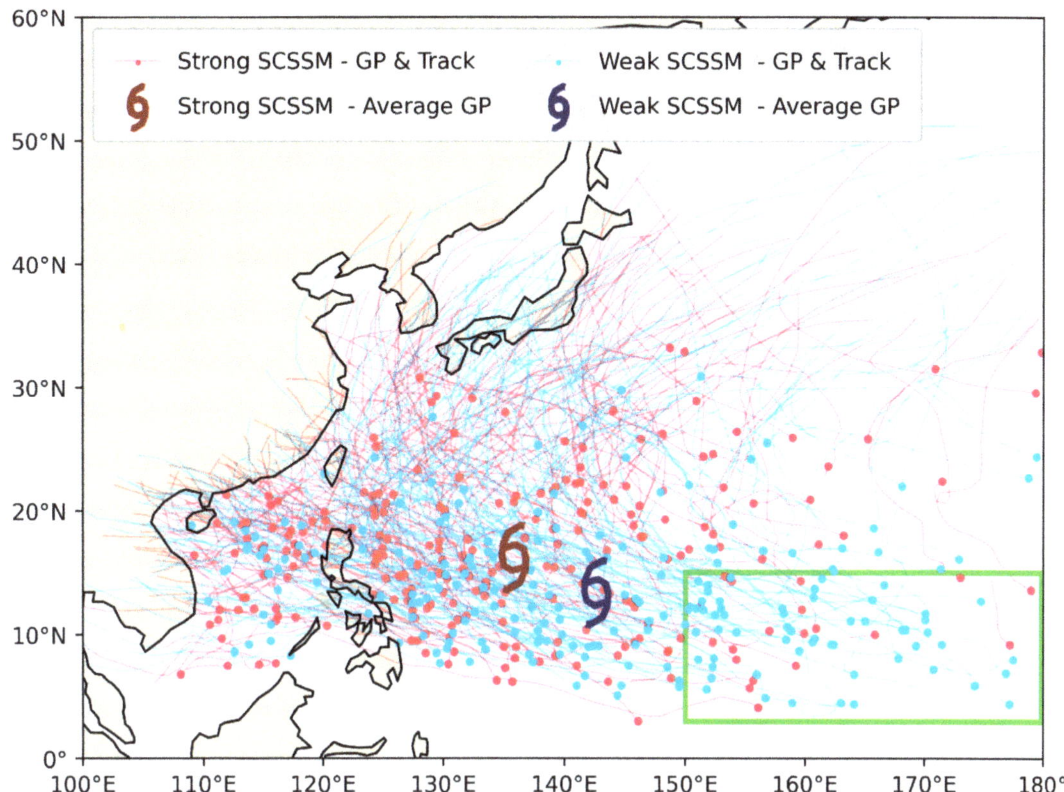

Figure 4. Genesis locations and activity ranges of TCs. Curves represent the TCs tracks, dots the TCs genesis locations, and cyclone symbols the TCs average genesis positions over WNP during TC season. The colors red and blue are used to distinguish between strong and weak SCSSM years, respectively. Within the green box, the number of TCs genesis per year in strong and weak SCSSM years differs by nearly three times.

The regional PDI spatial pattern (Figure 5c) can be well summarized as an integrated result of track density and average intensity characteristics, which can effectively represent the destructive potential of TCs in different areas. In weak SCSSM years, TCs are generated close to the tropical Central Pacific, and there are more and stronger TCs activities in the east of 130° E, resulting in significantly higher regional PDI. After the development of these TCs over the warm-pool region, they hit the Philippine coast with violent intensities. However, since TCs moved into the SCS, owing to the weak SCSSM, deep convection is weaker than that in strong years, and cyclone intensity gradually weakened, resulting in low regional PDI in the SCS.

In strong SCSSM years, the conditions are reversed and the opposite occurs. In these years, the East Asian continent, especially the southern coast of China, is required to defend against TCs that generate and intensify near the shore.

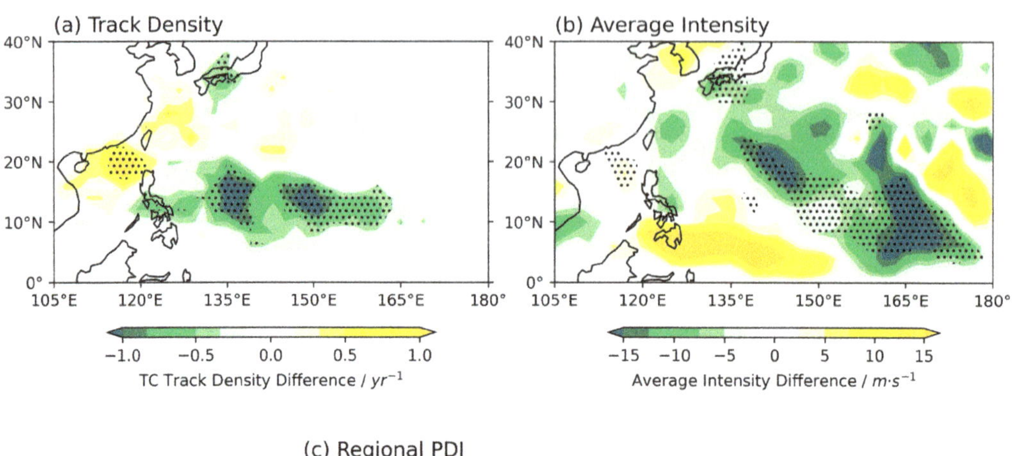

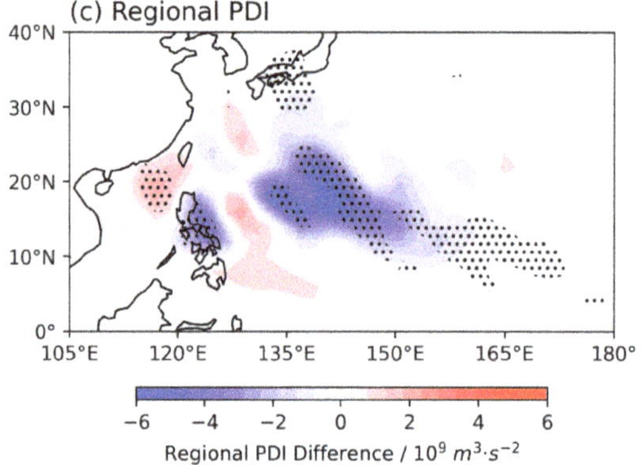

Figure 5. Nine-point-filtered spatial distribution differences (strong SCSSM years minus weak SCSSM years) of (**a**) TCs track density, (**b**) average intensity, and (**c**) regional PDI over WNP basin during TC season. Dots mark areas statistically significant at the 90% confidence level.

4. Possible Mechanisms

4.1. Monsoon Trough

Previous studies have shown that the location of the Monsoon Trough (MT) has a controlling effect on the activity distribution of WNPTC [20,42,43]. Wu et al. [18] found that more TCs were generated in the WNP when the MT extended eastward. To verify the relationship between SCSSM intensity and monsoon trough pattern over the WNP, this study conducted a regression analysis on the 850 hPa circulation and OLR in WNP during the period from May to December based on the annual I_{SM}. In both the strong and weak years, two MTs in the SCS and the Philippine Sea regions can be distinguished (as shown in Figure 6a,b). Compared to the weak years, the position of the SCS MT does not differ significantly in the strong years, but the intensity difference is remarkable, with stronger southwest winds and more intense convective activity. On the other hand, the intensity and pattern of the Philippine Sea MT undergo significant changes between the strong and weak years. In strong years, the Philippine Sea MT extends only to 140° E, with energy concentrated near the Philippine Islands due to the supplement of strong southwest moisture input from the SCS and the Western Pacific region, forming a block-shaped low OLR area. In the weak years of the SCSSM, the MT can extend eastward to nearly 155° E, at a lower latitude, with a more scattered distribution of low OLR values, and closer to the tropical central Pacific.

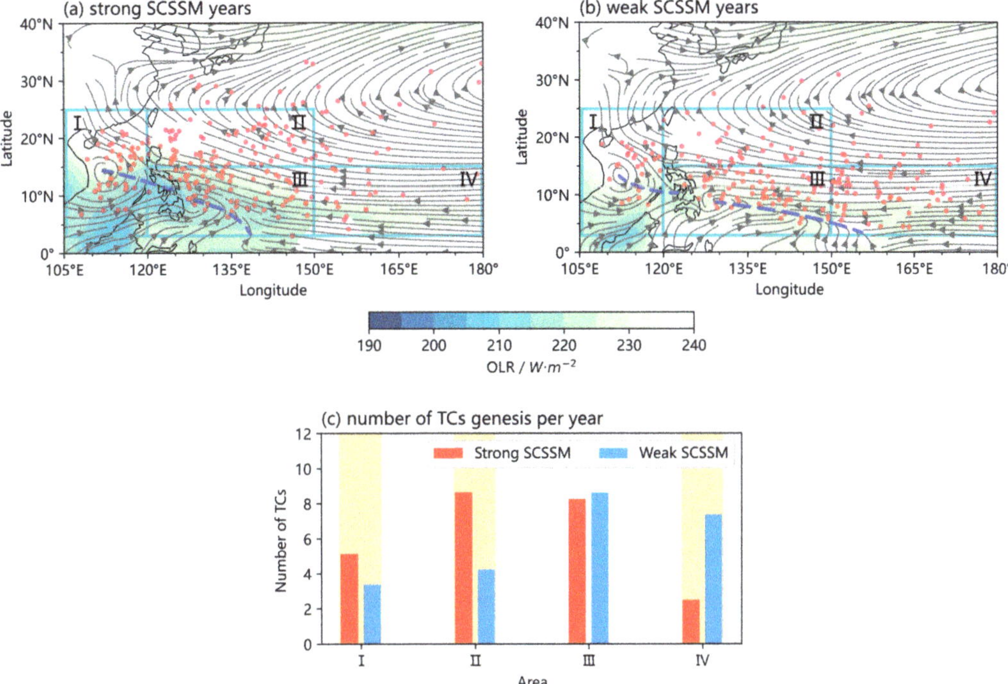

Figure 6. An 850 hPa horizontal wind field (streamline) and OLR (shade) during TC season (May to December) in (**a**) SCSSM strong years and (**b**) SCSSM weak years, respectively, regressed against the annual I_{SM}. The regression elements have statistical significance at the 90% confidence level. The MT is denoted by a purple, thick-dashed line. The solid red dots represent the locations of TC genesis. The blue boxes represent four sub-regions: Area I (i.e., SCS, 3°–25° N, 105°–120° E), Area II (15°–25° N, 120°–150° E), Area III (3°–15° N, 120°–150° E), and Area IV (3°–15° N, 150°–180° E). The bar charts (**c**) show the number of TCs genesis per year in each area during TC season, in which strong SCSSM years are in red and weak SCSSM years are in blue, with yellow shading indicating that the differences are significant at 90% confidence level.

As shown in Figure 6, based on the range of MT activity and TC genesis, four sub-regions are divided in the WNP: Area I (i.e., SCS, 3°–25° N, 105°–120° E), Area II (15°–25° N, 120°–150° E), Area III (3°–15° N, 120°–150° E), and Area IV (3°–15° N, 150°–180° E). The positions of TC genesis are closely related to the position of MT, and the number of TC genesis in the four regions differs significantly between the strong and weak years. In the SCSSM strong years, the SCS is affected by a stronger MT, resulting in a significantly higher number of TCs being generated. The same happens in Area II, the northern part of the Philippine Sea, where the number of TCs in strong years is more than twice as much as that in weak years, which is related to the northward MT. Conversely, in weak years, TCs tend to be generated at a lower latitude, in Areas III and IV. Especially in Area IV, during the weak years, the anomalous eastward expansion of MT results in more TCs generation, in contrast to the strong years. These composited patterns are consistent with previous studies. The MT is a background field for the growth of synoptic-scale wave disturbances [44], and it provides an environment for the enhancement of convective activity and TCs genesis [18–20], and thus anomalous MT activities controlling the variation in the frequency of TC formation in different regions.

4.2. Subtropical High

Climatologically, the MT over the Western Pacific extends along the southwestern edge of the subtropical high, which is well coupled with multi-scale systems including ENSO [45–47]. Before discussing the larger-scale ocean–atmosphere system, we examine the distribution of the Western Pacific subtropical high (WPSH) in anomalous SCSSM years. The patterns of the WPSH during TC season in the anomalous SCSSM years are shown in Figure 7. The WPSH intensity in strong SCSSM years is weak, the ridge latitude is northward, and the westward extension of the ridge point is eastward, which is in obvious contrast to the strong WPSH in the weak years, making it a contributor to the difference in destructive potential. The difference in ridge latitude is the most significant during the peak of the monsoon (JJA) (Figure 7b). In strong years, the WPSH ridge is northward, whereas the SCS MT and the convection over the northern part of the Philippine Sea are strong, which favors TC genesis over the SCS and Area II. On the contrary, in the weak years, when the WPSH ridge is southward, the TC formation condition of Area II turns worse due to the suppression of convection by the WPSH, so it can well explain the difference of Area II between the strong and weak years.

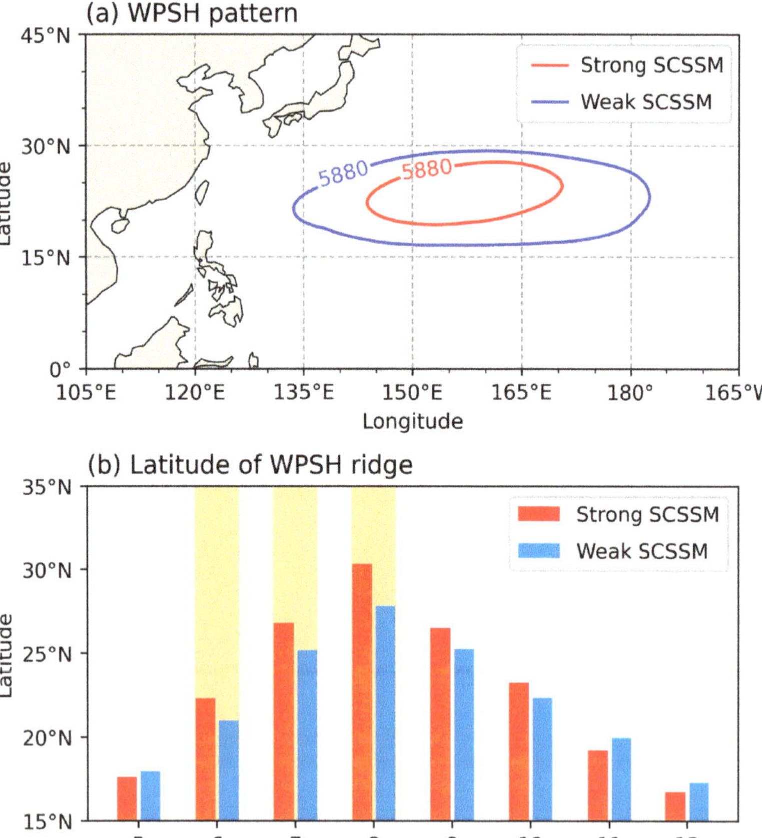

Figure 7. (**a**) The patterns (unit: gpm) and (**b**) the ridge latitudes of Western Pacific subtropical high (WPSH) at 500 hPa over the WNP basin during TC season (May to December), in which strong SCSSM years are in red and weak SCSSM years are in blue, with yellow shading in (**b**) indicating that the differences are significant at 90% confidence level.

The interannual variation of WPSH is modulated by tropical ocean and atmosphere interaction [48]. In order to understand the relationship between WPSH, SCSSM, and TCs activity, we need to turn our attention to a larger scale of the tropical Indo-Pacific Ocean.

4.3. Interaction of Tropical Indo-Pacific Ocean–Atmosphere System

The above analysis shows that the SCSSM intensity is coupled with the MT and the WPSH, which indicates that its interannual variability is not only a response to regional synoptic-scale changes. As a central system in the Asian–Australian monsoon region, the anomalies of the SCSSM are closely responsive to the large-scale tropical Indo-Pacific ocean–atmosphere system [49]. Studies have shown that there is a robust relationship between monsoon and sea surface temperature (SST) [23,50,51]. In terms of the SST anomaly (SSTA) during the TC season (May to December), the tropical Indo-Pacific Ocean shows tripolar distributions in anomalous SCSSM years (Figure 8), with a negative–positive–negative pattern in strong years and a positive–negative–positive pattern in weak years. In strong (weak) years, the Indian Ocean has an Indian Ocean Dipole (IOD) negative (positive) phase distribution, whereas the Central-Eastern Pacific Ocean has a La Niña-like (El Niño-like) pattern, and the Western Pacific warm-pool region is a large area of positive (negative) SSTA. This distribution further confirms the coupling relationship between the intensity of the SCSSM and Indo-Pacific pattern [52,53], such as El Niño-Southern Oscillation (ENSO) or IOD.

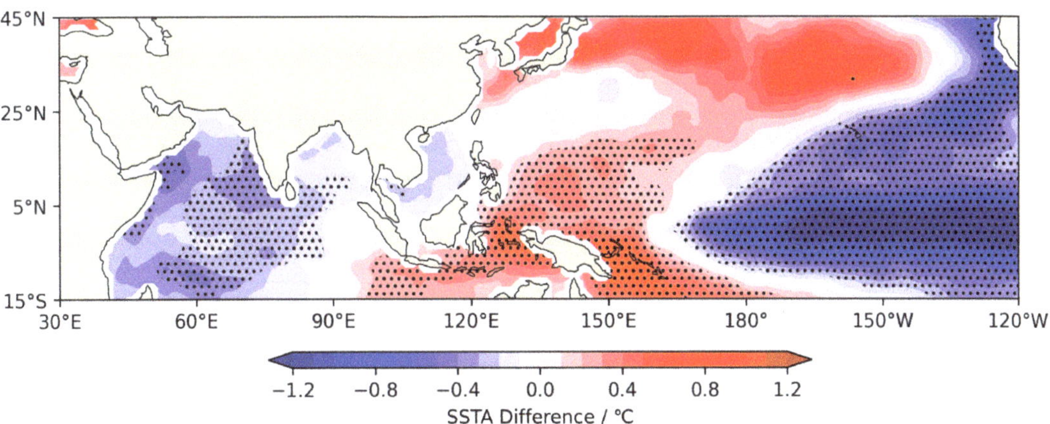

Figure 8. Distribution of SSTA differences (strong years minus weak years) in the tropical Indo-Pacific Ocean during TC season (May to December) in anomalous SCSSM years. Dots mark areas that are statistically significant at the 90% confidence level.

Numerous studies have described the activity of the WNPTC response to ENSO [23,54–60]: in El Niño years, there are more TCs in the eastern part of the WNP, and the TCs in El Niño years are stronger and longer-lived than in La Niña years. In addition to the influence of ENSO, the impact of the Indian Ocean mode on WNPTC has been emphasized in recent years. Zhan et al. [51] revealed that the influence of the East Indian Ocean SSTA on WNPTC has been significantly enhanced since the late 1970s. Zhou et al. [61] discovered a significant negative correlation between the number of TCs landfalling in China and the IOD. Liu et al. [62] proposed two joint modes of the Indo-Pacific SSTA, and the oscillations between them could reveal the interdecadal variation in the genesis frequency of WNPTC.

Furthermore, tropical Indo-Pacific SSTA exerts a critical influence on the WPSH [63]. Previous studies have highlighted the impact of tropical SSTA on WPSH and demonstrated that El Niño is a key source of its variability [48,63,64]. Additionally, some mechanisms related to the Indian Ocean and the Maritime Continent are also important contributors to the WPSH variability [65–68].

Therefore, through the action of WPSH, the tropical Indo-Pacific SSTA controls the spatial and temporal extent of TC activity and modulates the East Asian atmospheric circulation (including SCSSM), and plays a key role as a mediator in the strong relationship between the SCSSM and WNPTC.

As shown in Figure 9, the tripolar SSTA distribution leads to two anomalous Walker cells [69]. In strong (weak) SCSSM years, WPSH abnormally weakens (strengthens), and the Western Pacific is an anomalous ascending branch (descending branch) closely connected with the anomalous descending branches (ascending branches) over the Indian Ocean and Central-Eastern Pacific [66]. This means that in anomalous SCSSM years, the tropical Indo-Pacific air-sea systems are stably coupled. By combining the distributions of atmospheric circulation and related physical variables, the factors responsible for the differences in WNPTC activity in anomalous SCSSM years could be further understood.

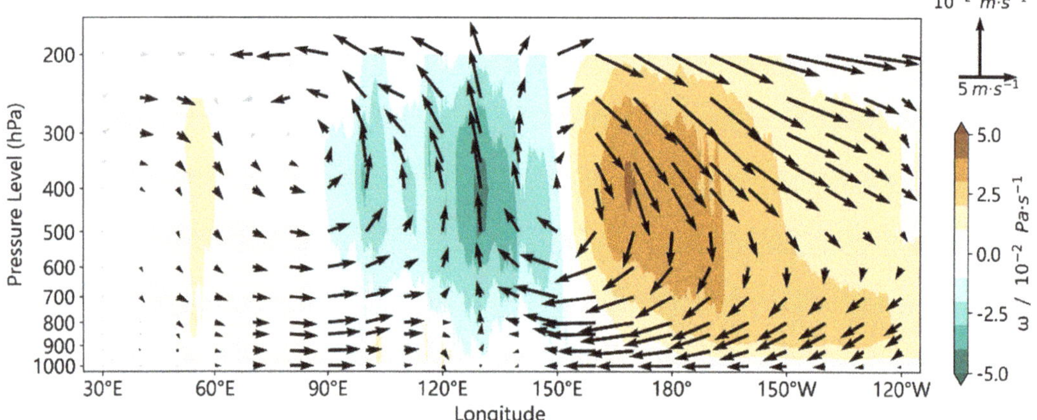

Figure 9. Vertical section differences (strong years minus weak years) of equatorial (average of 5° S–5° N) vertical velocity (shade) and latitudinal circulation (arrow) over the tropical Indo-Pacific Ocean during TC season (May to December) in anomalous SCSSM years. Shaded areas are statistically significant at the 90% confidence level.

In strong SCSSM years, anomalous eastward winds occur over the tropical Indian Ocean and anomalous westward winds occur over the tropical Pacific Ocean at the lower level, whereas the opposite is true at the upper level (Figure 10a,b), forming two Walker cells in their respective basins. They constitute a latitudinal wind anomalous convergence zone near the Western Pacific warm pool region (90–150° E), which also corresponds to the upward vertical motion anomalous zone (Figure 10c). This upward zone essentially controls the SCS to the Philippine Sea, which, together with the improved water vapor input from the SCSSM (Figure 10e), favors more TCs active in the region. Moreover, SCS is a zone of positive relative vorticity (Figure 10d), indicating active convection and enhanced monsoon trough, which also favors the generation and activity of the SCSTC. However, the opposite is true in the eastern part of WNP, with negative relative vorticity anomalies, anomalous downward vertical motion, and low water vapor, indicating that the dynamic and thermodynamic conditions are unfavorable for the generation of TCs in this region. Therefore, this pattern explains the seesaw relationship between the spatial activity characteristics of TCs over the SCS and the eastern part of WNP during the SCSSM anomaly years, as shown in Figure 5.

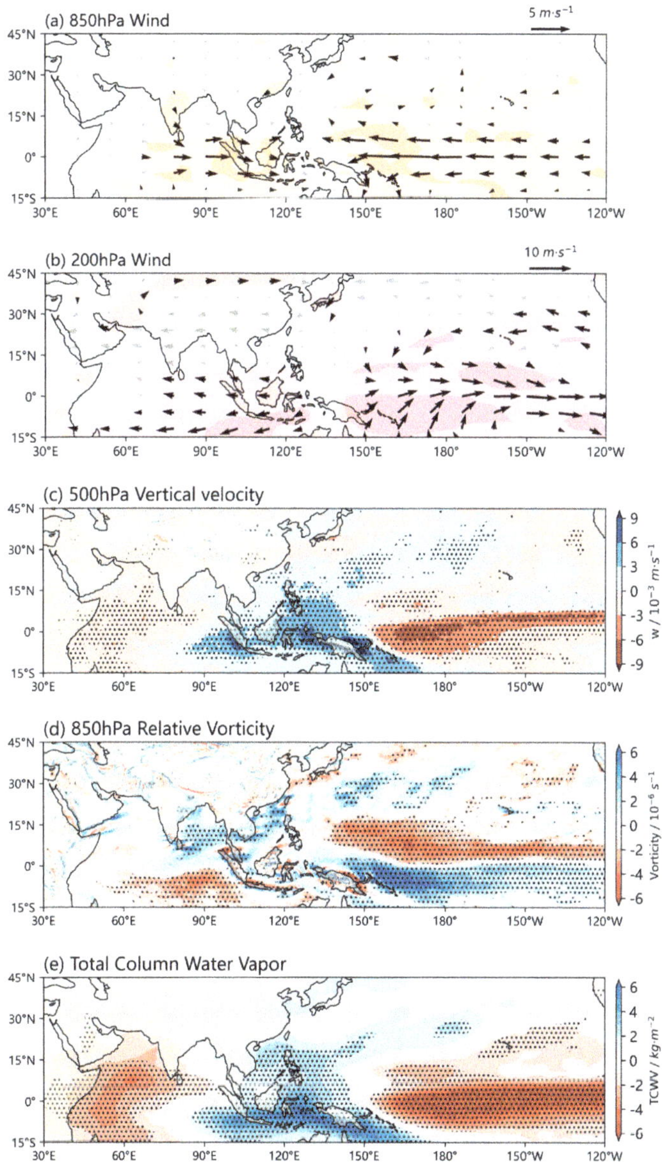

Figure 10. Distribution differences (strong years minus weak years) of (**a**) 850 hPa wind (arrow, unit: m/s); (**b**) 200 hPa wind (arrow, unit: m/s); (**c**) 500 hPa vertical velocity (shade, unit: 10^{-3} m/s); (**d**) 850 hPa relative vorticity (shade, unit: 10^{-6} s^{-1}); (**e**) total column water vapor (shade, unit: kg/m^2) over the tropical Indo-Pacific Ocean during TC season (May to December) in SCSSM anomalous years. Shaded areas in (**a**,**b**) and dotted areas in (**c**–**e**) are statistically significant at the 90% confidence level.

5. Conclusions and Discussion

5.1. Conclusions

In this study, we investigate differences in TCs activity indicators between strong and weak anomalous SCSSM years and explore potential mechanisms that explain these differences. Our analysis highlights the impact of the tropical Indo-Pacific ocean–atmosphere interaction on the relationship between SCSSM intensity and TCs activity. Here are the main points of the study:

1. Anomalously strong (weak) SCSSM years are associated with weak (strong) maximum intensities and short (long) lifetimes of WNPTC. SCSTC are more (less) frequent during strong (weak) SCSSM years.

2. During strong years, TCs are more frequent and powerful along the coast of East Asia, especially in Southern China. In weak years, TCs tend to reach their lifetime maximum intensities in the Philippine Sea, but after moving to coastal areas, their intensities decay more rapidly, and the threat to the East Asian continent is not as significant as in strong years.

3. A coupled status exists between the SCSSM and the tropical Indo-Pacific Ocean. During anomalously strong (weak) SCSSM years, the tropical Pacific Ocean SSTA tends to correspond to a La Niña-like (El Niño-like) distribution, whereas the tropical Indian Ocean shows an IOD-negative (positive) phase distribution, indicating a coupled relationship between the SCSSM and the tropical Indo-Pacific Ocean. Moreover, the Walker circulations in both the Indian and Pacific Oceans are coupled during these anomalous years.

4. During weak SCSSM years (Figure 11b), the tropical anomalous wind field corresponding to the SSTA distribution leads to a latitudinal wind anomaly divergence zone over the Western Pacific Ocean. The Western Pacific Ocean is dominated by a strong subtropical high, and anomalous easterly winds in the eastern Indian Ocean weaken moisture transport and reduce convective activity over the SCS. These conditions are unfavorable for SCSTC genesis and activity. At the same time, the Pacific Ocean east of 150° E has a positive SSTA, and both dynamic and thermodynamic conditions are favorable for the generation and development of TCs, so a high PDI zone appears in this region.

5. During strong SCSSM years (Figure 11a), the enhanced summer monsoon drives the enhanced SCS monsoon trough, which directly improves the convective environment over here. The positive vorticity anomaly in this region is favorable to TCs activity with high PDI. In contrast, the eastern part of the warm-pool region is covered by cold water and has worse conditions for TCs genesis, thus presenting a low PDI.

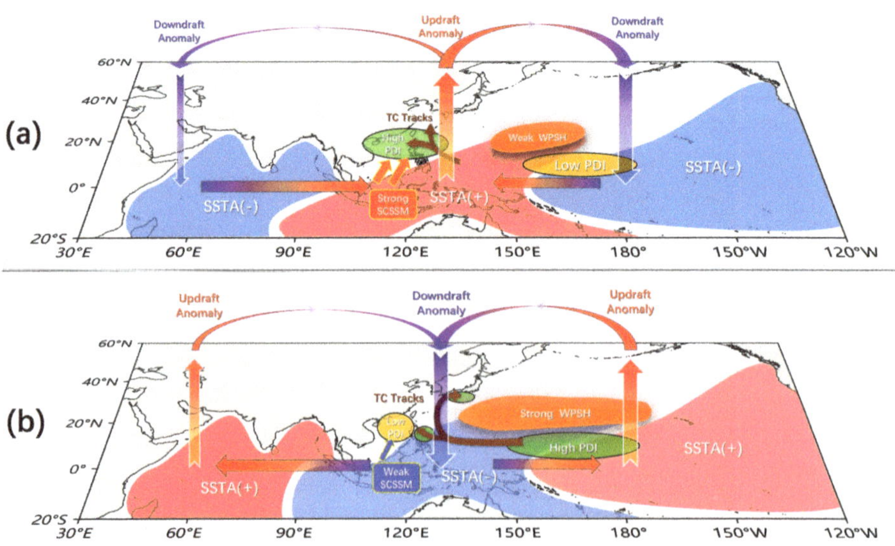

Figure 11. Mechanism diagram of the tropical Indo-Pacific ocean–atmosphere system and TCs activity characteristics in (**a**) strong and (**b**) weak SCSSM years.

5.2. Discussion

Although previous studies have investigated the impact of tropical oceans on WNPTC activity, there has been relatively little consideration of the role of the SCSSM. In this study, we have identified differences in WNPTC activity under different SCSSM intensity backgrounds, and have explained the relationship between the two through the coupled response of the Indo-Pacific SSTA pattern. This has improved our understanding of the influence of tropical ocean–atmosphere interactions, and emphasizes the crucial role of the SCSSM as an important signal amount. Although the Indo-Pacific system can explain the main co-variation between the SCSSM and WNPTC, the non-linearity of their connections makes it difficult to examine their direct relationship when controlling for Indo-Pacific SSTA effects. This is a limitation of our study, and we plan to validate the direct interaction between the two in a coupled ocean–atmosphere model in future research. With climate change causing uneven warming rates across ocean basins, it is possible that the relationship between SCSSM and TCs may also change [70]. Additionally, recent trends of TCs shifting towards land raise the question of whether the influence of the SCSSM on cyclones has increased [71]. These questions will be addressed in future research.

Author Contributions: Conceptualization, S.L.; methodology, S.L., J.X. and S.T.; formal analysis, S.L. and M.Z.; investigation, S.L.; writing—original draft, S.L.; writing—review & editing, S.L., J.X., S.T., M.Z. and Z.C.; visualization, S.L.; supervision, J.X. and S.T.; project administration, J.X. and Z.C. All authors have read and agreed to the published version of the manuscript.

Funding: This work is jointly supported by the National Key R&D Program of China (Grant No. 2019YFC1510002), the Major Program of the National Natural Science Foundation of China (Grant No. 72293604), and the Key Program of the National Natural Science Foundation of China (Grant No. 42130605).

Institutional Review Board Statement: Not applicable.

Informed Consent Statement: Not applicable.

Data Availability Statement: Data used in this study can be downloaded from the websites below: ERA5 reanalysis data: https://cds.climate.copernicus.eu/cdsapp#!/dataset/reanalysis-era5-pressure-levels-monthly-means (accessed on 15 March 2022); HadISST1: https://www.metoffice.gov.uk/hadobs/hadisst/data/download.html (accessed on 21 March 2022); Outgoing Longwave Radiation (OLR) Climate Data Record: https://www.ncei.noaa.gov/products/climate-data-records/outgoing-longwave-radiation-monthly (accessed on 25 March 2022); Best Track Archive for Climate Stewardship (IBTrACS): https://www.ncei.noaa.gov/products/international-best-track-archive?name=ib-v4-access (accessed on 7 March 2022).

Conflicts of Interest: The authors declare no conflict of interest.

References

1. Kajikawa, Y.; Yasunari, T.; Wang, B. Decadal Change in Intraseasonal Variability over the South China Sea. *Geophys. Res. Lett.* **2009**, *36*, L06810. [CrossRef]
2. Kajikawa, Y.; Wang, B. Interdecadal Change of the South China Sea Summer Monsoon Onset. *J. Clim.* **2012**, *25*, 3207–3218. [CrossRef]
3. Zhong, Z.; Hu, Y. Impacts of Tropical Cyclones on the Regional Climate: An East Asian Summer Monsoon Case. *Atmos. Sci. Lett.* **2007**, *8*, 93–99. [CrossRef]
4. Chan, J.C.L. Interannual and Interdecadal Variations of Tropical Cyclone Activity over the Western North Pacific. *Meteorol. Atmos. Phys.* **2005**, *89*, 143–152. [CrossRef]
5. Matsuura, T.; Yumoto, M.; Iizuka, S. A Mechanism of Interdecadal Variability of Tropical Cyclone Activity over the Western North Pacific. *Clim. Dyn.* **2003**, *21*, 105–117. [CrossRef]
6. Goh, A.Z.-C.; Chan, J.C.L. Interannual and Interdecadal Variations of Tropical Cyclone Activity in the South China Sea. *Int. J. Climatol.* **2010**, *30*, 827–843. [CrossRef]
7. Molinari, J.; Vollaro, D. What Percentage of Western North Pacific Tropical Cyclones Form within the Monsoon Trough? *Mon. Weather Rev.* **2013**, *141*, 499–505. [CrossRef]
8. Yoshida, R.; Ishikawa, H. Environmental Factors Contributing to Tropical Cyclone Genesis over the Western North Pacific. *Mon. Weather Rev.* **2013**, *141*, 451–467. [CrossRef]

70. Fosu, B.; He, J.; Liguori, G. Equatorial Pacific Warming Attenuated by SST Warming Patterns in the Tropical Atlantic and Indian Oceans. *Geophys. Res. Lett.* **2020**, *47*, e2020GL088231. [CrossRef]
71. Wang, S.; Toumi, R. Recent Migration of Tropical Cyclones toward Coasts. *Science* **2021**, *371*, 514–517. [CrossRef]

Article

Trends and Variability in Flood Magnitude: A Case Study of the Floods in the Qilian Mountains, Northwest China

Xueliang Wang [1,2,3,4], Rensheng Chen [1,*], Kailu Li [1,2], Yong Yang [1], Junfeng Liu [1], Zhangwen Liu [1] and Chuntan Han [1]

[1] Qilian Alpine Ecology and Hydrology Research Station, Key Laboratory of Ecohydrology of Inland River Basin, Northwest Institute of Eco-Environment and Resources, Chinese Academy of Sciences, Lanzhou 730000, China
[2] University of Chinese Academy of Sciences, Beijing 100049, China
[3] National Cryosphere Desert Data Center, Lanzhou 730000, China
[4] Hydrological Station of Gansu Province, Lanzhou 730000, China
* Correspondence: crs2008@lzb.ac.cn

Abstract: Analyzing trends in flood magnitude changes, and their underlying causes, under climate change, is a key challenge for the effective management of water resources in arid and semi-arid regions, particularly for inland rivers originating in the Qilian Mountains (QMs). Sen's slope estimator and the Mann–Kendall test were used to investigate the spatial and temporal trends in flood magnitude, based on the annual maximum peak discharge (AMPD) and Peaks Over Threshold magnitude (POT3M) flood series, of twelve typical rivers, from 1970 to 2021. The results showed that, in the AMPD series, 42% of the rivers had significantly decreasing trends, while 8% had significantly increasing trends; in the POT3M series, 25% of the rivers had significantly decreasing trends, while 8% had significantly increasing trends. The regional differences in the QMs from east to west were that, rivers in the eastern region (e.g., Gulang, Zamu, and Xiying rivers) showed significantly decreasing trends in the AMPD and POT3M series; most rivers in the central region had non-significant trends, while the Shule river in the western region showed a significantly increasing trend. Temperatures and precipitation showed a fluctuating increasing trend after 1987, which were the main factors contributing to the change in flood magnitude trends of the AMPD and POT3M flood series in the QMs. Regional differences in precipitation, precipitation intensity, and the ratio of glacial meltwater in the eastern, central and western regions, resulted in the differences in flood magnitude trends between the east and west.

Keywords: trends and variability; flood magnitude; climate change; Qilian Mountains

Citation: Wang, X.; Chen, R.; Li, K.; Yang, Y.; Liu, J.; Liu, Z.; Han, C. Trends and Variability in Flood Magnitude: A Case Study of the Floods in the Qilian Mountains, Northwest China. *Atmosphere* **2023**, *14*, 557. https://doi.org/10.3390/atmos14030557

Academic Editor: Haibo Liu

Received: 18 February 2023
Revised: 11 March 2023
Accepted: 13 March 2023
Published: 14 March 2023

1. Introduction

River flooding is one of the most concerning natural disaster issues. According to the United Nations Office for Disaster Risk Reduction, floods have been the most frequent of all recorded natural disasters worldwide, accounting for 43% of all disasters, with global average annual losses estimated at USD 104 billion [1]. It is estimated that the losses are likely to increase in the future, with climate change, economic acceleration, and urbanization development [2,3]. The IPCC [4] has shown that climate change is already an undeniable phenomenon and that the occurrence of extreme flood events is also associated with rising temperatures, heavy precipitation, and an accelerated hydrological cycle at global and regional scales [5–7]. Catastrophic floods, primarily associated with climate change, have also attracted public attention, and have been the focus of much research [8–11]. Many recent flood events around the world have led to growing concern that flood disasters are becoming more frequent and severe [5,12–14].

The flood variability induced by extreme climate change, has become a very focal area of research in the past two decades. Many studies worldwide have focused their attention on the issue of fluvial flooding, mainly on the impact of changes in magnitude, frequency, and timing of flood events on a regional, continental, or global scale [5–7,12–14]. Recent studies have shown that not only has the changing climate shifted the timing of flooding in Europe, but the increased precipitation in autumn and winter has also led to increased flooding in the northwest, while decreased precipitation, warmer temperatures, and increased evapotranspiration have led to decreased flooding in the southeast [6,7]. Significant trends in flood magnitudes were found in the time series based on the annual maximum and Peaks Over Threshold, in Canada [14]. The trend of increasing magnitude and extent of floods is due to the correlation among the magnitude, extent of precipitation, soil moisture, and the shift in flood generation processes [15]. Some studies have shown that more frequent heavy precipitation and increased moisture in the catchment area, in the context of climate change, are expected to result in a greater risk of flooding in the future [11,16,17].

In China, a nationwide characterization of flood hazards, based on a dataset of 1120 hydrographic stations, has been presented [18], and changes in flood characteristics in the Yangtze and Pearl River basins in southern China have also been studied [19,20]. Some studies have also been carried out in arid and semi-arid regions of northwest China, such as the increase in the magnitude of floods in the Tarim River basin after the 1990s, especially in the case of high-latitude rivers [21]. Typical basins in the Tien Shan Mountains, such as the Tuoshikan River and the Kumalak River, have shown a significant increase in flood magnitude in response to climate change over the last 50 years [22,23]. The QMs, which are located in northwest China, are the origin of inland rivers in the Hexi Corridor, and an important region for the economic development of the Belt and Road Initiative. The Hexi Corridor, which is a typical arid and semi-arid area in northwest China, is divided into three sub-basins from east to west, namely the Shiyang River basin (SYRB), the Hei River basin (HRB), and the Shule River basin (SLRB) [24]. Wang et al. [25] have analyzed the trends in the frequency of floods in the QMs using the flood series of twelve rivers from 1970–2019, and the results show that the frequency of floods, mainly small floods in summer, is increasing, and medium and large floods are generally decreasing. There are differences between the eastern and western regions, with decreases in the east and increases in the west. However, temperature and precipitation in the northwest region have been increasing since 1987 [26,27], and the trend in flood magnitude of the 12 major rivers in the QMs is still unknown. The impact of climate change on river flood magnitude remains an important issue, due to the complex changes in precipitation, topography, and the hydrological cycle.

In this study, the impact of meteorological variables on the variability in flood magnitude in the QMs was comprehensively assessed, through flood information and meteorological data. The main objectives of this paper are (1) to assess the characteristics of the variability in AMPD and POT3M magnitude series at temporal scales, for twelve rivers in the QMs; (2) to identify regional differences in AMPD and POT3M magnitude series from east to west; and (3) to explore the causes of flood magnitude variability, including the analysis of meteorological variables in the QMs. Regional differences in precipitation, topography, and the hydrologic cycle have led to complex changes in increasing and decreasing flood magnitudes. Analyzing and exploring the changes in flood magnitude characteristics over the historical period is not only necessary for the scientific management of water resources, but also important for the social development and safety of people's property, in the middle and lower reaches. This finding will raise new concerns about the changes in flood magnitude in the QMs, northwest China, under climate change.

2. Materials and Methods

2.1. Study Area

The QMs, whose geographical boundary is approximately 93.4°–103.4° E and 35.8°–40.0° N, are located in northwest China, and consist of several parallel mountains and broad valleys (Figure 1). The elevation ranges from below 3000 m to above 5000 m, with most peaks exceeding 4000 m [28]. There are large vertical differences in climate. Alpine areas are mainly subject to the coupling zone of three atmospheric circulation systems: westerly flow, the East Asian monsoon, and the Tibetan Plateau monsoon, and mean annual precipitation is 301.9 mm [29,30]. The northern part of the Qilian Mountains is the Hexi Corridor, which is geographically located between 92.4°–104.2° E and 37.3°–42.3° N, with a total area of 2.15×10^5 km^2. The Hexi Corridor is a typical arid and semi-arid region in northwest China (Figure 1). The region is located in a narrow corridor extending from the east to the west, for more than 1000 km, and from the south to the north for 100–200 km [24]. The Hexi Corridor is the most important passage from northern China to Central Asia, and is an important part of the historical Silk Road. In this study, twelve rivers, which originate from the northern slopes of the QMs and end at the northern oasis of the Hexi Corridor, were selected as the study subject, and divided into three sub-basins from east to west.

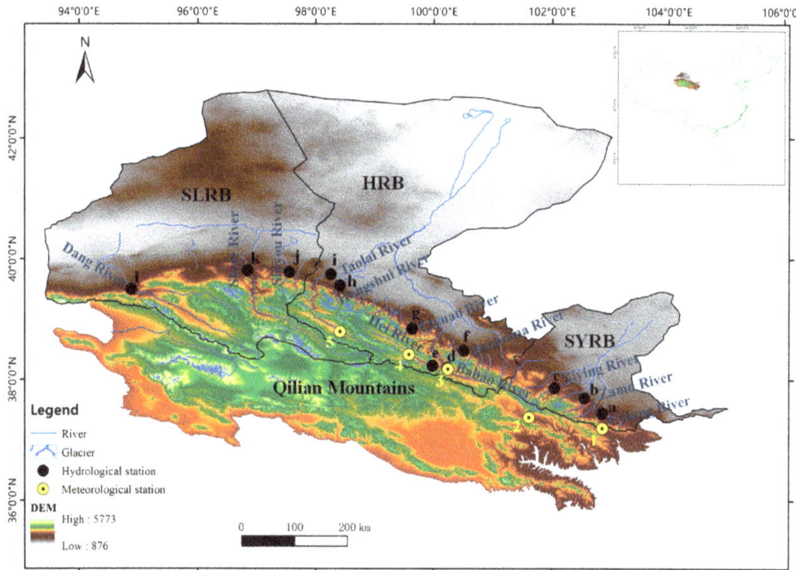

Figure 1. Location of the study area. (a–c) Denote Gulang, Zamu, and Xiying rivers in SYRB, (d–i) denote Babao, Hei, Dazhuma, Liyuan, Hongshui, and Taolai rivers in HRB, (j–l) denote Shiyou, Shule, and Dang rivers in SLRB, respectively. (1–5) Denote Wushaoling, Menyuan, Qilian, Yeniugou and Tuole meteorological stations, respectively.

2.2. Data

The dataset for this study includes meteorological data and historical flood discharges from 1970 to 2021. The discharge data were selected for analysis from twelve rivers originating from the QMs (Figure 1, Table 1), the upstream areas of which are mostly located in the high-altitude mountainous regions of the QMs, with little disturbance from human activities and no hydraulic projects, such as reservoirs and dykes. Peak flood discharges were collected from the Hydrological Yearbook of the People's Republic of China. Meteorological data were obtained from the National Meteorological Center of China's Meteorological Administration. All flood data and meteorological data have been subjected to strict quality control (e.g., extreme value test), reasonableness checks, and a standard normal homogeneity test [31–33], and missing meteorological data for individual years have been interpolated with a gap-filling method, after correlation check analysis [34]. The general information on the hydrological stations is shown in Table 1, and the basic information on the meteorological stations is shown in Table 2.

Table 1. Basic information on hydrological stations/catchments.

River Basin	Code	River Name	Discharge Station	Longitude	Latitude	Altitude (m)	Data Series
SYRB	a	Gulang	Gulang	102°52′	37°27′	2072	1970–2021
	b	Zamu	Zamusi	102°34′	37°42′	2010	1970–2021
	c	Xiying	Jiutiaoling	102°03′	37°52′	2270	1970–2021
HRB	d	Babao	Qilian	100°14′	38°12′	2710	1970–2021
	e	Hei	Zhamashike	99°59′	38°14′	2810	1970–2021
	f	Dazhuma	Wafangcheng	100°31′	38°29′	2440	1970–2021
	g	Liyuan	Sunan	99°38′	38°51′	2264	1970–2021
	h	Hongshui	Xindi	98°25′	39°34′	1880	1970–2021
SLRB	i	Taolai	Jiayuguan	98°16′	39°45′	1695	1970–2021
	j	Shiyou	Yumen	97°33′	39°47′	2300	1970–2021
	k	Shule	Changmabao	96°51′	39°49′	2080	1970–2021
	l	Danghe	Dangchengwan	94°53′	39°30′	2176	1970–2021

Table 2. Basic information on meteorological stations in the QMs.

Code	Meteorological Station	Longitude	Latitude	Altitude (m)	Data Series
1	Wushaoling	102°52′	37°12′	3045	1970–2021
2	Menyuan	101°37′	37°23′	2850	1970–2021
3	Qilian	100°15′	38°11′	2787	1970–2021
4	Yeniugou	99°35′	38°25′	3320	1970–2021
5	Tuole	98°25′	38°48′	3367	1970–2021

The continuously recorded AMPD and POT3M flood series were derived from hydrological stations originating from twelve rivers (i.e., three rivers in SYRB, six rivers in HRB, and three rivers in SLRB). Daily temperature and precipitation data from five meteorological stations were selected to analyze trends.

2.3. Methods

2.3.1. Determination of Flood Independence in the POT3M Flood Series

In this study, the POT3M sampling method was also used to supplement the AMPD sampling, with information such as flood magnitude, and this method has a good application in inland river basins in arid regions. The independence of the flood peak was determined by Lang et al. [35].

$$\begin{cases} D > 5 + log(A) \\ Q_{min} < \frac{3}{4}min(Q_1, Q_2) \end{cases} \tag{1}$$

where D denotes the flood duration between the two flood peaks; A denotes the catchment area in km^2; and Q_1 and Q_2 denote the magnitude of the two flood flows in m^3/s, respectively.

2.3.2. Test Methods for Trend and Abrupt Change Analysis

The non-parametric method is well suited to detecting trends and significance levels in hydrometeorological time series. In this study, the Sen's slope estimator [36] and the Mann–Kendall test (M–K test) [37,38], which are widely used to identify trends in hydrometeorological variables, were employed to explore trends in the flood and climate variables [39–42].

Sen's slope estimator is as follows.

$$\beta = median(\frac{x_j - x_k}{j - k}) \quad j > k \tag{2}$$

where the β symbol indicates whether a trend is positive or negative, while its value reflects the magnitude of the steepness of the trend, and x_j and x_k are the data values at times j and k $(j > k)$.

Additionally, The M–K test statistic S, is calculated as:

$$S = \sum_{i=1}^{n-1} \sum_{j=i+1}^{n} sgn(x(j) - x(i)) \tag{3}$$

where

$$sgn(x(j) - x(i)) = \begin{cases} 1 & if \quad x(j) - x(i) > 0 \\ 0 & if \quad x(j) - x(i) = 0 \\ -1 & if \quad x(j) - x(i) < 0 \end{cases} \tag{4}$$

A positive (negative) value of S indicates an increasing (decreasing) trend. When $n > 8$, the statistic S is approximately normally distributed, and its mean $E(S)$ and variance $Var(S)$ are identified as follows:

$$E(S) = 0 \tag{5}$$

$$Var(S) = \frac{1}{18} \left[n(n-1)(2n+5) - \sum_{i=1}^{m} t_i(t_i - 1)(2t_i + 5) \right] \tag{6}$$

$$Z = \begin{cases} \frac{S-1}{\sqrt{Var(S)}} & if \quad S > 0 \\ 0 & if \quad S = 0 \\ \frac{S+1}{\sqrt{Var(S)}} & if \quad S < 0 \end{cases} \tag{7}$$

where n is the number of data points; m is the number of tied groups, and t_i denotes the number of ties of extent i. The null hypothesis is rejected if the absolute value of Z is greater than the theoretical value $Z_{1-\alpha/2}$, where α is the statistical significance level. In this study, all trend results were evaluated at 90%, 95%, and 99% significance levels, respectively.

The sequential Mann–Kendall (SQ–MK) [43,44] test constructs the order series of the time series X.

$$s_k = \sum_{i=1}^{k} r_i \quad (k = 2, 3, \cdots, n) \tag{8}$$

where n is the length of the time series X.

$$r_i = \begin{cases} +1 & x_i - x_j \\ 0 \end{cases} \quad (j = 1, 2, \cdots, i) \tag{9}$$

Assuming that the time series X is random, define the statistic:

$$UF = \frac{[s_k - E(s_k)]}{\sqrt{Var(s_k)}} \quad (k = 1, 2, \cdots, n) \tag{10}$$

where $UF_1 = 0$, $E(s_k)$ and $Var(s_k)$ are the mean and variance of s_k, and $x_1, x_2, \cdots, x_n$ are independent of each other, when they have the same continuous distribution, which can be deduced from the following equation:

$$E(s_k) = \frac{n(n-1)}{4} \quad (2 \leq k \leq n) \tag{11}$$

$$Var(s_k) = \frac{n(n-1)(2n+5)}{72} \quad (2 \leq k \leq n) \tag{12}$$

where UF is the standard normal distribution, which is the sequence calculated for the order of the time series X $(x_1, x_2, \cdots, x_n)$.

The inverse order of the time series X $(x_n, x_{n-1}, \cdots, x_1)$ is used to calculate the statistic UB, for the inverse series of the time series X.

In this paper, given a significance level of $\alpha = 0.05$ (95% significance level), then the critical value $U_{0.05} = \pm 1.96$, and the two statistic series curves of UF and UB, and the two straight lines of ± 1.96, are plotted on a single graph. If UF and UB intersect between the critical values, then the intersection point corresponds to the time at which the mutation of time series X begins.

2.3.3. Meteorological Variables

Six temperature and precipitation indices were employed to analyze the meteorological variables and to explore the changes in temperature and precipitation (Table 3).

Table 3. Definitions of the six temperature and precipitation indices used in this study.

Abbreviations	Meteorological Indices	Unit
AMT	Annual mean temperature	°C
TXX	Annual maximum value of daily maximum temperature	°C
TXN	Annual minimum value of daily maximum temperature	°C
P1	Maximum daily precipitation in 1 year	mm
P3	3-day mean maximum precipitation in 1 year	mm
P7	7-day mean maximum precipitation in 1 year	mm

3. Results

3.1. Flood Independence Analysis

Flood independence analysis is a prerequisite for determining trends in the AMPD and POT3M flood series. In order to make the assumption of independence of the sampled flood peaks reasonable, the flood independence criterion method proposed by Lang et al. [35] was used, while the POT3M sampling method, applied in the arid and semi-arid regions of northwest China, was considered [21,23]. For the AMPD and POT3M flood series of the 12 rivers in the study, all sampled flood peaks comply with the two conditions of the flood independence criterion method, in terms of duration, D, and intermediate minimum flow, $Q_{\min}$, and thus flood independence was valid.

3.2. Trends in AMPD Series

The AMPD series of twelve rivers in the QMs were tested using Sen's slope estimator and the M–K test, and the results are shown in Figure 2 and Table 4. Five rivers exhibited significantly decreasing trends, while one river showed a significantly increasing trend, and six rivers had non-significant trends. In SYRB (a–c rivers), Gulang, Zamu, and Xiying rivers showed decreasing trends in AMPD at 99%, 95%, and 95% significance levels, respectively. In HRB (d–i rivers), five rivers (Babao, Hei, Liyuan, Hongshui, and Taolai rivers) showed non-significant trends, while Dazhuma River showed a decreasing trend, at the significance level of 99%. However, in SLRB (j–l rivers), Shiyou River showed a decreasing trend at the significance level of 90% and Shule River exhibited an increasing trend at the significance level of 99%, while Dang River had a non-significant trend. The trend line of AMPD for the twelve rivers is shown in Figure 2, where the evolutionary trend in historical floods can be seen. Although the AMPD trends are different among the eastern, central, and western parts of the QMs, most of the rivers presented obvious decreasing trends from 1970 to 2021, especially in the eastern and central regions.

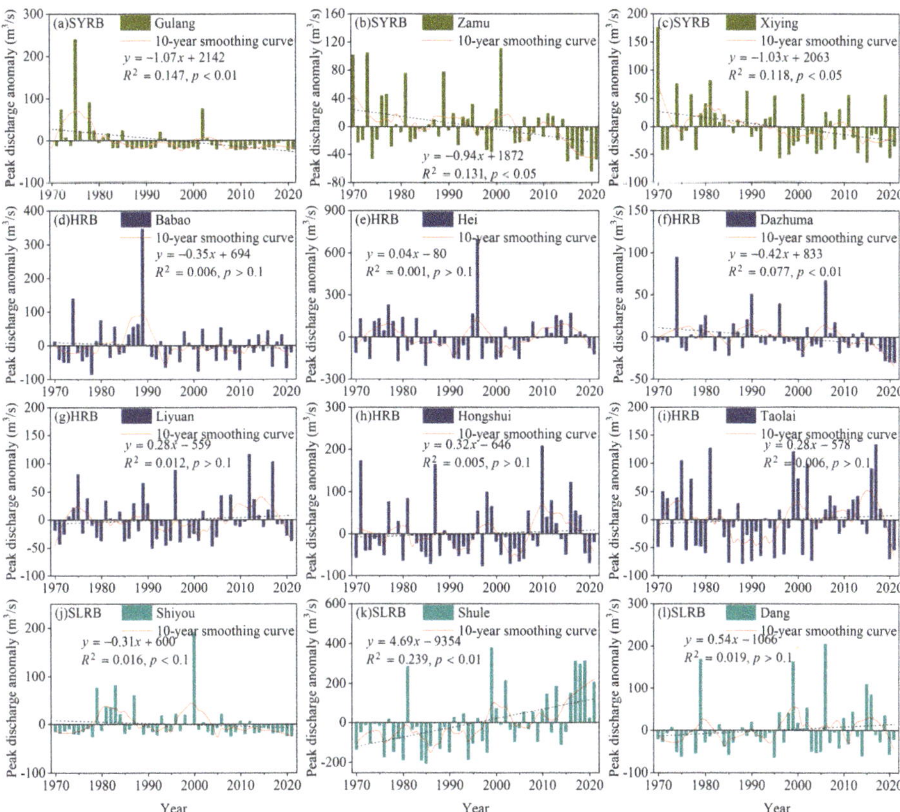

Figure 2. Trends of AMPD for 12 representative rivers in the QM region, northwest China. (**a–c**) Gulang, Zamu, and Xiying rivers in SYRB, (**d–i**) Babao, Hei, Dazhuma, Liyuan, Hongshui, and Taolai rivers in HRB, (**j–l**) Shiyou, Shule, and Dang rivers in SLRB, respectively.

Table 4. Results of trend analysis of AMPD and POT3M series, in the QM region, northwest China.

River Basin	Code	AMPD Series	Mean Value (m^3/s)	POT3M Series	Threshold Value (m^3/s)
SYRB	a	Z(−3.71) S(0.01)↓	27	Z(−2.59) S(0.01)↓	5
	b	Z(−2.04) S(0.05)↓	101	Z(−1.04) NS	55
	c	Z(−2.28) S(0.05)↓	129	Z(−2.35) S(0.05)↓	79
HRB	d	Z(0.11) NS	143	Z(−0.14) NS	75
	e	Z(0.54) NS	314	Z(0.61) NS	158
	f	Z(−2.37) S(0.01)↓	43	Z(−0.17) NS	16
	g	Z(0.49) NS	99	Z(−0.49) NS	56
	h	Z(0.58) NS	145	Z(−0.01) NS	69
	i	Z(0.72) NS	171	Z(0.25) NS	121
SLRB	j	Z(−1.33) S(0.1)↓	34	Z(−1.61) S(0.1)↓	13
	k	Z(3.44) S(0.01)↑	338	Z(3.14) S(0.01)↑	251
	l	Z(0.73) NS	90	Z(0.77) NS	39

In Table 4, the M–K test for the AMPD and POT3M series detected trends at 90%, 95%, and 99% significance levels, respectively; S means statistically significant at the significance level shown in parentheses; NS means not significant; "↑" and "↓" mean increasing and decreasing trend. (a–l) Denote: Gulang, Zamu, Xiying, Babao, Hei, Dazhuma, Liyuan, Hongshui, Taolai, Shiyou, Shule, and Dang rivers, respectively. The corresponding hydrological stations for the (a–l) rivers are Gulang, Zamusi, Jiutiaoling, Qilian, Zhamashike, Wafangcheng, Sunan, Xindi, Jiayuguan, Yumen, Changmabao, and Dangchengwan, respectively.

3.3. Abrupt Behavior for Changes in AMPD Series

The twelve rivers, that originate at high-altitude in the QMs, are mainly recharged by precipitation, snow, and glacial meltwater and have specific, typical flood generation mechanisms. Heavy precipitation and the melting of snow and ice water, due to increased temperatures, are the main factors in flood generation. As a result, the magnitude of floods has fluctuated over the 50 years of observation, but the structure is stable. The results of the SQ-MK test for twelve representative rivers are shown in Figure 3. Six of the twelve rivers exceeded the critical value of 1.96, for which $\alpha = 0.05$ in the significance test. Thus, the AMPD series showed abrupt changes. Among them, in the eastern region of QMs, the SQ-MK test values of the Gulang and Xiying rivers continued to decrease and break the critical values during the period 1970–2021, but the SQ-MK test values of the Zamu River fluctuated steadily. In the central region, the change points of the Babao and Dazhuma rivers occurred in 1989 and 2020, with the significance level of 95%. However, the Hei, Liyuan, Hongshui, and Taolai rivers all showed stable fluctuation from 1970 to 2021, without breaking the critical values. In the western region of the QMs, the SQ-MK test values of the Shiyou, Shule, and Dang rivers behaved differently. Shiyou River showed an increasing trend from 1970 to 1984 and broke the critical value in 1983, followed by a continuous decreasing trend from 1985–2021. Though it had a decreasing trend from 1970 to 1986, Shule River showed a continuously increasing trend after 1987, bypassing the critical value of 1.96, to a value of 3.47. From 1970 to 2021, Dang River did not break through the critical value and presented a slightly increasing trend. In Figure 3, most of the change points can be seen after the mid-1990s, whereas significant trends occurred in SYRB (a and c), HRB (f), and SLRB (k).

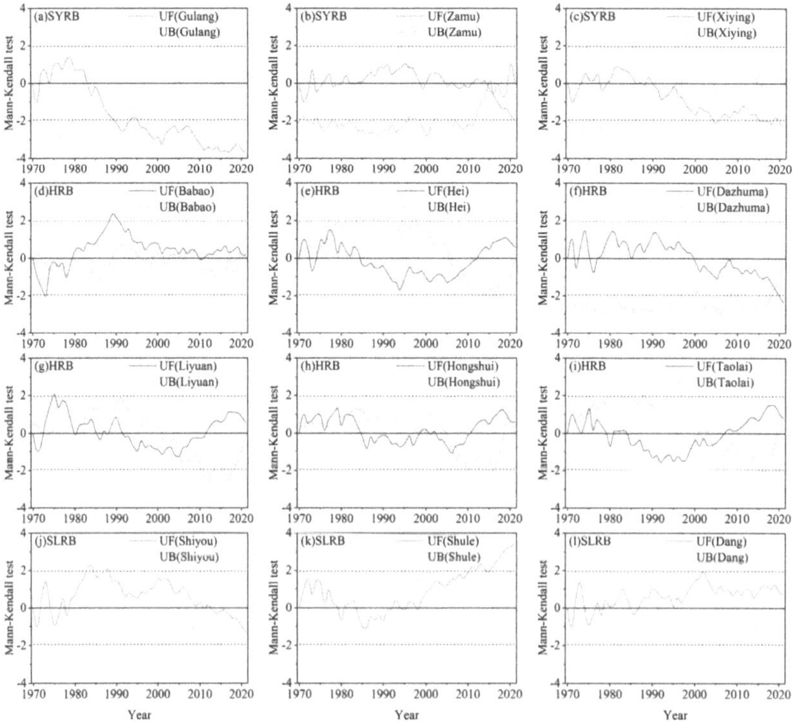

Figure 3. Trends of SQ–MK test in AMPD for 12 representative rivers in the QM region, northwest China. (**a–c**) Denote Gulang, Zamu, and Xiying rivers in SYRB, (**d–i**) denote Babao, Hei, Dazhuma, Liyuan, Hongshui, and Taolai rivers in HRB, (**j–l**) denote Shiyou, Shule, and Dang rivers in SLRB, respectively.

3.4. Trends in POT3M Flood Series

Twelve rivers in the QMs were selected to calculate the trends of the POT3M flood series and the statistical results are presented in Figure 4 and Table 4. Significantly increasing and decreasing trends of the POT3M series were detected at 90%, 95%, and 99% significance levels, respectively. The results demonstrate that Shule River in SLRB displayed a significantly increasing trend, at the significance level of 99%, while the Gulang and Xiying rivers in SYRB, and Shiyou River in SLRB showed decreasing trends, at the significance levels of 99%, 95%, and 90%, respectively. It should be noted again, that non-significant trends were found in the Zamu, Babao, Hei, Dazhuma, Liyuan, Hongshui, Taolai, and Dang rivers. Compared to the AMPD series, more reliable conclusions could be drawn from these series, because samplings of the POT3M series provided more flood information. By comparing the trends of the POT3M series and AMPD series in Table 4 in parallel, the trends of increasing and decreasing are consistent for most of the rivers, with some differences in the Zamu and Dazhuma rivers, which are mainly caused by the different flood information of the sample selection.

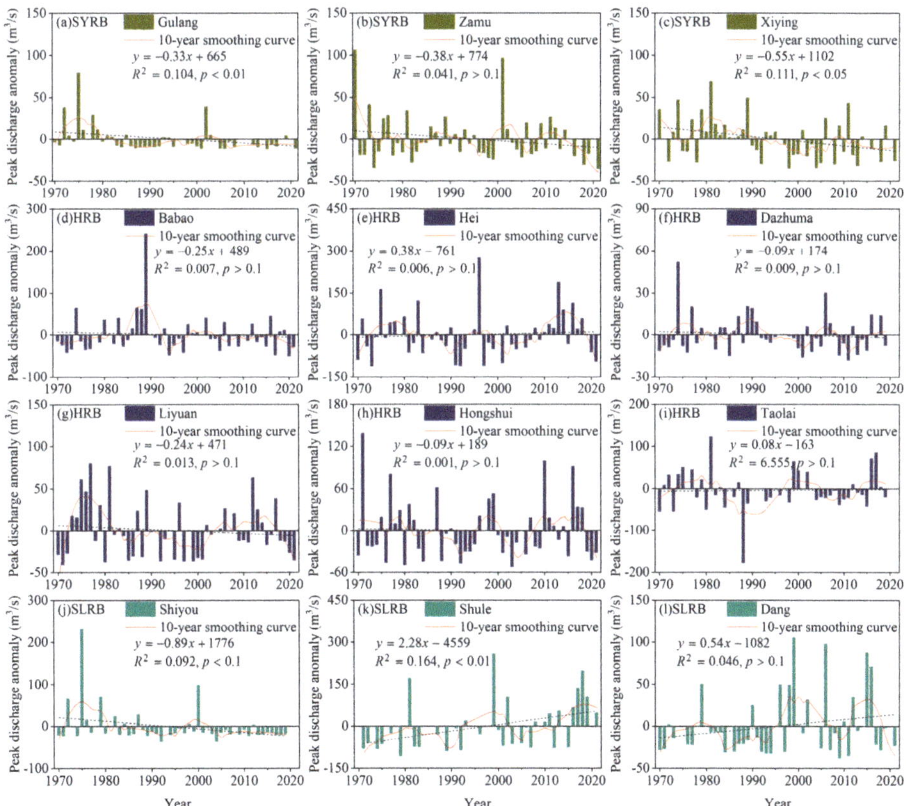

Figure 4. Trends of POT3M flood series for 12 representative rivers in the QM region, northwest China. (**a–c**) Denote Gulang, Zamu, and Xiying rivers in SYRB, (**d–i**) denote Babao, Hei, Dazhuma, Liyuan, Hongshui, and Taolai rivers in HRB, (**j–l**) denote Shiyou, Shule, and Dang rivers in SLRB, respectively.

3.5. Trends of Flood Magnitude in the Eastern, Central, and Western Regions

According to the regional comparative analysis of the different flood series from the flood observation dataset, the flood magnitude trends for twelve rivers in the QMs are both increasing and decreasing in the eastern, central, and western regions, as shown in Figure 5 and Table 4. In the eastern SYRB (a–c rivers), the results of the AMPD and POT3M series showed an overall decreasing trend from 1970 to 2021, with different significance levels. In the central HRB (d–i rivers), there were minor differences in the trends of the AMPD and POT3M series. In the trend in the AMPD series, five of the six rivers were non-significant trends, except for the Dazhuma River, which exhibited a decreasing trend at the significance level of 99%. However, in the trend in the POT3M series, all six rivers showed non-significant trends. Among the western SLRB (j–l rivers), Shiyou River displayed a decreasing trend, at the significance level of 90%, in both the AMPD and POT3M series, Shule River showed an increasing trend at the significance level of 99% in both the AMPD and POT3M series, whereas Dang River had a non-significant trend during the period of 1970–2021.

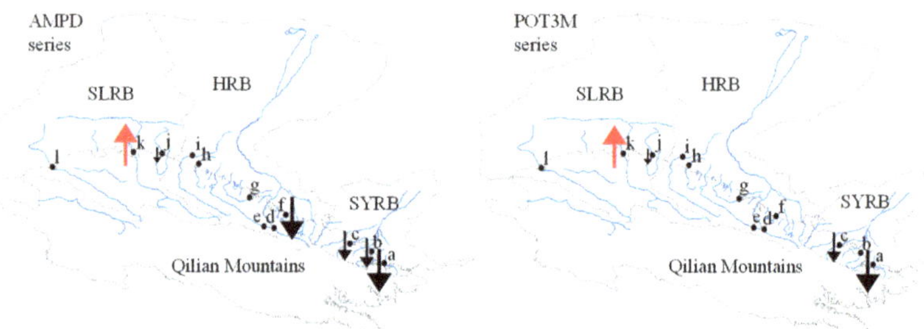

Figure 5. Trends in AMPD and POT3M flood series for 12 representative rivers in the QM region, northwest China. (a–c) Denote Gulang, Zamu, and Xiying rivers in SYRB, (d–i) denote Babao, Hei, Dazhuma, Liyuan, Hongshui, and Taolai rivers in HRB, (j–l) denote Shiyou, Shule, and Dang rivers in SLRB, respectively. A big arrow indicates a statistically increasing trend (red) and decreasing trend (black), at the significance level of 99%, a medium arrow indicates a statistically increasing trend (red) and decreasing trend (black), at the significance level of 95%, a small arrow indicates a statistically increasing trend (red) and decreasing trend (black), at the significance level of 90%. No marked symbol at stations represents a non-significant trend.

4. Discussion

4.1. Attribution of Climate Change to Variability in Flood Magnitude

The results of this study indicate that the flood magnitude of the QMs changed significantly from east to west during the period 1970–2021. The linear trends of the AMPD and POT3M series show a decreasing trend in flood magnitude for SYRB in the east (e.g., Gulang, Zamu, and Xiying rivers), a slight fluctuation for six rivers in HRB in the central part, and an increasing trend, with larger values, for SLRB in the west (e.g., Shule River). The AMPD and POT3M series differ slightly, in that the AMPD series shows more rivers with a decreasing trend (e.g., Zamu and Dazhuma rivers), as shown in Figures 2, 4 and 5, and Table 4. A recent study has shown that climate change in the Qilian Mountains region has led to variations in the frequency of floods of different levels, with a decrease in SYRB in the east, a slight fluctuation in HRB in the center, and an increase in SLRB in the west, and the main factors causing those changes in flood frequency in the QMs are heavy rainfall, abnormal warming, and accelerated glacial melting [25]. Even changes in river discharge in arid and semi-arid regions of northwest China are associated with climate change [21–23,45]. In the past decades, e.g., Shi, et al. [27] have confirmed that the climate of arid and semi-arid regions in northwest China changed from warm-dry to warm-wet during 1961–2003, and Chen, et al. [26] also found an increasing trend in temperature and precipitation in arid northwest China during 1960–2015, and precipitation started to rise sharply after 1987. Climate change may cause high temperatures and heavy precipitation, which in turn lead to changes in flooding [46–48].

Twelve rivers, originating in the high-altitude mountainous regions of the QMs, have a special and typical flooding mechanism, mainly influenced by heavy precipitation and the melting of snow and ice water, due to rising temperatures. Owing to the complex topography of the QMs, the high altitude, and the inconvenient access, only five national meteorological stations are located at high-altitude in the mountains, while the rest of the meteorological stations are located in the piedmont and oasis plain areas. In this study, five meteorological stations in high-altitude mountainous areas, were selected to calculate trends using daily precipitation and temperature data for the period 1970–2021. To better explore the meteorological factors affecting the variation in flood magnitude in the QMs, temperature and precipitation trends at five meteorological stations were analyzed, by defining six temperature and precipitation indices (Figures 6 and 7). Over the past 52 years, temperature and precipitation in SYRB, HRB, and SLRB have generally increased to different degrees (Figures 6 and 7). Although the only five meteorological stations in the high-altitude mountainous regions do not correspond to analyzing the temperature and precipitation change patterns of the twelve rivers one by one, it can be found that the trend varies from region to region. In terms of the causes of variation in flood magnitude generated in the twelve rivers, it is mainly influenced by temperature, heavy precipitation, substratum, and topography, but in high-altitude areas of the QMs, heavy precipitation is the main factor causing the variation in flood magnitude. According to the analysis of the data from five meteorological stations, temperature and precipitation have generally increased at different rates in SYRB, HRB, and SLRB over the past 52 years (Figures 6 and 7), however, the magnitudes of the AMPD and POT3M series have increased or decreased in different rivers (e.g., Gulang, Zamu, Xiying, Dazhuma, Shiyou, and Shule rivers), the possible causes are the regional differences of heavy precipitation in the upper reaches of the rivers, which are responsible for the changes in flood magnitude.

Among high-latitude rivers in northwest and southwest China, similar results of increasing and decreasing trends in the magnitude of river floods due to climate change, dominated by changes in temperature and precipitation, have been obtained in the Tien Shan Mountains, Aksu, Tarim, and Lancang–Mekong river basins [21–23,49]. In Central Asia, climate change is leading to an increased risk of flooding and landslides, and the likelihood of glacial lake outburst flooding is expected to increase with rising temperatures and an increase in the number of glacial lakes [50]. Similar studies can be compared in Europe, increases and decreases in temperature, evaporation, and precipitation in different regions have led to increased flooding in the northwest and decreased flooding in the south and east. Flood magnitude trends in different regions of Europe range from an increase of 11% to a decrease of 23%, per decade [6]. Even in Canada and North America, the trend of increasing and decreasing flood magnitude in different regions over the past decades has been identified by researchers [14]. Despite the spatial heterogeneity of the observed records of temperature, precipitation, and flooding in different regions, it is reasonable that the variation in flood magnitudes in the high-altitude mountains of the QMs found in our study, is mainly attributed to climate change.

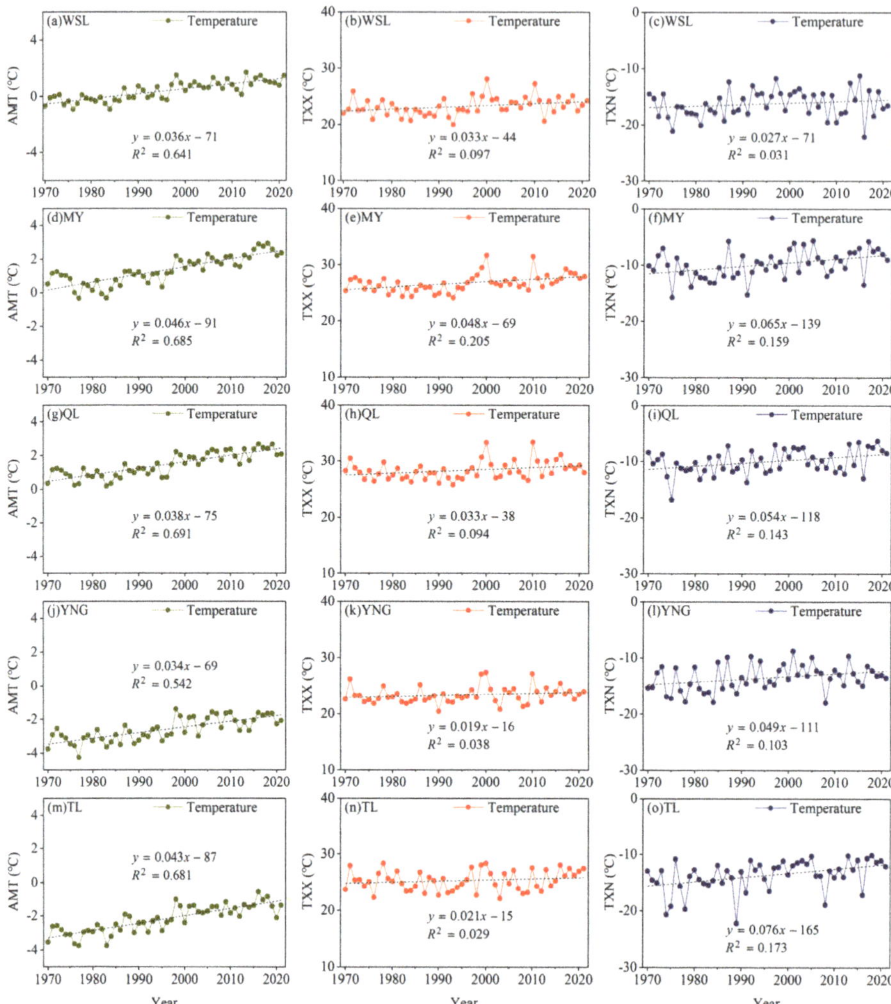

Figure 6. Temporal variations in temperature indices at meteorological stations in the QM region. The stations are, WSL: Wushaoling; MY: Menyuan; QL: Qilian; YNG: Yeniugou; TL: Tuole. (**a–c**) WSL; (**d–f**) MY; (**g–i**) QL; (**j–l**) YNG; (**m–o**) TL. Definitions of these temperature indices are given in Table 3.

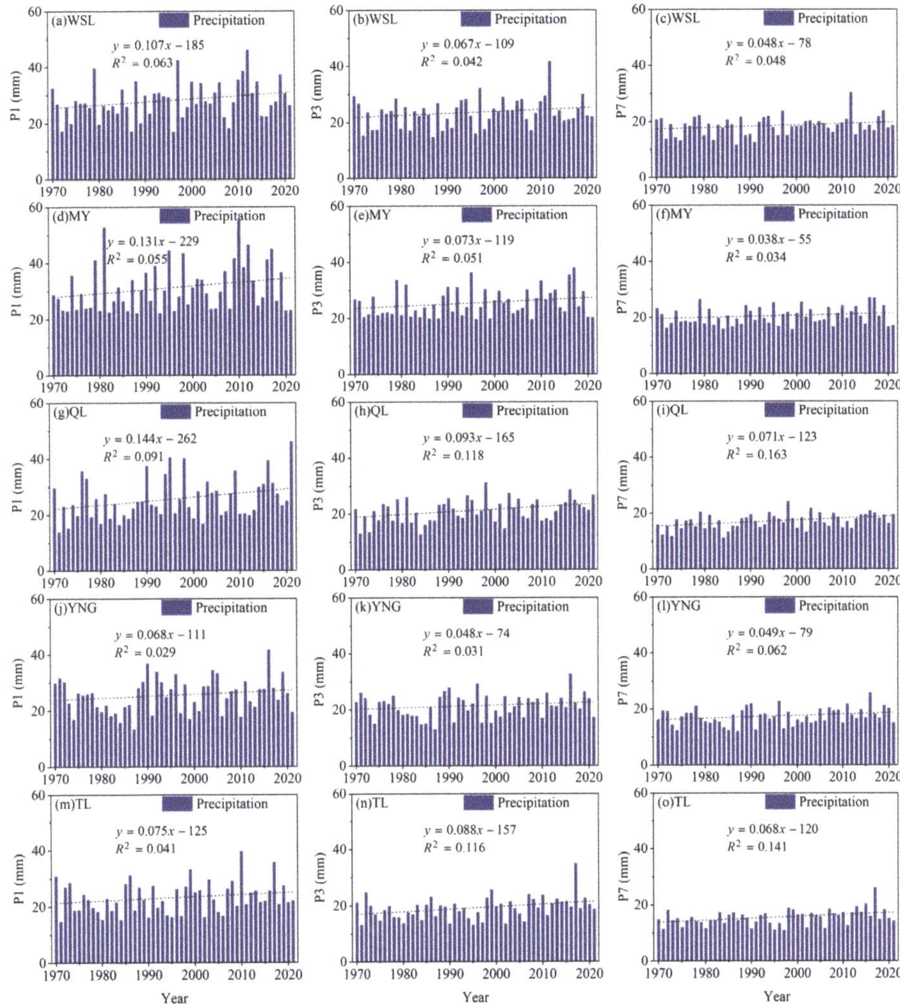

Figure 7. Temporal variations in precipitation indices at meteorological stations in the QM region. The stations are, WSL: Wushaoling; MY: Menyuan; QL: Qilian; YNG: Yeniugou; TL: Tuole. (**a–c**) WSL; (**d–f**) MY; (**g–i**) QL; (**j–l**) YNG; (**m–o**) TL. Definitions of these precipitation indices are given in Table 3.

4.2. Regional Differences in the Evolution of Flood Magnitude

The long-term variability in the AMPD and POT3M flood series highlights the regional differences in understanding the evolution of flood magnitude in the QMs. Analysis of the datasets shows a clear regional pattern of flooding trends across the QMs (Figures 2, 4 and 5), with a decreasing trend in SYRB in the east, a slight fluctuation for six rivers in HRB in the central part, and an increasing trend with larger values in SLRB in the west (e.g., Shule River). In arid and semi-arid regions, for the twelve inland rivers originating in the QMs, regional variations in flood magnitudes are mainly caused by heavy precipitation, abnormal warming, and accelerated glacial melting. Regional differences in precipitation and precipitation intensity, river catchment, underlying surface, and topography are the main drivers of flood magnitude variability. Nevertheless, the climate shows great variability with altitude, and most precipitation events in mountainous regions are influenced by three main sources of water vapor; namely, the East Asian monsoon, the Tibetan Plateau monsoon, and the westerly circulation [29]. The SYRB in the east, HRB in the center, and SLRB in the west are mainly influenced by the combination of the East Asian monsoon, the Tibetan Plateau monsoon, and the westerly circulation, respectively. The enhanced westerly circulation and increased water vapor, are the main factors leading to the increased precipitation in the QMs [30,51]. Different climate types result in different regional precipitation and precipitation intensities, causing differences in flood magnitude trends in the eastern, central, and western regions. In addition, due to the different glacier coverage areas and reserves in the eastern, central and western regions of the QMs, (Table 5), the eastern part is the smallest, the central part is larger, and the western part is the largest. The glaciers in the east are retreating rapidly, while the central and western regions are retreating more slowly in turn [52]. Due to the differences in glacier area and glacial meltwater ratios in the upper reaches of the twelve rivers in three sub-basins, the three rivers in the eastern SYRB have the smallest glacial meltwater ratios, whereas the three rivers in the western SLRB have the largest (Table 5) [53,54]. Differences in precipitation and the proportion of glacial meltwater, due to temperature increases in the east and west regions are the main reasons for the spatial patterns of flood magnitude variation in those three main regions (Figures 2, 4 and 5).

Table 5. Glacier area and percentage of glacier meltwater runoff in this study [25,54].

River Basin	River Name	Glacier Area (km^2)	Glacial Meltwater Ratio (%)	Periods
SYRB	Gulang	–	–	1960s–2010s
	Zamu	3.75	1.3	1960s–2010s
	Xiying	19.77	5.3	1960s–2010s
HRB	Babao	–	–	1960s–2010s
	Hei	58.90	2.7	1960s–2010s
	Dazhuma	5.94	11.4	1960s–2010s
	Liyuan	16.28	7.1	1960s–2010s
	Hongshui	125.62	44.0	1960s–2010s
	Taolai	137.89	17.2	1960s–2010s
SLRB	Shiyou	6.38	19.3	1960s–2010s
	Shule	469.52	42.2	1960s–2010s
	Danghe	233.83	46.8	1960s–2010s

The trend comparison of the AMPD and POT3M series, shows that in the eastern part of the QMs, three rivers (e.g., Gulang, Zamu, and Xiying rivers) showed an overall decreasing trend, at the significance level of 99% or 95%. In the central region, most of the rivers showed non-significant trends, except for minor differences in individual rivers. For example, in the AMPD series, Dazhuma River exhibited a decreasing trend at the significance level of 99%. In the western region, Shule River displayed a significant increasing trend at the significance level of 99%, whereas Shiyou River was decreasing at the significance level of 90%, and Dang River had a non-significant trend. Similar results for regional differences in flood magnitude trends can also be found for the Tien Shan Mountains, and Aksu and Tarim rivers, in northwest China, as well as in Europe and North America [6,14,21–23].

5. Conclusions

In this study, flood events of twelve inland rivers in the QMs, northwest China, were examined, using Sen's slope estimator and the M–K test, for the continuously recorded AMPD and POT3M flood series, from 1970 to 2021. Through analyzing the trends, abrupt changes, and causes of flood magnitude changes, the main findings can be summarized as follows:

1. The evolution of the flood magnitude of the twelve rivers originating in the QMs over the last 52 years, has been mainly influenced by the gradual increase in temperature

and precipitation. The main factors causing abrupt changes in flood magnitude are heavy precipitation events and anomalous warming.

2. The trend analysis of the AMPD and POT3M magnitude series of the twelve rivers, for the period of 1970–2021, shows that the spatial distribution of flood magnitude changes is different. The AMPD and POT3M series show a decreasing trend in SYRB in the east (e.g., Gulang, Zamu, and Xiying rivers), a slight fluctuation for six rivers in HRB in the central region, and an increasing trend, with larger significance values, in SLRB in the west (e.g., Shule River), especially since 1987.

In the future, research on the formation process of flood events in arid and semi-arid regions should be strengthened, including elements such as meteorological variables related to flood frequency, the timing of flood peak occurrence, and flood magnitude. These findings can suggest solutions for water resources management and influence decisions for adaptation to climate change.

Author Contributions: Conceptualization, X.W.; data curation, C.H.; funding acquisition, R.C.; investigation, K.L.; methodology, X.W. and Y.Y.; project administration, J.L.; resources, R.C.; software, X.W. and Z.L.; supervision, R.C.; validation, K.L.; writing—original draft, X.W.; Writing—review and editing, Z.L. All authors have read and agreed to the published version of the manuscript.

Funding: This research was funded by the National Key Research and Development Program of China, grant number 2019YFC1510505, the National Natural Science Foundation of China, grant number 42171145, the Natural Science Foundation of Gansu Province, China, grant number 21JR7RA043, the Qinghai Key R&D and Transformation Program, grant number 2020-SF-146, and the open research fund of the National Cryosphere Desert Data Center, grant number 2021kf09.

Institutional Review Board Statement: Not applicable.

Informed Consent Statement: Not applicable.

Data Availability Statement: Meteorological data were collected by the National Climate Center of the China Meteorological Administration and are available on http://data.cma.cn/ under request (accessed on 10 October 2022). Flood data were collected by the Bureau of Hydrology and Water Resources of Gansu Province, China, and are available under request.

Acknowledgments: We are very grateful for the support of the National Meteorological Center of China's Meteorological Administration on meteorological data and the Hydrological Station of Gansu Province regarding the experimental flood data. We would like to thank the editor and four anonymous reviewers for their insightful and constructive comments that greatly improved the paper.

Conflicts of Interest: The authors declare no conflict of interest.

References

1. Desai, B.; Maskrey, A.; Peduzzi, P.; De Bono, A.; Herold, C. *Making Development Sustainable: The Future of Disaster Risk Management, Global Assessment Report on Disaster Risk Reduction*; United Nations Office for Disaster Risk Reduction (UNISDR): Geneva, Switzerland, 2015; Available online: https://archive-ouverte.unige.ch/unige:78299 (accessed on 10 October 2022).
2. Winsemius, H.C.; Aerts, J.; van Beek, L.P.H.; Bierkens, M.F.P.; Bouwman, A.; Jongman, B.; Kwadijk, J.C.J.; Ligtvoet, W.; Lucas, P.L.; van Vuuren, D.P.; et al. Global drivers of future river flood risk. *Nat. Clim. Chang.* **2016**, *6*, 381–385. [CrossRef]
3. Hirabayashi, Y.; Mahendran, R.; Koirala, S.; Konoshima, L.; Yamazaki, D.; Watanabe, S.; Kim, H.; Kanae, S. Global flood risk under climate change. *Nat. Clim. Chang.* **2013**, *3*, 816–821. [CrossRef]
4. IPCC. *Intergovernmental Panel on Climate Change, Climate Change 2021: Impacts, Adaptation, and Vulnerability*; Cambridge Univ. Press: Cambridge, UK, 2021.
5. Kundzewicz, Z.W.; Su, B.; Wang, Y.J.; Wang, G.J.; Wang, G.F.; Huang, J.L.; Jiang, T. Flood risk in a range of spatial perspectives—From global to local scales. *Nat. Hazards Earth Syst.* **2019**, *19*, 1319–1328. [CrossRef]
6. Bloeschl, G.; Hall, J.; Viglione, A.; Perdigao, R.A.P.; Parajka, J.; Merz, B.; Lun, D.; Arheimer, B.; Aronica, G.T.; Bilibashi, A.; et al. Changing climate both increases and decreases European river floods. *Nature* **2019**, *573*, 108. [CrossRef]
7. Bloschl, G.; Hall, J.; Parajka, J.; Perdigao, R.A.P.; Merz, B.; Arheimer, B.; Aronica, G.T.; Bilibashi, A.; Bonacci, O.; Borga, M.; et al. Changing climate shifts timing of European floods. *Science* **2017**, *357*, 588–590. [CrossRef]
8. Wilby, R.L.; Keenan, R. Adapting to flood risk under climate change. *Prog. Phys. Geogr.* **2012**, *36*, 348–378. [CrossRef]
9. Nigel, W.; Gosling, S.N. The impacts of climate change on river flood risk at the global scale. *Clim. Chang.* **2014**, *134*, 387–401. [CrossRef]

10. Cornwall, W. Europe's deadly floods leave scientists stunned. *Science* **2021**, *373*, 372–373. [CrossRef]
11. Alfieri, L.; Bisselink, B.; Dottori, F.; Naumann, G.; de Roo, A.; Salamon, P.; Wyser, K.; Feyen, L. Global projections of river flood risk in a warmer world. *Earths Future* **2017**, *5*, 171–182. [CrossRef]
12. Hodgkins, G.A.; Whitfield, P.H.; Burn, D.H.; Hannaford, J.; Renard, B.; Stahl, K.; Fleig, A.K.; Madsen, H.; Mediero, L.; Korhonen, J.; et al. Climate-driven variability in the occurrence of major floods across North America and Europe. *J. Hydrol.* **2017**, *552*, 704–717. [CrossRef]
13. Mediero, L.; Santillan, D.; Garrote, L.; Granados, A. Detection and attribution of trends in magnitude, frequency and timing of floods in Spain. *J. Hydrol.* **2014**, *517*, 1072–1088. [CrossRef]
14. Zadeh, S.M.; Burn, D.H.; O'Brien, N. Detection of trends in flood magnitude and frequency in Canada. *J. Hydrol.-Reg. Stud.* **2020**, *28*, 13. [CrossRef]
15. Kemter, M.; Merz, B.; Marwan, N.; Vorogushyn, S.; Bloschl, G. Joint Trends in Flood Magnitudes and Spatial Extents Across Europe. *Geophys. Res. Lett.* **2020**, *47*, e2020GL087464. [CrossRef]
16. Archfield, S.; Hirsch, R.; Viglione, A.; Blöschl, G. Using a peaks-over-threshold approach to assess changes in flood magnitude, volume, duration, and frequency across the United States. In Proceedings of the EGU General Assembly Conference, Vienna, Austria, 27 April–2 May 2014.
17. Murray, V.; Ebi, K.L. IPCC Special Report on Managing the Risks of Extreme Events and Disasters to Advance Climate Change Adaptation (SREX). *J. Epidemiol. Community Health* **2012**, *66*, 759–760. [CrossRef]
18. Yang, L.; Wang, L.C.; Li, X.; Gao, J. On the flood peak distributions over China. *Hydrol. Earth Syst. Sci.* **2019**, *23*, 5133–5149. [CrossRef]
19. Xiong, B.; Xiong, L.H.; Guo, S.L.; Xu, C.Y.; Xia, J.; Zhong, Y.X.; Yang, H. Nonstationary Frequency Analysis of Censored Data: A Case Study of the Floods in the Yangtze River From 1470 to 2017. *Water Resour. Res.* **2020**, *56*, 20. [CrossRef]
20. Gu, X.; Zhang, Q.; Wang, Z. Evaluation on Stationarity Assumption of Annual Maximum Peak Flows during 1951-2010 in the Pearl River Basin. *J. Nat. Resour.* **2015**, *30*, 824–835, (In Chinese with English Abstract).
21. Zhang, Q.; Gu, X.H.; Singh, V.P.; Sun, P.; Chen, X.H.; Kong, D.D. Magnitude, frequency and timing of floods in the Tarim River basin, China: Changes, causes and implications. *Glob. Planet. Chang.* **2016**, *139*, 44–55. [CrossRef]
22. Mao, W.; Fan, J.; Shen, Y.; Yang, Q.; Gao, Q.; Wang, G.; Wang, S.; Wu, S. Variations of Extreme Flood of the Rivers in Xinjiang Region and Some Typical Watersheds from Tianshan Mountains and Their Response to Climate Change in Recent 50 Years. *J. Glaciol. Geocryol.* **2012**, *34*, 1037–1046, (In Chinese with English Abstract).
23. Jiang, J.X.; Cai, M.; Xu, Y.J.; Fang, G.H. The changing trend of flooding in the Aksu River basin. *J. Glaciol. Geocryol.* **2021**, *43*, 1200–1209, (In Chinese with English Abstract).
24. Zhang, Y.Y.; Fu, G.B.; Sun, B.Y.; Zhang, S.F.; Men, B.H. Simulation and classification of the impacts of projected climate change on flow regimes in the arid Hexi Corridor of Northwest China. *J. Geophys. Res.-Atmos.* **2015**, *120*, 7429–7453. [CrossRef]
25. Wang, X.L.; Chen, R.S.; Li, H.Y.; Li, K.L.; Liu, J.F.; Liu, G.H. Detection and attribution of trends in flood frequency under climate change in the Qilian Mountains, Northwest China. *J. Hydrol.-Reg. Stud.* **2022**, *42*, 101153. [CrossRef]
26. Chen, Y.N.; Li, Z.; Fan, Y.T.; Wang, H.J.; Deng, H.J. Progress and prospects of climate change impacts on hydrology in the arid region of northwest China. *Environ. Res.* **2015**, *139*, 11–19. [CrossRef] [PubMed]
27. Shi, Y.F.; Shen, Y.; Li, D.L.; Zhang, G.W.; Ding, Y.; Hu, R.J.; Kang, E.S. Discussion on the present climate change from warm-dry to warm-wet in northwest china. *Quatern. Sci.* **2003**, *23*, 152–164, (In Chinese with English Abstract).
28. Wang, X.; Chen, R.; Han, C.; Yang, Y.; Liu, J.; Liu, Z.; Song, Y. Changes in river discharge in typical mountain permafrost catchments, northwestern China. *Quatern. Int.* **2019**, *519*, 32–41. [CrossRef]
29. Zhang, Q.; Yu, Y.X.; Zhang, J. Characteristics of Water Cycle in the Qilian Mountains and the Oases in Hexi Inland River Basins. *J. Glaciol. Geocryol.* **2008**, *30*, 907–913, (In Chinese with English Abstract).
30. Wang, L.; Chen, R.S.; Han, C.T.; Wang, X.Q.; Liu, G.H.; Song, Y.X.; Yang, Y.; Liu, J.F.; Liu, Z.W.; Liu, X.J.; et al. Change characteristics of precipitation and temperature in the Qilian Mountains and Hexi Oasis, Northwestern China. *Environ. Earth Sci.* **2019**, *78*, 13. [CrossRef]
31. Guan, Y.H.; Zheng, F.L.; Zhang, X.C.; Wang, B. Trends and variability of daily precipitation and extremes during 1960–2012 in the Yangtze River Basin, China. *Int. J. Climatol.* **2017**, *37*, 1282–1298. [CrossRef]
32. Hussain, A.; Cao, J.H.; Ali, S.; Muhammad, S.; Ullah, W.; Hussain, I.; Akhtar, M.; Wu, X.Q.; Guan, Y.H.; Zhou, J.X. Observed trends and variability of seasonal and annual precipitation in Pakistan during 1960-2016. *Int. J. Climatol.* **2022**, *42*, 8313–8332. [CrossRef]
33. Hussain, A.; Cao, J.H.; Hussain, I.; Begum, S.; Akhtar, M.; Wu, X.Q.; Guan, Y.H.; Zhou, J.X. Observed Trends and Variability of Temperature and Precipitation and Their Global Teleconnections in the Upper Indus Basin, Hindukush-Karakoram-Himalaya. *Atmosphere* **2021**, *12*, 973. [CrossRef]
34. Zhang, Q.; Singh, V.P.; Li, J.F.; Chen, X.H. Analysis of the periods of maximum consecutive wet days in China. *J. Geophys. Res.-Atmos.* **2011**, *116*, D23106. [CrossRef]
35. Lang, M.; Ouarda, T.; Bobee, B. Towards operational guidelines for over-threshold modeling. *J. Hydrol.* **1999**, *225*, 103–117. [CrossRef]
36. Sen, P.K. Estimates of the regression coefficient based on Kendall's tau. *J. Am. Stat. Assoc.* **1968**, *63*, 1379–1389. [CrossRef]
37. Mann, H.B. Nonparametric test against trend. *Econometrica* **1945**, *13*, 245–259. [CrossRef]

38. Kendall, M.G. Rank Correlation Methods. *Brit. J. Psychol.* **1975**, *25*, 86–91. [CrossRef]
39. Sang, Y.F.; Wang, Z.; Liu, C. Comparison of the MK test and EMD method for trend identification in hydrological time series. *J. Hydrol.* **2014**, *510*, 293–298. [CrossRef]
40. Yang, Y.; Chen, R.S.; Liu, G.H.; Liu, Z.W.; Wang, X.Q. Trends and variability in snowmelt in China under climate change. *Hydrol. Earth Syst. Sci.* **2022**, *26*, 305–329. [CrossRef]
41. Gocic, M.; Trajkovic, S. Analysis of changes in meteorological variables using Mann-Kendall and Sen's slope estimator statistical tests in Serbia. *Glob. Planet. Chang.* **2013**, *100*, 172–182. [CrossRef]
42. Li, Z.L.; Xu, Z.X.; Li, J.Y.; Li, Z.J. Shift trend and step changes for runoff time series in the Shiyang River basin, northwest China. *Hydro. Process* **2008**, *22*, 4639–4646. [CrossRef]
43. Mosmann, V.; Castro, A.; Fraile, R.; Dessens, J.; Sanchez, J.L. Detection of statistically significant trends in the summer precipitation of mainland Spain. *Atmos. Res.* **2004**, *70*, 43–53. [CrossRef]
44. Bari, S.H.; Rahman, M.T.U.; Hoque, M.A.; Hussain, M.M. Analysis of seasonal and annual rainfall trends in the northern region of Bangladesh. *Atmos. Res.* **2016**, *176*, 148–158. [CrossRef]
45. Wang, Y.; Qin, D. Influence of Climate Change and Human Activity on Water Resources in Arid Region of Northwest China: An Overview. *Clim. Chang. Res.* **2017**, *13*, 483–493, (In Chinese with English Abstract). [CrossRef]
46. Cao, B.; Pan, B.; Wang, J.; Shangguan, D.; Wen, Z.; Qi, W.; Cui, H.; Lu, Y. Changes in the glacier extent and surface elevation along the Ningchan and Shuiguan river source, eastern Qilian Mountains, China. *Quatern. Res.* **2014**, *81*, 531–537. [CrossRef]
47. Tian, H.; Yang, T.; Liu, Q. Climate change and glacier area shrinkage in the Qilian mountains, China, from 1956 to 2010. *Ann. Glaciol.* **2014**, *55*, 187–197. [CrossRef]
48. Zhang, Y.; Liu, S.; Shangguan, D.; Li, J.; Zhao, J. Thinning and shrinkage of Laohugou No. 12 glacier in the Western Qilian Mountains, China, from 1957 to 2007. *J. Mt. Sci.* **2012**, *9*, 343–350. [CrossRef]
49. Wang, S.; Zhang, L.; She, D.; Wang, G.; Zhang, Q. Future projections of flooding characteristics in the Lancang-Mekong River Basin under climate change. *J. Hydrol.* **2021**, *602*, 126778. [CrossRef]
50. Reyer, C.P.O.; Otto, I.M.; Adams, S.; Albrecht, T.; Baarsch, F.; Cartsburg, M.; Coumou, D.; Eden, A.; Ludi, E.; Marcus, R.; et al. Climate change impacts in Central Asia and their implications for development. *Reg. Environ. Chang.* **2017**, *17*, 1639–1650. [CrossRef]
51. Chen, R.S.; Han, C.T.; Liu, J.F.; Yang, Y.; Liu, Z.W.; Wang, L.; Kang, E.S. Maximum precipitation altitude on the northern flank of the Qilian Mountains, northwest China. *Hydrol. Res.* **2018**, *49*, 1696–1710. [CrossRef]
52. Wang, P.; Li, Z.Q.; Gao, W.Y. Rapid Shrinking of Glaciers in the Middle Qilian Mountain Region of Northwest China during the Last similar to 50 Years. *J. Earth Sci.* **2011**, *22*, 539–548. [CrossRef]
53. Sun, M.P.; Liu, S.Y.; Yao, X.J.; Guo, W.Q.; Xu, J.L. Glacier changes in the Qilian Mountains in the past half-century: Based on the revised First and Second Chinese Glacier Inventory. *J. Geogr. Sci.* **2018**, *28*, 206–220. [CrossRef]
54. Liu, G.H.; Chen, R.S.; Li, K.L. Glacial Change and Its Hydrological Response in Three Inland River Basins in the Qilian Mountains, Western China. *Water* **2021**, *13*, 2213. [CrossRef]

Article

Climate Change, Land Use, and Vegetation Evolution in the Upper Huai River Basin

Abel Girma [1,2,*], Denghua Yan [3,4,*], Kun Wang [3,4], Hailu Birara [2], Mohammed Gedefaw [2], Dorjsuren Batsuren [1,5], Asaminew Abiyu [1], Tianlin Qin [3], Temesgen Mekonen [2] and Amanuel Abate [2]

[1] College of Environmental Science and Engineering, Donghua University, Shanghai 201620, China
[2] Department of Natural Resource Management, University of Gondar, Gondar P.O. Box 196, Ethiopia
[3] State Key Laboratory of Simulation and Regulation of Water Cycle in River Basin, China Institute of Water Resources and Hydropower Research (IWHR), Beijing 100038, China
[4] Water Resources Department, China Institute of Water Resources and Hydropower Research (IWHR), Beijing 100038, China
[5] Department of Environment and Forest Engineering, School of Engineering and Applied Sciences, National University of Mongolia, Ulaanbaatar 210646, Mongolia
* Correspondence: abelethiopia@yahoo.com (A.G.); yandh@iwhr.com (D.Y.); Tel.: +251-969146028 (A.G.)

Abstract: Land-use/land-cover change and climate change have changed the spatial–temporal distribution of water resources. The Huai River Basin shows the spatial and temporal changes of climate from 1960 to 2016 and land-use/land-cover changes from 1995 to 2014. Thus, this study aims to investigate climate change, land use, and vegetation evolution in the Upper Huai River Basin. The Mann–Kendall test (MK), Innovative Trend Analysis Method (ITAM), and Sen's slope estimator test were used to detect climate change trends. The land-use/land-cover change was also examined using a transformation matrix and Normalized Difference Vegetation Index (NDVI). The results of this study revealed that precipitation has shown a slightly decreasing trend during the past 56 years. However, the air temperature has increased by 1.2 °C. The artificial and natural vegetation and wetland were decreased by 12,097 km^2, 3207 km^2, and 641 km^2, respectively. On the other hand, resident construction land and artificial water bodies increased by 2277 km^2 and 3691 km^2, respectively. This indicates that the land cover has significantly changed during the past 30 years. The findings of this study will have implications for predicting the water resources safety and eco-environment of The Huai River Basin. The spatial distribution showed an uneven change in the Huai River Basin. Together, we suggested that the variability of water resources availability in the Huai River Basin was mainly attributed to climate variability, while land use change plays a key role in the sub-basins, which experienced dramatic changes in land use.

Keywords: climate change; land cover/land use; vegetation; NDVI; Huai River Basin

Citation: Girma, A.; Yan, D.; Wang, K.; Birara, H.; Gedefaw, M.; Batsuren, D.; Abiyu, A.; Qin, T.; Mekonen, T.; Abate, A. Climate Change, Land Use, and Vegetation Evolution in the Upper Huai River Basin. *Atmosphere* **2023**, *14*, 512. https://doi.org/10.3390/atmos14030512

Academic Editor: Haibo Liu

Received: 4 February 2023
Revised: 27 February 2023
Accepted: 28 February 2023
Published: 7 March 2023

1. Introduction

Climate change has a significant influence on vegetation and water resources in river basins across the globe [1–3]. The production of large amounts of greenhouse gases will exacerbate global temperature rise [4]. Land-use/land-cover (LULC) change is linked with climate change because of its diverse environmental impacts [5]. It is also an important driving force of environmental changes across all spatial and temporal scales. LULC change contributes significantly to earth atmosphere interactions, forest fragmentations, and loss of biodiversity [6]. It is also one of the factors contributing to climate change. However, the disorderly expansion of urban construction land and the massive loss of ecological land has restricted the sustainable development of the overall ecological environment [4].

In the past decades, water resource shortages have become increasingly severe in many parts of the world, associated with the impacts of climate variability and population expansion. To study the impacts of climate change, the Haui River Basin is a suitable area.

Climate variables include precipitation, sunshine hours, air temperature, cloudiness, pressure, rainfall/snow, and wind velocity humidity. These metrological variables interrelate directly or indirectly and significantly influence the environment and living entities [6–8]. Land surface temperature is an imperative ecological factor, and its warming tendency affects the topsoil [9].

The rate of biochemical and plant growth is significantly influenced by soil temperature [6]. Temperature and precipitation influences other variables that have a direct impact on vegetation growth during the concurrent year [10]. Climate change can greatly affect land cover by altering vegetation [5–9]. Changes in vegetation cover alter the strength of the sink and may turn it into a source of green house gasses as a result of land processses such as deforestation [11]. The most used vegetation quantity index derived from remote-sensing resources is the normalized difference vegetation index (NDVI) [11]. The NDVI is a compelling index for monitoring vegetation, soil, and water [1,12].

Land-use/land-cover change (LUCC) is the major content of research on world climate and environmental change [13]. It has a huge impact on resources, environment, biodiversity, artificial water environment changes, and economic and social development [14]. Land-use change serves as the ideal starting point for coupled systems about human–environment research in land systems, and the core content of the LUCC dynamic change process has gradually become the focus of scholars. In the past, land-use investigation has merely recognized various land-use scales and structures, however, this method is unable to detect the evolution of the different vegetation types. Lousia employed the TM imagery to analyze land-use/land-cover changes between 1990 to 2004 in Manica province, Mozambique. Ma Xiaoxue et al. [9] explicitly analyzed the characteristics and driving mechanism of land use of the Qinhuai river basin; they proposed the main driving mechanisms of land use in this basin were climate change, economic development, population growth, and the improvement of people's living standards [15]. Wang Xiulan and Zhang Jin et al. organized relevant methods to research the impacts of change of different land use/cover; they presented the meaning and significance of these models of land resources quantity change, land-use degree change, and land demand forecast in the study of land-use change [16–18].

Changes in land cover may occur as a result of the shift in vegetation cover. The determination of land-use and land-cover change is a potent tool for comprehending and evaluating the crucial connection between socioeconomic activities and natural processes [19]. Specifically, it is essential to examine the changes in the different land use/land cover in naturally sensitive areas [20]. The arid and semi-arid regions of central Asia are the most sensitive areas to climate change [21–23]. The Huai River Basin could be the largest demonstrative of these regions [24,25]. Climate change has a significant impact on the Huai River, with water quality being the most sensitive parameter [26]. In recent decades, water resources shortages have become increasingly severe in many parts of the world, associated with the impacts of climate variability and population expansion. Additionally, changes in hydrological cycles may attribute to the change in climate and land uses [27].

The Huai River Basin is spatially all-embracing, with prominent environmental gradients principally determined by temperature and precipitation on wide-ranging scales. Hence, the Huai River Basin is an ideal place to observe the changing aspects of landscape structures and biological communities [21]. Changes in the hydroclimate can also alter lake conditions, land cover, and vegetation cover [22–29]. The main aim of the study is to investigate spatiotemporal trends of climate changes during 1960 to 2016 and examine the vegetation change as well as land-use/land-cover changes during 1985 to 2014 in the Huai River Basin. The main objectives of this are (i) to identify historical climatic trends with land use change, (ii) to determine the extent to which observed trends in land-use/cover change can be reproduced, and (iii) to determine the extent to which trends in vegetation cover change in the region under study.

2. Study Area

The study basin is found between 111°56′ E~109°15′ E and 30°57′ N~37°50′ N with area coverage of 207 × 105 km [24]. The mean annual precipitation and temperature of the basin is 883 mm and 11~16 °C, respectively. The average annual surface evaporation is between 600~1500 mm [23]. The basin is known to experience water shortages (Figure 1).

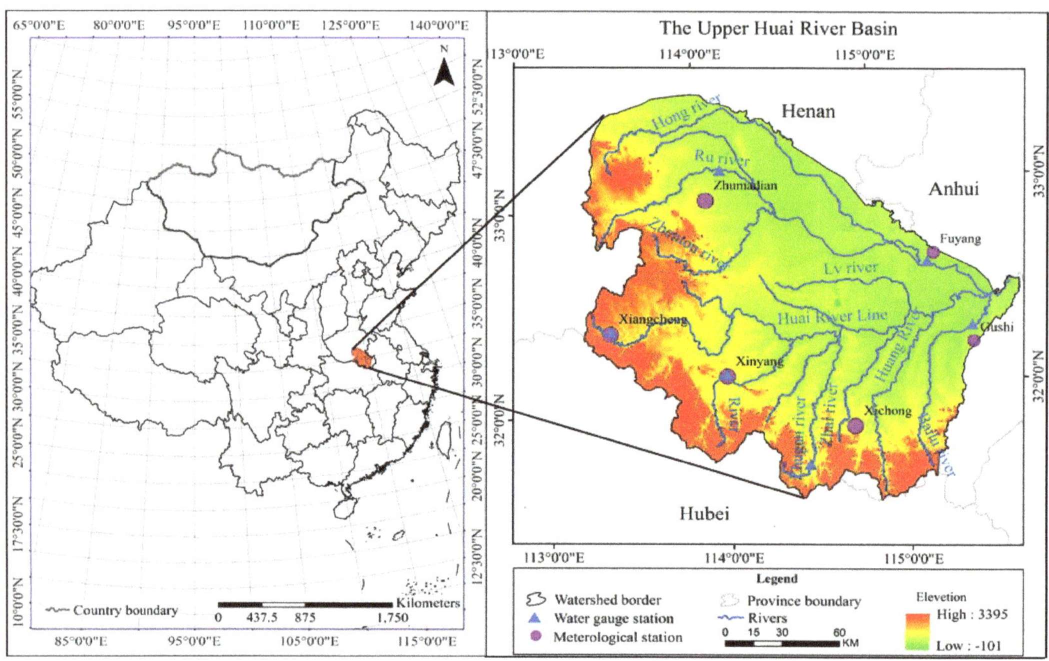

Figure 1. Location map of the Huai River Basin and drainage.

3. Data Source

3.1. Meteorological Data

Climate data were gathered for this study from six distinct representatives meteorological and water gauge stations located throughout the Huai River Basin (Table 1). The duration of the study was from 1960 to 2016. The basic information of the representative stations are presented in (Table 1).

Table 1. List of representative Meteorological stations' information.

Stations	Altitude (m)	Latitude (N)	Longitude (E)	Annual Mean Precipitation (mm)	Annual Mean Temperature (°C)
Xiangcheng	149.1	32.383333	113.416667	1124.69	15.37
Zhumadian	82.7	33.533333	114.016667	953.42	15.12
Gushi	42.9	32.163333	115.616667″	1064.62	15.64
Fuyang	60.5	31.733333	116.516667	910.07	15.34
Xinyang	68.1	31.413333	116.316667″	839.09	15.52
Xichong	71.5	31.563333	114.116667	1089.32	15.77

3.2. Data on Land Cover

For validation, the topographic map, the soil map, the ecological landscape potential map, the forest map, and the vegetation map were selected. From 1982 to 2014, Landsat Thematic Mapper (TM) and Landsat Enhanced Thematic Mapper (ETM+) maps with a 30 m resolution were used to classify land cover. The National Geomatics Center of China in 2014 described the Global Land covers dataset (GlobeLand30) product, from which the study area's satellite data for land covers were derived. The name, code, and definition of land cover classification are presented in Table 2. Over the past 30 years, the land cover spatial pattern data were interpreted using all Landsat TM and ETM + images downloaded from USGS (http://landsat.usgs.gov/ accessed on 1 September 2019).

Table 2. Gray scale value and vegetation coverage equivalent distribution table.

Rating	Gray Scale Value Division	The Percentage of Vegetation Cover	Land-Use/Landcover Type	Vegetation Coverage Evaluation
Level 1	191~255	>60%	Forest land, dense shrub land, and shrub land.	Excellent
Level 2	156~190	30%~60%	Potential degraded land, good farmland, high coverage grassland, and forest land.	Good
Level 3	139~155	15%~30%	Grassland, the middle plain has fixed sand, and beach.	Medium
Level 4	128~138	5%~15%	Forest land, desert grassland, and scattered vegetation.	Subalternation
Level 5	below 128	below 5%	Artificial water areas, desert, residential areas, etc.	Inferior

3.3. Vegetation Data

The MODIS NDVI product (MOD13Q1), which was obtained from NASA's land processes distributed active archive center, served as the source for the normalized difference vegetation index (NDVI) data. The MODIS surface reflectance values from the red band (610–680 nm) and the near-infrared band (780–890 nm) was used to calculate the NDVI after they were adjusted for molecular scattering, ozone absorption, and aerosols [30]. The MOD13 Q1 products, which include 12 scientific data sets with a 16-day temporal sampling period and a spatial resolution of 250 m × 250 m, were derived from the most recent version.

4. Methods

4.1. Climate Variable Analysis

4.1.1. Innovative Trend Analysis Method (ITAM)

ITAM divides dataseries into two equivalent sub-series and categorizes them in increasing order [6,15,31–34]. The X-axis (x_i:i = 1, 2, 3, . . . , n/2) and the Y-axis (x_j:j = n/2 +1, n/2 + 2, . . . , n) were then used to arrange the two sections on a coordinate system. There is no trend if time-series data are collected on a scattered plot along a 1:1 (45°) straight line. When data points accumulate above the 1:1 straight line, the trend is upward, and when data points accumulate below it, the trend is downward. The difference in mean values between x I and x j might reflect how far a data series is trending (Figure 2).

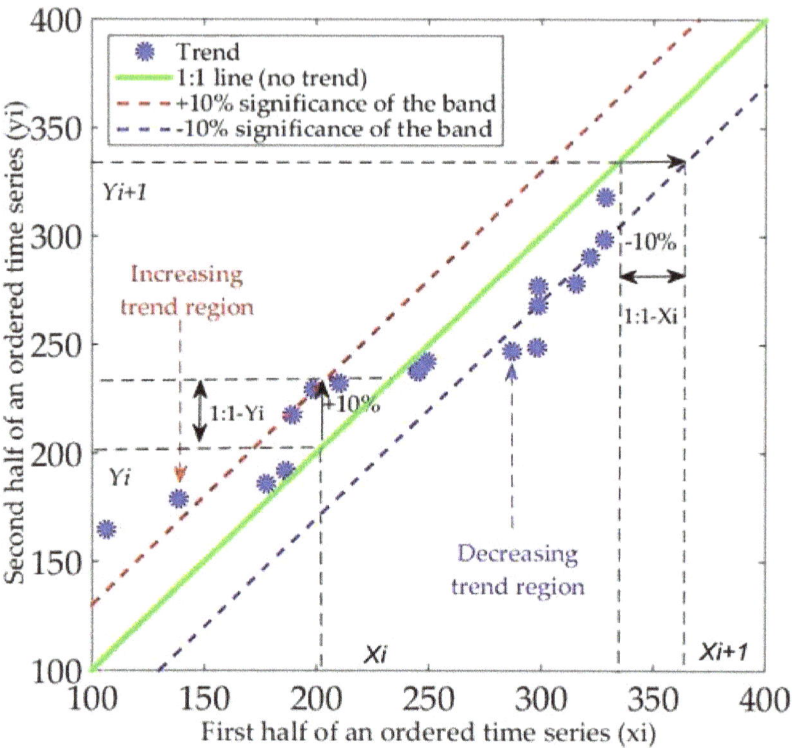

Figure 2. Paradigmatic illustrations of the ITA method.

4.1.2. Mann–Kendall (M–K) Test Method

The Mann–Kendall (M–K) test is used to detect the trends of time series data [2,35]. The test statistics "S" is equated as follows:

$$S = \sum_{i=1}^{n-1} \sum_{j=i+1}^{n} sgn\left(x_j - x_i\right) \tag{1}$$

$$sgn(x_j - x_i) = \begin{cases} +1\ if\ \left(x_j - x_i\right) > 0 \\ 0\ if\ \left(x_j - x_i\right) = 0 \\ -1\ if\ \left(x_j - x_i\right) < 0 \end{cases} \tag{2}$$

where x_j and x_i represents the data points in periods j and i. While the amount of data series is larger than or equivalent to 10 ($n \geq 10$), the M–K test is then categorized by a standard distribution with the mean $E(S) = 0$ and variance $Var(S)$, which is given as follows [35]:

$$E(S) = 0 \tag{3}$$

$$Var(S) = \frac{n(n-1)(2n+5) - \sum_{k=1}^{m} t_k(t_k - 1)(2t_k + 5)}{18} \tag{4}$$

where m is the number of the tied groups in the time series, and t_k is the number of ties in the kth tied group. From this the test, the Z statistics is obtained using approximation as follows:

$$Z = \begin{cases} \frac{s-1}{\delta} & if\ S > 0 \\ 0, & if\ S = 0 \\ \frac{s+1}{\delta} & if\ S < 0 \end{cases} \tag{5}$$

4.1.3. Sen's Slope Estimator Test

The slope Q_i between two data points is given by the equation [2,35].

$$Q_i = \frac{x_j - x_k}{j - k}, \; for \; i = 1, 2, \ldots N \tag{6}$$

where x_j and x_k are data points at time j and ($j > k$), respectively.

When there is only single datum in each time, then $N = \frac{n(n-1)}{2}$; n are a number of time periods. However, if there are many data in each year, then $N < \frac{n(n-1)}{2}$; n total number of observations [36]. The N values of slope estimator are arranged from smallest to biggest. Then, the median of slope (β) is computed as follows:

$$\beta = \begin{cases} Q[(N+1)/2] & \text{when } N \text{ is odd} \\ Q[(N/2) + Q(N+2)/(2)/(2)] & \text{when } N \text{ is even} \end{cases} \tag{7}$$

4.2. Analysis of Land Use and Land Cover

The overlaying operation was used to perform spatial analysis, which revealed how land use and land cover changed over time and established a connection between the two. A land cover transformation map was obtained and utilized for transformation matrix analysis by intersecting the two land cover/land use maps (1985 and 2014). The magnitudes of land cover shifts were calculated as [12].

$$CA = TA(t_2) - TA(t_1), \tag{8}$$

$$CE = [CA/TA(t_1)] * 100, \tag{9}$$

where: t_1 and t_2 represent the beginning and end, while TA, CA, and CE represent the total area, changed area, and extent of change, respectively. The Kappa coefficient (Kappa) was also calculated. The agreement between user-assigned ratings and the predefined producer ratings is measured by kappa. The difference between the amount of agreement that is present (the "observed" agreement) and the amount of agreement that would be expected to be present by chance alone (the "expected" agreement) is the basis of the calculation [37].

$$K = P(A) - P(E)/1 - P(E) \tag{10}$$

$$P(A) = \frac{(A + D)}{N}, \tag{11}$$

$$P(E) = \left(\frac{A1}{N}\right) * \left(\frac{B1}{N}\right) + \left(\frac{A2}{N}\right) * \left(\frac{B2}{N}\right), \tag{12}$$

where K is the Kappa coefficient, $P(A)$ is the number of times the K raters agree, and is the number of times the K rates are expected to agree only by chance, A and D are unchanged categories, $A1$ and $B1$ are subject's categories, and N is the change of results. ArcGIS 10.2 and the land-use transfer matrix were used to investigate a comprehensive land-use dynamics degree [38] of the Huai River Basin.

$$LC = \frac{\sum_{i=1}^{n} VLU_{i-j}}{\sum_{i=1}^{n} LU_i \, T} \times 100\% \tag{13}$$

where LC is the total land-use dynamic degree, Lui is the type of area, Lui−j is the total value of an area in land-use type I to J, and T is the amount of time that has passed since the last monitoring. LC is the land-use dynamic change rate when T is set for years.

We can generate five groups of transfer matrixes for various land-use types using five years of land-use data—1985–1990, 1990–2000, 2000–2005, 2005–2014, and 1985–2014—to comprehend the basin's transfer situation.

4.3. Vegetation Analysis

One of the most widely used vegetation indices of plant biomass and activity is the normalized difference vegetation index (NDVI) which is used to denote a shift in an area's vegetative greenness [1,39].

$$NDVI = \frac{(NIR - RED)}{(NIR + RED)},$$

(14)

where the electromagnetic spectrum's near-infrared (NIR) and red (RED) channels correspond to bands 2 and 1 of the MODIS (MOD13Q1) product, respectively. We used a linear regression model based on MODIS data to spread GIMMS data until 2014 due to the different spatial resolutions of GIMMS and MODIS. In addition, we utilized the same time data to check outspread data. Afterward, the NDVI average correlation coefficient reached greater than 90% in the Huai River Basin. The land-use type and the NDVI gray-scale value were additionally examined (Table 1) to determine vegetation coverage evolution tendency.

5. Results and Discussions

5.1. Trends of Observed Climate Changes

In the study region, positive and negative tendencies are present by the M–K test estimator, and annual mean precipitation, and display temporal variations (Figure 3). This is in good agreement with studies and annotations from various parts of China [8,24]. Since precipitation tends to decrease across the study basin, the anticipated decrease in precipitation determination will probably result in a decrease in water accessibility in the years to come [8].

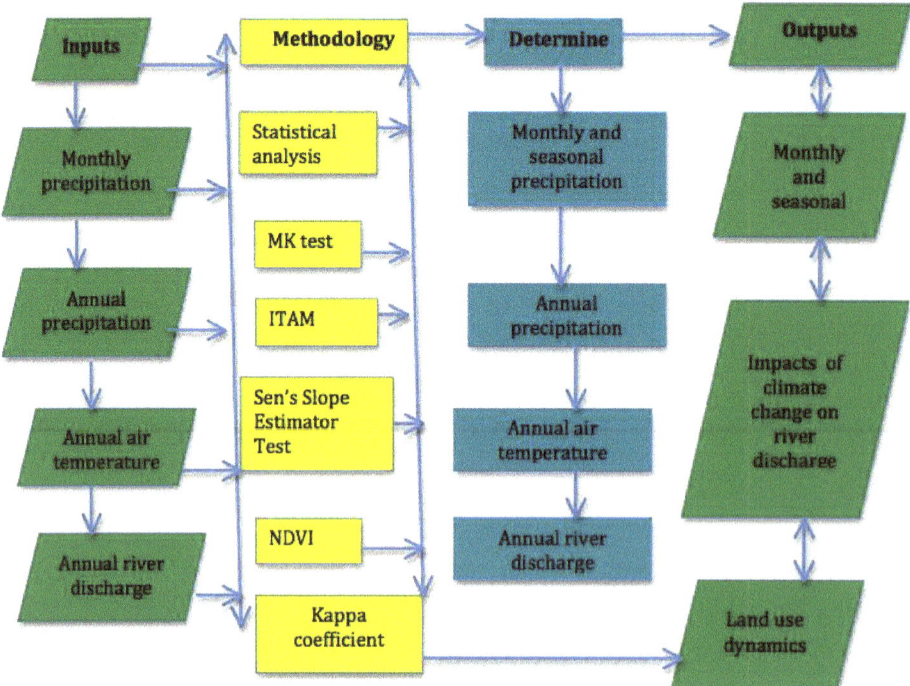

Figure 3. Spatially distribution land-use/land-cover map of the Huaihe River Basin.

The hydrological series and the supply of water resources for ecological units and society can be disrupted by a decrease in precipitation during the wet season [40]. The summer is the most precipitous time of year in the study area, contributing nearly 49.3% of the total precipitation, indicating a high intensity of precipitation and snowfall. Additionally, stream flow is at its best during this time of year. The squat rainy season, which begins in December and lasts through February (winter), accounts for approximately 7.3% of the total precipitation. Winter is more susceptible to the occurrence of persistent drought events, as revealed by the consequences of light precipitation concentration [30]. In China, a variety of trend exploration studies have been piloted at various spatiotemporal measures, yielding a wide range of results with various trend test parameters. Although they discovered a statistically significant upward trend in temperature, the situation regarding precipitation varied.

The rise in temperature is amongst the indices of global climate transformation. Globally, mean air temperature has increased by 0.85 °C from 1880 onwards, which is expected to move along in the near future [4,5,7,40]. The global large inland water bodies temperature has been promptly heating ever since 1980, at the rate of 0.05 ± 0.012 °C/year and by the maximum rate of 0.1 ± 0.011 °C/year [41]. An abrupt upward trend of average annual air temperature was detected in the upper reaches of the Huai River Basin via 1.2 °C or 0.021 °C/year during the deliberated chronological period from 1960 to 2016 (Figure 4b). It is nearly twice as fast as the global average heating rate (0.012 °C/year) [42]. The basin's average annual temperature was found to be 15.5 °C. From 1990 onwards, a significant rise in temperature was observed.

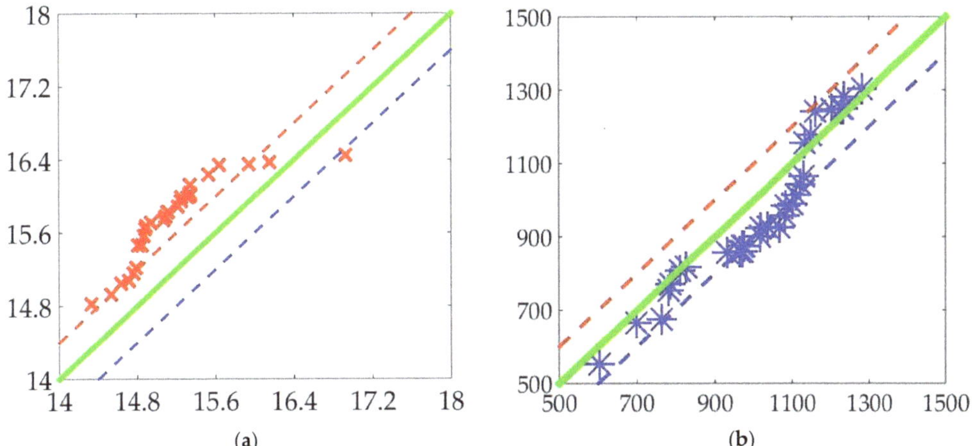

Figure 4. (a) Trends of annual temperature. (b) Trends of annual precipitation.

The cumulative upward trend of air temperature could be the outcome of global warming, which is due to the "built-up heat island", greenhouse influence, and long-standing climate inconsistency [6–9]. In the present study, the temperature during the summer season was higher. Similarly, Gu et al. [43]. found a higher magnitude of air temperature with an increasing trend in the summer seasons, which was superior to the other periods (Figure 5).

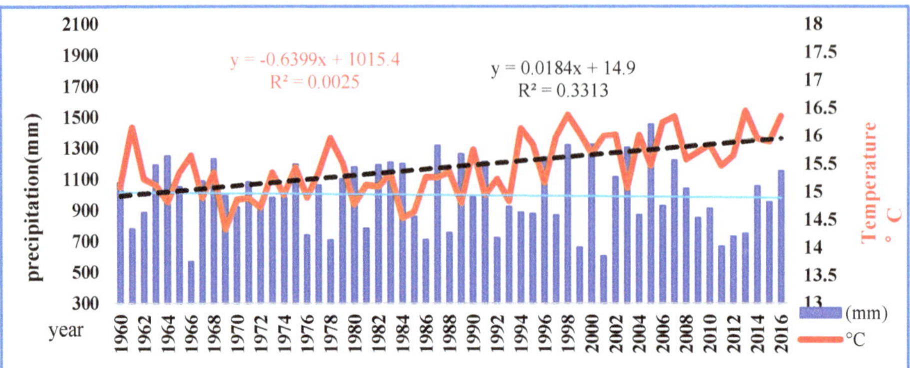

Figure 5. Mean annual precipitation and temperature of the Huai River Basin (1960–2016).

The M–K trend detection test result discovered that the annual mean temperatures have been considerably increasing throughout the period. The general increment in mean annual temperature in the study region then is principally indorsed to a rise in the minimum temperature. Increased temperatures in winter and spring will affect the precipitation phase and, as a consequence, the snow/precipitation ratio and the volume of water stored in snow cover will be changed. Therefore, the hydrology of rivers in the Northern Hemisphere is sensitive to climate change [44].

5.2. Change Detection and Classification of Land Cover

We may be able to comprehend the synoptic temporal change in the study area's land cover types through the use of satellite data, which can make concurrent, synoptic, and repetitive annotations [45].

A maximum-likelihood classifier and the classification scheme in Table 2, have made five classification maps of the Huai River Basin. With substantial precision, the study region landscape was classified into 65 land-use classes (Figure 6).

In the present study, the artificial and natural vegetation, as well as the wetland, showed a decreasing trend of 12,097 km^2, 3207 km^2, and 641 km^2, respectively, during the period 1985–2014. Meanwhile, the area of resident construction land and artificial water showed an increasing trend by 2277 km^2 and 3691 km^2, respectively. However, the other areas only displayed a small change (Figure 7).

It should be noted that the artificial water bodies experienced the majority of the area's greatest changes and urban land during the past 30 years. The dynamic degree was 9.2% and 6.4%, respectively. The area for wetland artificial vegetation and natural vegetation was diminished, and the shifting ratio was inferior. From the comprehensive land-use dynamic degree of the whole study years, the degree of intervention to land use by human activities was prevalent from 2005–2014. The dynamic degree is approximately 0.5% from 1985–2014, which implies that about 4.5% of the land use type has been transformed throughout this period every year (Table 3). In the course of the early 21st century, the population of the Huai River Basin had a high rate of growth; throughout this phase, the per capita possession of land resources was small, which caused a shift in land use to some extent (Figure 7).

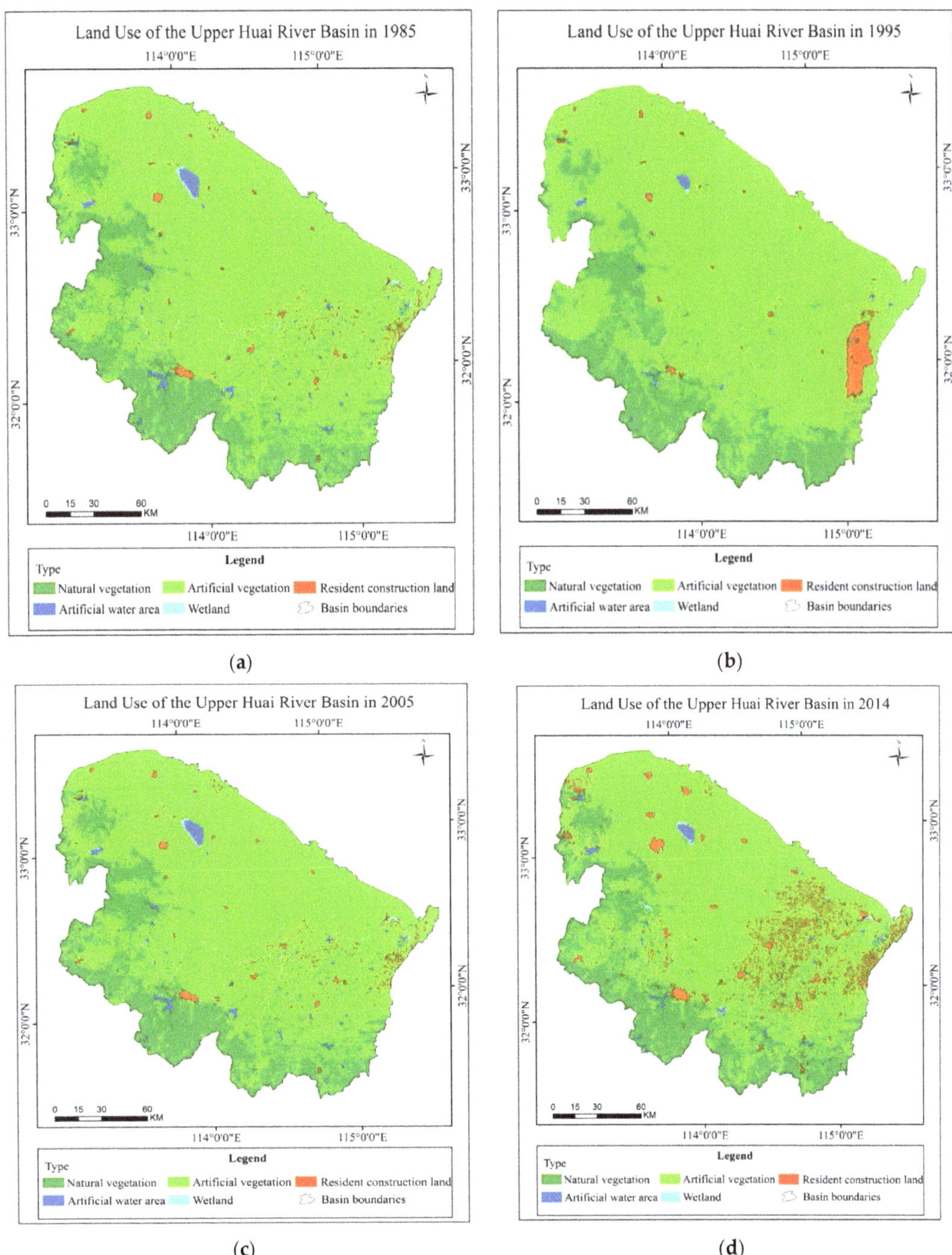

Figure 6. Study area map of land-use type by spatial transformation variation.

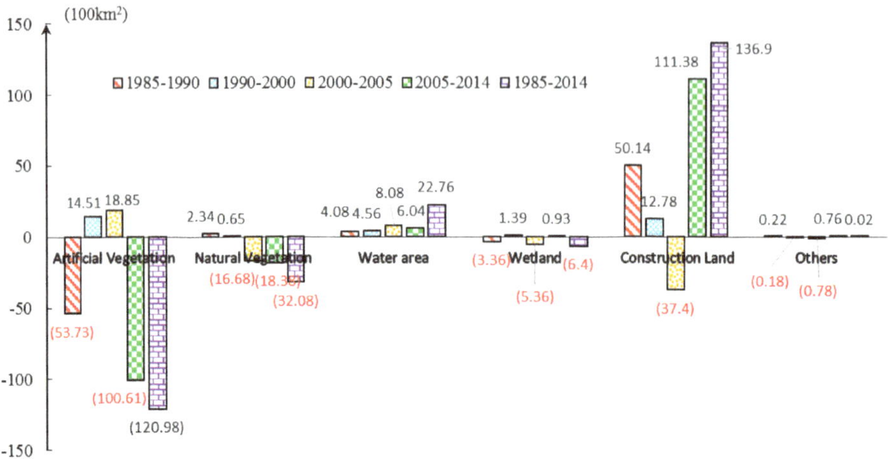

Figure 7. Land-use/land-cover transformations from 1985 to 2014.

Table 3. Land-use/land-cover dynamics of the Huai River Basin from 1985 to 2014 (%).

Land-Use/Cover Type	1985–1990	1990–2000	2000–2005	2005–2014	1985–2014
Artificial Vegetation (AV)	−0.56	0.16	0.21	−1.04	−1.32
Natural Vegetation (NV)	0.16	0.05	−1.09	−1.28	−2.38
Artificial water area (AW)	3.03	2.94	4.54	2.78	9.15
Wetland (W)	−0.71	0.31	−1.15	0.23	−1.44
Resident construction land (RCL)	3.45	0.76	−2.13	7.04	6.41
Others (O)	2.18	−1.57	−7.47	11.55	0.21
(LC) Comprehensive land-use dynamic degree	0.44	0.14	0.33	0.57	0.46

RCL and AW bodies are formed from more than one third of artificial vegetation. More than 30% of Nevada is converted into additional land uses, such as empty land and sand. The leading causes for the alterations are hasty population growth, urbanization, ecological deterioration, and high intensity of socio-economic activities in the study area [46].

Natural factors are yet another significant cause of land-cover shift. The components of the land cover, for instance, may be impacted directly or indirectly by the effects of climate change. The overall pattern of regional vegetation was characterized by the shifting of various forest types, whereas the distribution of forest boundaries may not have changed as clearly [25]. When the climate changes rapidly, fast-growing pioneer communities expand rapidly and require a significant amount of migration distance. As a result, it is essential to define the size of changes in land cover (Table 4).

Table 4. The Huai River Basin's land-use and land-cover transformation matrix from 1985 to 2014 (in percent).

Land Use Type	AV	NV	AW	W	RCL	Others
Natural Vegetation	7.44	91.82	0.208	0.264	0.202	0.065
Artificial Vegetation	93.541	2.269	0.121	0.447	3.588	0.034
Artificial water area	31.086	6.071	42.135	15.643	5.03	0.035
Wetland	8.222	1.434	2.195	87.362	0.775	0.012
Resident construction land	45.984	1.682	0.299	0.305	51.668	0.062
Others	11.843	36.978	0.055	0.079	4.709	46.337

5.3. NDVI

MODIS vegetation indices produced on 16-day intervals and at multiple spatial resolutions provide consistent spatial and temporal comparisons of vegetation canopy greenness, a composite property of leaf area, chlorophyll, and canopy structure. Two vegetation indices were derived from atmospherically corrected reflectance in the red, near-infrared, and blue wave bands—the normalized difference vegetation index (NDVI). The strength of global NDVI data is their high temporal information content [1,47]. The common compositing time of 8–14 days provides at least 25–30 global NDVI datasets per year [2]. This ensures consistency with the historical and climate applications of NOAA's AVHRR NDVI time series record, reducing variations in the canopy and soil and increasing sensitivity in dense vegetation conditions. The two products better describe the global range of states and processes of vegetation [47]. As a result, the NDVI was used to determine the vegetation cover (Figure 8).

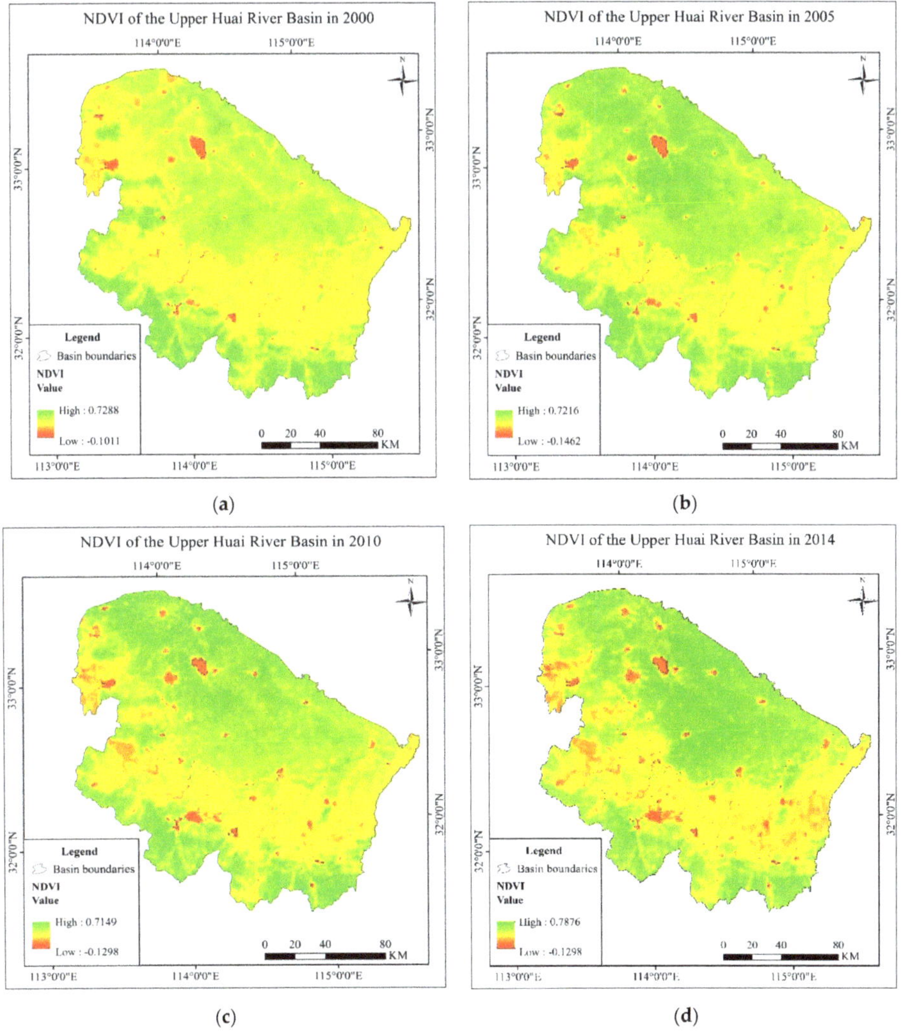

Figure 8. NDVI—the spatial distribution map of the Huai River Basin vegetation.

The pixel's NDVI values ranged from -1 to 1. The highest NDVI values were greater than 0.6, indicating the healthiest or richest vegetation. The normalized difference vegetation index had a significant difference in the study region. This is due to the prevalence of sufficient precipitation during the period from 1985 onwards. Contrarily, during 2000, the vegetation coverage had significantly declined, and the percentage of preeminent coverage area accounted for only about 8.7% of the study region. This was principally because the total mean annual precipitation significantly declined [44]. For the period from 2005–2014, the area of preeminent vegetation coverage declined by 30%; on the other hand, the area of lower vegetation coverage NDVI < 0.4 increased by about 75%. The major reason for the vegetation boost was the prevalence of adequate precipitation throughout the summer season [44], and to some extent, harsh climatic circumstances were reduced during this period that generated advantageous situations for vegetation growth. On the contrary, urbanization leads to a scattered configuration of vegetation coverage in the study region.

Temperature and precipitation may influence the dynamics of vegetation in the Huai River Basin. Changes in the global carbon and hydrology cycle, as well as feedback on climate change, could be caused by vegetation dynamics [25]. One of the factors that alter the ecosystem, water system circulation, and surface water supply balance is the cover of vegetation. There may be a number of factors connected to the change in the Huai River's vegetation cover. This indicates a significant degree of climate change throughout the basin due to the coincidence of overlaps with climatic variations of the current weather. It is obvious that cultivated land areas are growing bigger and areas of vegetation are becoming smaller. This suggests that human activity may have altered the river basin's land cover.

5.4. Implication of Climate Change over Land Cover/Use of Huai River Basin

In the Huai Basin, precipitation has decreased significantly over the past 56 years. The basin's temperature has varied by approximately 1.2 °C over the same period (Figure 9). The most changed parts of the land covers in the Huai River Basin are the artificial vegetation, natural vegetation, and wetland, which showed a decrease of 12,098, 3028, and 640 km^2, respectively. This could be due to the increasing temperature and an overall decline pattern of precipitation in the study basin. On the other hand, the rest of the other land transformation type is related to human activities.

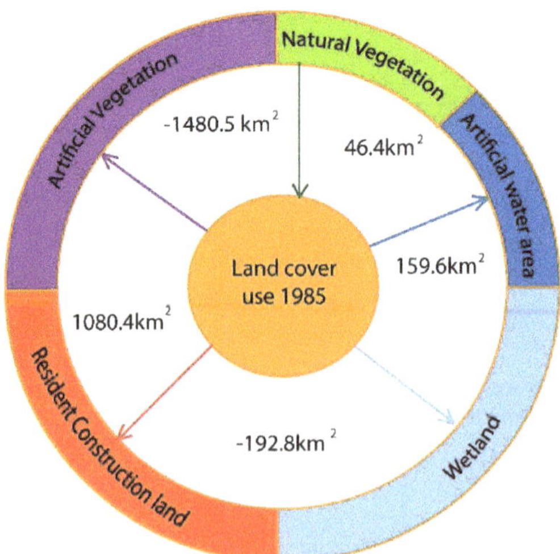

Figure 9. The changes in land cover from 1985 to 2014 are depicted in the diagram. Gains and losses are indicated by arrows. Herein: Pink is a result of human activity, red is a result of temperature rise, yellow is a result of humans and nature, and blue is a result of other natural factors.

In this region, climate change may alter the land cover [48]. Large mountains and river basins have different levels of vegetation cover. This demonstrates that the mountain and river valley vegetation cover changed between 1985 and 2014 [49]. Additionally, the 2007 average annual temperature overlap was the highest (Figure 10). This is a result of shifting land cover as a result of intense human activity and global warming. Along the river valley, there is also a large amount of agricultural land. In addition, extensive human activity in the region has altered the vegetation cover of the Huai River Basin (Figure 10). The vegetation cover in the river basin will change if there are many people working there [49]. Artificial surfaces and cultivated land have increased as a result of intense human activity. Thus, this region's vegetation and land cover may be significantly affected by climate change and human activity.

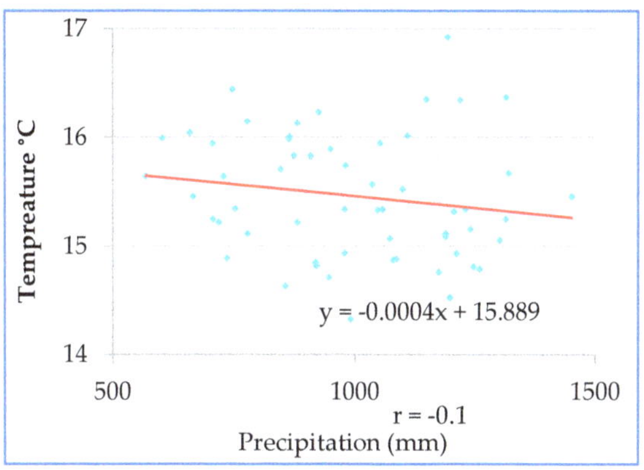

Figure 10. Correlation between precipitation and temperature.

6. Conclusions

We assessed the Huai River Basin's climate changes between 1960 and 2016, land use and land cover, and vegetation between 1985 and 2014 using remote sensing, classification of land use and land cover, and detection of vegetation.

The study area's precipitation slightly changed between 1960 and 2016. On the other hand, the temperature often went up by 1.2 °C, leading to a significant increase over the past 40 years in the semi-arid central Asian region.

Artificial vegetation, natural vegetation, and wetland changed the most in the Huai River Basin with a decreasing trend. Construction land for homes significantly increased. The Huai River Basin's land cover and land use have changed slightly over time, both in quantity and quality.

There was a clear shift in the vegetation cover in the river basin and mountainous regions. Between 1985 and 2014, the vegetation cover and land cover changed significantly. In the study area, these changes coincide with the process of climate change and human activity. Furthermore, the increase in run-off may also have a wider implication for increasing soil erosion and sedimentation. Thus, curving the trends of LULC towards increasing vegetation cover is very important to reduce wet season flow and surface run-off. Therefore, the concerned entities should take appropriates measures for the sustainable management of land as well as water resources.

Scientific research on the causes of land-cover change and its potential effects on the hydrological and ecological systems of the Huai River Basin is essential in the near future. Due to the limitation of data, this study only considers six stations for this analysis. Therefore, detail study should be conducted in the future by considering more sample stations.

Author Contributions: Conceptualization, A.G., D.Y., H.B., K.W., D.B.; M.G and T.Q., methodology, A.G., D.Y., D.B. and M.G., associated software, D.Y., D.B. and M.G., investigation, D.Y., H.B., K.W., T.M., A.A. (Asa-minew Abiyu) and A.A. (Amanuel Abate) data curation, A.G. and D.Y.; Formal analysis, A.G., D.Y., T.M., D.B., T.Q. and A.A., writing original draft preparation, A.G., D.Y., K.W. and M.G., writing review and editing, H.B., T.M. and A.A. All authors have read and agreed to the published version of the manuscript.

Funding: This research was supported by the National Science Fund Project (Grant No. 52130907; 52109043) and the Five Major Excellent Talent Programs of IWHR (WR0199A012021).

Institutional Review Board Statement: Not applicable.

Informed Consent Statement: Not applicable.

Data Availability Statement: Data will be obtained uopon the request of the corresponding authors.

Acknowledgments: The researchers thank the National Key Research and Development Project (Grant No. 2016YFA0601503). The individuals who contributed data to this study are appreciated by the authors. We are also grateful to the China Institute of Water Resources and Hydropower Research (IWHR) for their financial assistance.

Conflicts of Interest: The authors declared no conflict of interest.

References

1. Dastour, H.; Ghaderpour, E.; Zaghloul, M.S.; Farjad, B.; Gupta, A.; Eum, H.; Achari, G.; Hassan, Q.K. Wavelet-based spatiotemporal analyses of climate and vegetation for the Athabasca river basin in Canada. *Int. J. Appl. Earth Obs. Geoinf.* **2022**, *114*, 103044. [CrossRef]
2. Afzal, A.; Bhutto, J.K.; Alrobaian, A.; Kaladgi, A.R.; Khan, S.A. Modelling and Computational Experiment to Obtain Optimized Neural Network for Battery Thermal Management Data. *Energies* **2022**, *14*, 7370. [CrossRef]
3. Gedefaw, M. Assessment of changes in climate extremes of temperature over Ethiopia. *Cogent Eng.* **2023**, *10*, 2178117. [CrossRef]
4. Zhang, M.; Kafy, A.-A.; Ren, B.; Zhang, Y.; Tan, S.; Li, J. Application of the Optimal Parameter Geographic Detector Model in the Identification of Influencing Factors of Ecological Quality in Guangzhou, China. *Land* **2022**, *11*, 1303. [CrossRef]
5. Feng, J.; Yan, D.; Li, C.; Bao, S.; Gao, Y. Assessing the impact of climate variability on potential evapotranspiration during the past 50 years in North China. *J. Food Agric. Environ.* **2013**, *11*, 1025–1031.
6. Conway, D. A water balance model of the Upper Blue Nile in Ethiopia. *Hydrol. Sci. J.* **2009**, *42*, 265–286. [CrossRef]
7. Zhang, M.; Al Kafy, A.; Xiao, P.; Han, S.; Zou, S.; Saha, M.; Zhang, C.; Tan, S. Impact of urban expansion on land surface temperature and carbon emissions using machine learning algorithms in Wuhan, China. *Urban Clim.* **2023**, *47*, 101347. [CrossRef]
8. Guidigan, M.L.G.; Sanou, C.L.; Ragatoa, D.S.; Fafa, C.O.; Mishra, V.N. Assessing Land Use/Land Cover Dynamic and Its Impact in Benin Republic Using Land Change Model and CCI-LC Products. *Earth Syst. Environ.* **2019**, *3*, 127–137. [CrossRef]
9. Yates, B.D.N.; Strzepek, K.M. Modeling the Nile basin under climatic change. *J. Hydrol. Eng.* **1998**, *3*, 98–108. [CrossRef]
10. Liang, L.; Zhao, L.A.; Gong, Y.F. Probability distribution of summer daily precipitation in the Huaihe Basin of China based on the gamma distribution. *Acta Meteorol. Sin.* **2012**, *26*, 72. [CrossRef]
11. Wurbs, R.A.; Muttiah, R.S.; Felden, F. Incorporation of Climate Change in Water Availability Modeling. *J. Hydrol. Eng.* **2006**, *10*, 375–385. [CrossRef]
12. Duveiller, G.; Hooker, J.; Cescatti, A. The mark of vegetation change on Earth's surface energy balance. *Nat. Commun.* **2018**, *9*, 1–12. [CrossRef]
13. Novillo, C.J.; Arrogante-Funes, P.; Romero-Calcerrada, R. Recent NDVI Trends in Mainland Spain: Land-Cover and Phytoclimatic-Type Implications. *ISPRS Int. J. Geo-Information* **2019**, *8*, 43. [CrossRef]
14. Abate, T.; Angassa, A. Conversion of savanna rangelands to bush dominated landscape in Borana, Southern Ethiopia. *Ecol. Process.* **2016**, *5*, 1–18. [CrossRef] [PubMed]
15. Bičík, I.; Himiyama, Y.; Feranec, J.; Kupková, L. *Land Use/Cover Changes in Selected Regions in the World–IX*; IGU Commission on LUCC, Charles University in Prague: Prague, Czech Republic, 2014.
16. Lawler, J.J.; Lewis, D.J.; Nelson, E.; Plantinga, A.J.; Polasky, S.; Withey, J.C.; Helmers, D.P.; Martinuzzi, S.; Pennington, D.; Radeloff, V.C. Projected land-use change impacts on ecosystem services in the United States. *Proc. Natl. Acad. Sci. USA* **2014**, *111*, 7492–7497. [CrossRef]
17. Pozen, D.E. The mosaic theory, national security, and the freedom of information act. *Yale Law J.* **2005**, *115*, 630–681.
18. Defries, R.S.; Rudel., T.; Uriarte, M.; Hansen, M. Deforestation driven by urban population growth and agricultural trade in the twenty-first century. *Nat. Geosci.* **2010**, *3*, 178–181. [CrossRef]
19. Li, G.; Zhang, F.; Jing, Y.; Liu, Y.; Sun, G. Science of the Total Environment Response of evapotranspiration to changes in land use and land cover and climate in China during 2001–2013. *Sci. Total Environ.* **2017**, *596–597*, 256–265. [CrossRef]
20. Getachew. Uc Santa Cruz. Powered by Calif. *Digit. Libr. Univ. Calif.* **2013**, 1–168.

21. Haregeweyn, N.; Tsunekawa, A.; Tsubo, M.; Meshesha, D.; Adgo, E.; Poesen, J.; Schütt, B. Analyzing the hydrologic effects of region-wide land and water development interventions: A case study of the Upper Blue Nile basin. *Reg. Environ. Chang.* **2015**, *16*, 951–966. [CrossRef]

22. Kearney, J. Food consumption trends and drivers. *Philos. Trans. R. Soc. B Biol. Sci.* **2010**, *365*, 2793–2807. [CrossRef] [PubMed]

23. Blumstein, M.; Thompson, J.R. Land-use impacts on the quantity and configuration of ecosystem service provisioning in Massachusetts, USA. *J. Appl. Ecol.* **2015**, *52*, 1009–1019. [CrossRef]

24. Getachew, H.E.; Melesse, A.M. The Impact of Land Use Change on the Hydrology of the Angereb Watershed, Ethiopia. *Int. J. Water Sci.* **2014**, *1*, 1–7. [CrossRef]

25. She, D.X.; Xia, J.; Zhang, Y.Y.; Du, H. The trend analysis and statistical distribution of extreme rainfall events in the Huaihe River basin in the past 50 years. *Acta Geogr. Sin.* **2011**, *66*, 1200.

26. Wang, K.Q.; Zeng, Y.; Xie, Z.Q. Chang trend of temperature and precipitation in Huaihe river basin from 1961–2008. *J. Meteorol. Sci.* **2012**, *32*, 671.

27. Yilma, A.D.; Awulachew, S.B. Characterization and Atlas of the Blue Nile Basin and its Sub basins. *Int. Water Manag. Inst.* **2009**.

28. Tekleab, S.; Mohamed, Y.; Uhlenbrook, S. Hydro-climatic trends in the Abay/Upper Blue Nile basin, Ethiopia. *Phys. Chem. Earth* **2013**, *61–62*, 32–42. [CrossRef]

29. Kim, U.; Kaluarachchi, J.J. Application of parameter estimation and regionalization methodologies to ungauged basins of the Upper Blue Nile River Basin, Ethiopia. *J. Hydrol.* **2008**, *362*, 39–56. [CrossRef]

30. Running, S.W.; Loveland, T.R.; Pierce, L.L.; Nemani, R.; Hunt, E. A remote sensing based vegetation classification logic for global land cover analysis. *Remote. Sens. Environ.* **1995**, *51*, 39–48. [CrossRef]

31. Kim, U.; Kaluarachchi, J.J. Climate Change Impacts on Water Resources in the Upper Blue Nile River Basin, Ethiopia[1]. *JAWRA J. Am. Water Resour. Assoc.* **2009**, *45*, 1361–1378. [CrossRef]

32. Seleshi, Y.; Zanke, U. Recent changes in rainfall and rainy days in Ethiopia. *Int. J. Clim.* **2004**, *24*, 973–983. [CrossRef]

33. Gedefaw, M.; Wang, H.; Yan, D.; Song, X.; Yan, D.; Dong, G.; Wang, J.; Girma, A.; Ali, B.A.; Batsuren, D.; et al. Trend Analysis of Climatic and Hydrological Variables in the Awash River Basin, Ethiopia. *Water* **2018**, *10*, 1554. [CrossRef]

34. Yan, D.; Ludwig, F.; Huang, H.Q.; Werners, S.E. Many-objective robust decision making for water allocation under climate change. *Sci. Total. Environ.* **2017**, *607–608*, 294–303. [CrossRef]

35. Gedefaw, M.; Yan, D.; Wang, H.; Qin, T.; Girma, A.; Abiyu, A.; Batsuren, D. Innovative Trend Analysis of Annual and Seasonal Rainfall Variability in Amhara Regional State, Ethiopia. *Atmosphere* **2018**, *9*, 326. [CrossRef]

36. Butt, A.; Shabbir, R.; Ahmad, S.S.; Aziz, N. Land use change mapping and analysis using Remote Sensing and GIS: A case study of Simly watershed, Islamabad, Pakistan. *Egypt. J. Remote. Sens. Space Sci.* **2015**, *18*, 251–259. [CrossRef]

37. Schmidt, E.; Kedir, M. Urbanization and Spatial Connectivity in Ethiopia: Urban Growth Analysis Using GIS. ESSP II Work 2009, p. 3.

38. Goward, S.N.; Markham, B.; Dye, D.G.; Dulaney, W.; Yang, J. Normalized difference vegetation index measurements from the advanced very high resolution radiometer. *Remote. Sens. Environ.* **1991**, *35*, 257–277. [CrossRef]

39. Li, X.; Zhang, L.; Yang, G.; Li, H.; He, B.; Chen, Y.; Tang, X. Impacts of human activities and climate change on the water environment of Lake Poyang Basin, China. *Geoenviron. Disasters* **2015**, *2*, 1–2. [CrossRef]

40. Yan, D.H.; Wang, H.; Li, H.H.; Wang, G.; Qin, T.L.; Wang, D.Y.; Wang, L.H. Quantitative analysis on the environmental impact of large-scale water transfer project on water resource area in a changing environment. *Hydrol. Earth Syst. Sci.* **2012**, *16*, 2685–2702. [CrossRef]

41. IPCC. Who is Who in the IPCC? *Wmo* **2004**, 96–97.

42. Gu, X.; Zhang, Q.; Singh, V.P.; Shi, P. Changes in magnitude and frequency of heavy precipitation across China and its po-tential links to summer temperature. *J. Hydrol.* **2017**, *547*, 718–731. [CrossRef]

43. Wagena, M.B.; Sommerlot, A.; Abiy, A.Z.; Collick, A.S.; Langan, S.; Fuka, D.R.; Easton, Z.M. Climate change in the Blue Nile Basin Ethiopia: Implications for water resources and sediment transport. *Clim. Chang.* **2016**, *139*, 229–243. [CrossRef]

44. Geremew, A.A. Assessing the Impacts of Land Use and Land Cover Change on Hydrology of Watershed: A Case Study on Gilgel–Abbay Watershed, Lake Tana. Ph.D. Thesis, Bahir Dar University, Bahir Dar, Ethiopia, 2013; p. 82.

45. Gebrehiwot, S.G.; Bewket, W.; Gärdenäs, A.I.; Bishop, K. Forest cover change over four decades in the Blue Nile Basin, Ethiopia: Comparison of three watersheds. *Reg. Environ. Chang.* **2013**, *14*, 253–266. [CrossRef]

46. Liu, H.; Yin, Y. Response of forest distribution to past climate change: An insight. *Chin. Sci. Bull.* **2013**, *58*, 4426–4436. [CrossRef]

47. Rouse, W.; Haas, H.; Deering, W. 20 monitoring vegetation systems in the great plains with arts. 2018.

48. Li, Z.; Liu, W.Z.; Zhang, X.C.; Zheng, F.L. Impacts of land use change and climate variability on hydrology in an agri-cultural catchment on the Loess Plateau of China. *J. Hydrol.* **2009**, *377*, 35–42. [CrossRef]

49. Liu, F.; Qin, T.; Girma, A.; Wang, H.; Weng, B.; Yu, Z.; Wang, Z. Dynamics of Land-Use and Vegetation Change Using NDVI and Transfer Matrix: A Case Study of the Huaihe River Basin. *Pol. J. Environ. Stud.* **2018**, *28*, 213–223. [CrossRef]

Article

Characteristics of Rainstorm Intensity and Its Future Risk Estimation in the Upstream of Yellow River Basin

Wanzhi Li [1,2], Ruishan Chen [2,3,*], Shao Sun [4], Di Yu [1], Min Wang [5], Caihong Liu [1] and Menziyi Qi [1]

1   Qinghai Climate Center, Xining 810001, China
2   Institute of Plateau Science and Sustainable Development, Xining 810001, China
3   School of Design, Shanghai Jiaotong University, Shanghai 200030, China
4   State Key Laboratory of Severe Weather (LASW), Chinese Academy of Meteorological Sciences, Beijing 100081, China
5   Huangnan Prefecture Meteorological Bureau, Tongren 811300, China
*   Correspondence: rschen@sjtu.edu.cn

Abstract: Under the background of climate warming, the occurrence of extreme events upstream of the Yellow River Basin has increased significantly. Extreme precipitation tends to be even more intense, and occurs more frequently. The impacts of various extreme weather and climate events in the basin have become increasingly complex, which is increasingly difficult to cope with and affects the basin's long-term stability and ecological security. Based on the daily precipitation data of 33 meteorological stations in the upper reaches of the Yellow River Basin from 1961 to 2021, this paper analyzes the characteristics of rainstorm intensity. Moreover, combined with the simulation results of 10 global climate models of the Coupled Model Intercomparison Project (CMIP6) and the social and economic prediction data from SSPs, it analyzes the possible changes of rainstorm disaster risk in the upper reaches of the Yellow River Basin in the 21st century, under the three emission scenarios of SSP126, SSP245, and SSP370. The results show that the precipitation in the upstream area of the Yellow River Basin is increasing at a rate of 8.1 mm per 10 years, and the number of rainstorm processes and their indicators is increasing, which indicates an increase in the extremeness of precipitation; the rainstorm process intensity index shows an increasing trend, especially in the northeast region with a concentrated population and economy, where the rainstorm process intensity index is high; it is estimated that the number of rainstorm days in low-, medium-, and high-risk scenarios will increase, which leads to an increase in the social risk by at least 60% by around 2050 (2036–2065); with the increasing disaster risk, the population exposure to rainstorm disasters is also on the rise. If no measures are taken, the population exposure will increase to 7.316 million people per day by around 2050, increasing by more than double, especially in the northeast. This study shows that, with the increasing rainstorm disaster risk and population exposure in the upper reaches of the Yellow River Basin, relevant measures need to be taken to ensure the safety of people's lives and property.

Keywords: rainstorm process; rainstorm intensity; risk estimation and mapping; CMIP6; risk prediction

Citation: Li, W.; Chen, R.; Sun, S.; Yu, D.; Wang, M.; Liu, C.; Qi, M. Characteristics of Rainstorm Intensity and Its Future Risk Estimation in the Upstream of Yellow River Basin. *Atmosphere* **2022**, *13*, 2082. https://doi.org/10.3390/atmos13122082

Academic Editor: Haibo Liu

Received: 5 November 2022
Accepted: 8 December 2022
Published: 10 December 2022

Publisher's Note: MDPI stays neutral with regard to jurisdictional claims in published maps and institutional affiliations.

1. Introduction

Under the background of climate warming, the frequency, intensity, and duration of extreme climate events are increasing, which will magnify and exacerbate the risk of future climate disasters [1–3] and have huge impacts on human life, agricultural production, and the social economy, etc. [4,5]. According to statistics, in the past 30 years, 86% of the world's major natural disasters, 59% of deaths, 84% of economic losses, and 91% of insurance losses were caused by meteorological and derived disasters. China has become one of the world's most seriously affected countries by meteorological disasters [6], where the disasters account for more than 70% of natural disasters in recent years [7]. It is estimated that the possibility of flooding, high temperatures, and other disaster risks in China in the future will also be further increased [8]. On the regional and global scales, precipitation changes are more localized than temperature changes. Therefore, extreme precipitation usually shows high regional complexity and spatial-temporal variation [9]. Local heavy rainfall is more likely to lead to severe meteorological disasters. For example, the "7.21" torrential rain in Beijing in 2012 [10], the extremely heavy rain in North China from 19 to 20 July in 2016 [11], the extremely heavy rain in Hubei Province from 18 to 20 July in 2016 [12], the "5.7" extremely heavy rain in Guangzhou in 2017 [13], and the "7.20" extremely heavy rain in Zhengzhou in 2021 [14] all caused severe floods and their derivative disasters, high casualties, and economic and property losses. Therefore, analyzing the characteristics of rainstorm disasters and predicting their future risks is very meaningful.

Human-induced global warming has increased the frequency, intensity and number of heavy precipitation events worldwide. As the atmosphere warms, it contains more moisture, which means more rain during storms [15,16]. Climate change research advances have helped improve predictions of hydrologic extremes, with potentially severe implications for human societies and natural landscapes. The impact of precipitation is mainly reflected by the overall changes in the precipitation amount, intensity, spatial and temporal distribution, and the increase in extreme precipitation events, the most direct impact of which is the change in the intensity of precipitation erosion forces and the occurrence of extreme precipitation due to the changes in precipitation [17–20]. At the same time, the increased uncertainty of extreme precipitation and the influence of human activities lead to certain types of extreme precipitation patterns that can cause significant stress on river hydrology, and thus increase the risk of regional flood control [21,22].

In the rainstorm hazard census, the intensity of the rainstorm process is usually calculated to assess the risk level of the rainstorm process. Scholars have conducted related studies. Wang et al. established a comprehensive intensity assessment model and classification criteria for precipitation processes in four rainstorm-hazard-sensitive areas in China using three indicators: rainfall intensity, coverage area, and duration [23]. Zou et al. selected four indicators of maximum daily rainfall, maximum process rainfall, and rainstorm extent and duration, and constructed a comprehensive intensity assessment model of the regional rainstorm process and the comprehensive intensity classification criteria in Fujian Province [24]. Han et al. established a comprehensive rainstorm disaster assessment index for Liaoning Province, based on the average amount of rainfall, rainfall intensity extremes, and coverage size [25]. Moderate rain, heavy rain, and rainstorm events are estimated to increase significantly nationwide at the end of the 21st century [26]. The population exposure to extreme precipitation in China is also expected to increase by nearly 22% [27], aggravating the potential risk of disasters caused by rainstorms [28–31]. Adequate measures need to be taken in advance [32].

Existing studies have analyzed the runoff characteristics, precipitation characteristics, and their causes of formation in the Yellow River Basin [33–35]. However, few have concentrated on the characteristics of rainstorm intensity in the basin and the direction of future risk estimation. Currently, the Coupled Model Intercomparison Project is at a sixth stage (CMIP6), and has improved resolution, physical processes, and parameterization [36]. Moreover, compared with CMIP5, CMIP6 can enable the simulation results to be closer to the observations [37,38]. Based on CMIP6, studies have simulated and predicted the future temperature in the upper reaches of the Yellow River. This paper aims to solve the following key questions, by analyzing the characteristics of rainstorm intensity and estimating the future risk in the upper reaches of the Yellow River Basin: (1) Are the extremes of precipitation in the upper reaches of the Yellow River Basin increasing? How about their distribution? (2) Under the background of climate change, what about the future rainstorm disaster risk in the upper reaches of the Yellow River Basin? Where is the highest risk?

2. Data and Methods

2.1. Study Area

Qinghai Province is located in northwestern China, between latitude 31°36′–39°19′ north and longitude 89°35′–103°04′ east, as the headstream and mainstream area of the Yellow River; it has a drainage area of 152,300 km² (accounting for 21% of the Yellow River Basin), a drainage length of 1983 km (taking up 31% of the Yellow River), and the annual runoff out of the province accounts for 49.2% of the total flow of the Yellow River [39]. It impacts the sustainable development and utilization of water resources in the Yellow River Basin. The upper reaches of the Yellow River Basin are the population and economic agglomeration area of Qinghai Province, as well as the high precipitation and rainstorm-disaster-prone areas. For example, the flash flood disaster in Datong on 18 August 2022 caused heavy casualties and economic and property losses. Extreme rainstorms and the complex geographical environment are the main reasons for the frequent occurrence of rainstorm disasters in the upper reaches of the Yellow River Basin.

The upstream area of the Yellow River Basin is the source and mainstream area of the Yellow River, and the part of its watershed overlapping with the administrative region of Qinghai Province is located in the eastern part of Qinghai Province (Figure 1). The area of this region is 277,800 km², with a total population of 5,252,100, and a GDP of 226.44 billion CNY as of the end of 2019, accounting for 39.9%, 86.4%, and 76.3% of the province, respectively. The region is a significant ecological security barrier and a core area for high-quality development. The environmental quality in the study area significantly impacts the Sanjiangyuan Basin's ecological functions, the north and south slopes of the Qilian Mountains, and the arid eastern mountains. The upper reaches of the Yellow River Basin belong to the highland continental climate, with a low temperature and low precipitation. The annual average temperature is −3.8~9.5°C, which is high in the east and low in the south and north; annual precipitation is 256.7~742.0 mm, with more in the south and less in the north; annual rainstorm days last 9~69 days, with a greater number in the north and lower in the south. The distribution of rainstorm days is similar to the distribution of the annual average temperature, which is different from the distribution of the annual precipitation.

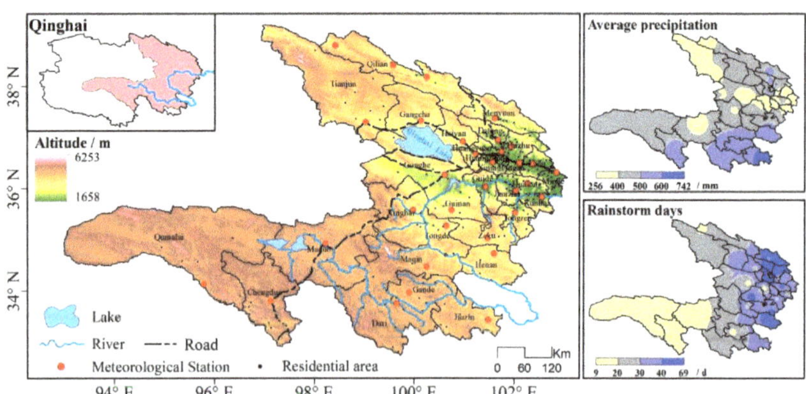

Figure 1. Map of the location of the upper Yellow River Basin.

2.2. Data Sources

The meteorological data were obtained from the Qinghai Meteorological Information Center, including the daily precipitation observations from 33 national meteorological observation stations (including national benchmark stations, basic stations, and general stations). The time that the meteorological data was obtained was selected from 1961 to 2021, and all data were subjected to strict quality control (Figure 2). The mean values in the paper are adopted from the standard climatic mean (mean values for 1991–2020).

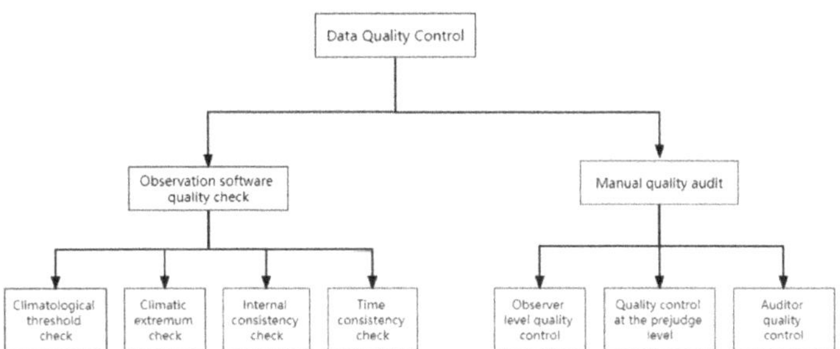

Figure 2. Data quality control flow chart.

The simulations from 10 climate models in the CMIP6 project were selected through the availability of daily precipitation for the historical run and future scenario runs (i.e., SSP1-2.6, SSP2-4.5, and SSP5-8.5) [40]. The multimodal ensemble mean (MME) method can effectively reduce the uncertainty of the simulations and presents the spatial pattern of precipitation extremes, which has been verified in numerous studies [41,42]. Only the first simulation member (r1i1p1f1) of each model is adopted. We resample all outputs of the 10 models to a common 0.25 × 0.25 resolution by using bilinear interpolation and only the data in the upper reaches of the Yellow River Basin.

Shared Socioeconomic Pathways (SSPs) are coupled with representative concentration pathways (RCPs), often applied to future project scenarios of socioeconomic development [43]. The projected population datasets are available at Socioeconomic Data and Applications Center (SEDAC), with ten-year intervals of 2010–2100 and a precision of 1 km on global land [44]. SSP1, SSP2, and SSP5 scenarios were selected to present population changes corresponding to low-, moderate-, and high-emission scenarios (i.e., RCP2.6, RCP4.5, and RCP8.5), respectively.

2.3. Methods

Since the rainfall in Qinghai province is small, the standard of rainstorm disaster is reached when the daily rainfall reaches ≥25 mm [45]. Therefore, in this paper, the rainfall process with ≥10 mm rainfall at a single station lasting ≥1 day and reaching ≥25 mm rainfall occurring during the process is called the rainstorm process.

The weighted comprehensive evaluation method is used to calculate the intensity of the rainstorm process [46]. It integrates the degree of influence of each indicator on the overall object. The effects of each specific indicator are combined and a quantitative indicator is used to focus on the degree of influence of the overall evaluation object. The calculation process begins with the normalization of each evaluation indicator [47]. Normalization is transforming a quantified value into a dimensionless value, which eliminates the quantified differences of each indicator. The calculation formula is:

$$X_{ij} = \frac{R_{ij} - R_{imin}}{R_{imax} - R_{imin}},$$ (1)

where X_{ij} is the normalized value of the ith indicator at site j; R_{ij} is the ith indicator value at site j; R_{imin} and R_{imax} are the minimum and maximum values of the ith indicator value, respectively.

In this study, the maximum daily rainfall of the rainstorm process, the accumulated rainfall of the process, and the duration of the process are selected as the index factors for calculating the intensity of the rainstorm process. The intensity index of the rainstorm process is obtained by weighted summation. The calculation formula is:

$$IR = A \times R_{24pre} + B \times R_{pre} + C \times R_{day},$$ (2)

where IR is the intensity index of the rainstorm process, R_{24pre} is the process daily maximum rainfall, R_{pre} is the process cumulative rainfall, and R_{day} is the process duration days. A, B, and C are the weight coefficients of each index, which are determined by the information entropy assignment method. This method is based on the actual data of each sample used to find the optimal weights, avoiding artificial influencing factors; therefore, the index weights given are more objective and thus have higher reproducibility and credibility. The specific calculation method is described in reference [48].

A linear trend equation, a significance t-test [49], and the Yue–Pilon [50] method were used in performing the statistical analysis. The linear trend equation is:

$$x_i = a + b \times t_i. \tag{3}$$

The climate variable of a sample size is denoted by x_i, the corresponding event is denoted by t_i, a univariate linear regression is created between x_i and t_i, a is the regression constant, and b is the regression coefficient.

3. Characteristics of the Intensity Change of the Rainstorm Process

3.1. Precipitation Variation Characteristics

From 1961 to 2021, the annual average precipitation upstream of the Yellow River Basin was 330.5 mm, with an overall increase of 8.1 mm/10a(years abbreviated as a). In terms of chronological changes, the annual precipitation fluctuated greatly in the early 1960s, and did not change significantly from the late 1960s to the 1990s. Moreover, the 1990s was the period with the lowest annual precipitation, which was only 308.8 mm. At the beginning of the 21st century, the annual precipitation began to increase significantly. Since the 2010s, it has been the period with the highest annual precipitation, reaching 346.2 mm. The year with the highest annual precipitation was 2018, which reached 429.4 mm (Figure 3a). From the perspective of spatial distribution, the annual precipitation in Minhe, Henan, and Huzhu, showed a decreasing trend, while the annual precipitation in the rest of the regions mainly increased, with an increase of 2.3~22.1 mm/10a. Additionally, the middle Yellow River Basin is the region with the most significant annual precipitation increase, with Guinan having the most significant increase, reaching 22.1 mm/10a (Figure 3b).

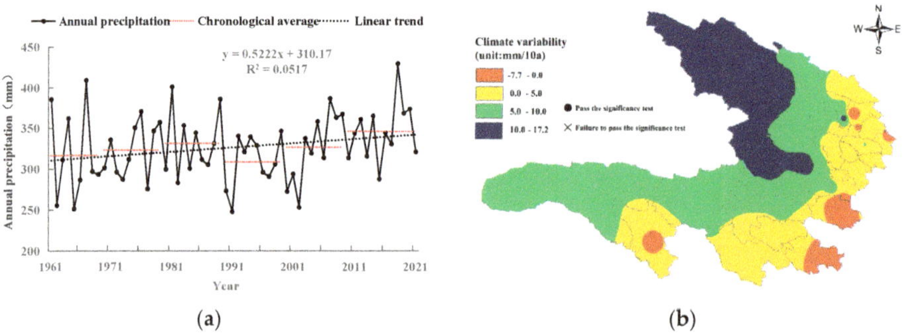

Figure 3. Variation curve (**a**) and spatial distribution of variability (**b**) of annual precipitation in the upstream of Yellow River Basin from 1961 to 2021.

3.2. Index Characteristics of the Rainstorm Process

From 1961 to 2021, the annual average number of a rainstorm upstream of the Yellow River Basin was 35.5 stations, showing an overall increasing trend, with an increase of 2.3 stations/10a. The number of rainstorms was relatively few from the 1960s to the 1980s, and began to increase in the 1990s. The period of the 2010s had the heaviest precipitation processes, with an average of 39.2 stations per year, of which the highest frequency was in 2018, with 77 stations (Figure 4a).

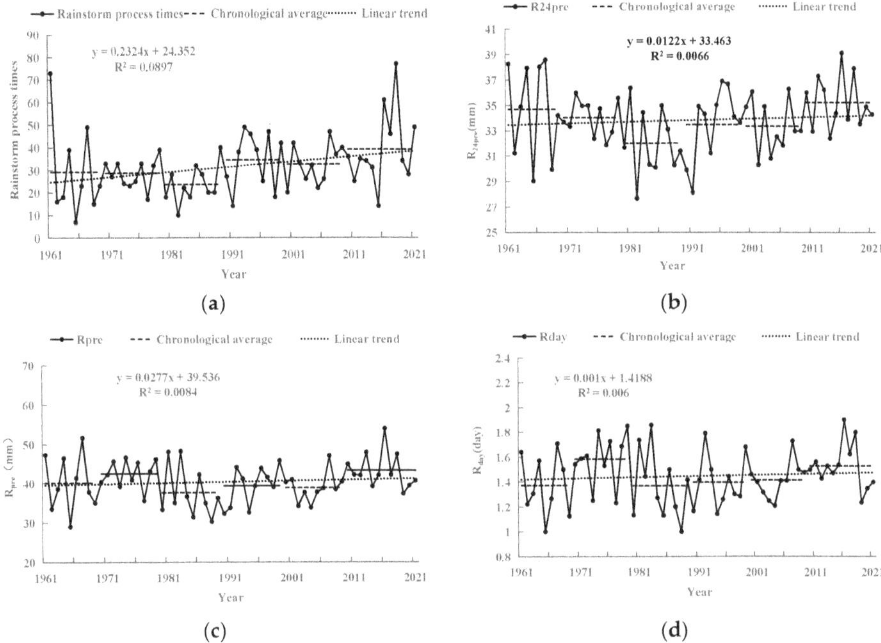

Figure 4. Variation curves of the number of rainstorm processes (**a**), maximum daily precipitation (**b**), accumulated precipitation (**c**), and duration of precipitation (**d**) in the upstream of Yellow River Basin from 1961 to 2021.

During the rainstorm, the average daily maximum precipitation, cumulative precipitation, and duration of precipitation were 34.4 mm, 64.7 mm, and 2 days, respectively, therefore all show an increasing trend. The maximum daily precipitation showed a trend of first decreasing and then increasing, with a minimum in the 1980s, and it continued to increase, reaching a maximum in the 1910s (Figure 4b). The accumulated precipitation fluctuated, and the minimum and maximum values also appeared in the 1980s and after the 21st century (Figure 4c), respectively. Except for the high value in the 1970s, the duration of precipitation showed a steady upward trend in other years (Figure 4d). Judging from the changes of various indices, not only the number of rainstorm processes in the upstream of Yellow River Basin increased, but also the various indices of the rainstorm processes showed an increasing trend, indicating that the extremeness of precipitation was also increasing.

From the perspective of spatial distribution, rainstorms tended to occur more in the east and less in the west. The number of rainstorms in Xining, Huangzhong, Datong, and Huzhu in the northeastern basin, and Henan in the southeast are all over 50. Huangzhong had the most frequent precipitation events, reaching a total of 69 times, which occurred more than 2 times a year on average (Figure 5a). The distribution trend of the average daily maximum precipitation during the rainfall process is generally consistent with the distribution trend of the number of rainfall processes. The high-value areas are mainly distributed in Datong, Xining, Minhe, Huzhu, and Huang in the northeast of the Yellow River Basin. In the high-value area, Datong had the largest average daily precipitation, with a value of 41.3 mm (Figure 5b). The distribution characteristics of the accumulated precipitation and the duration of the precipitation process are essentially the same, and the overall distribution trend is high in the north and south and low in the middle areas. Among them, the high-value areas of accumulated precipitation include Gangcha, Datong, Menyuan, Tianjun, and Huzhu, and the accumulated precipitation is above 45 mm. Furthermore, there are two high-value areas in the southeast, Henan and Jiuzhi (Figure 5c). The duration of the rainstorm process lasts longer in the high-value southern regions, including in Jiuzhi, Qingshuihe, Maqin, Qumalai, Dari, Henan, Zeku, and Guinan, among other places. The northern high-value areas include Datong, Qilian, Menyuan, Gangcha, Haiyan, and Tianjun, and the average duration of precipitation is more than 1.5 days (Figure 5d). Henan County is the region with the largest average accumulated precipitation and the longest duration of precipitation, with values of 52.5 mm and 2 d, respectively. The maximum historical daily precipitation was 119.9 mm in Datong in 2013, the maximum accumulated precipitation was 154.3 mm in Henan County in 2016, and the maximum duration of the process was 6 days in Jiuzhi in 1979.

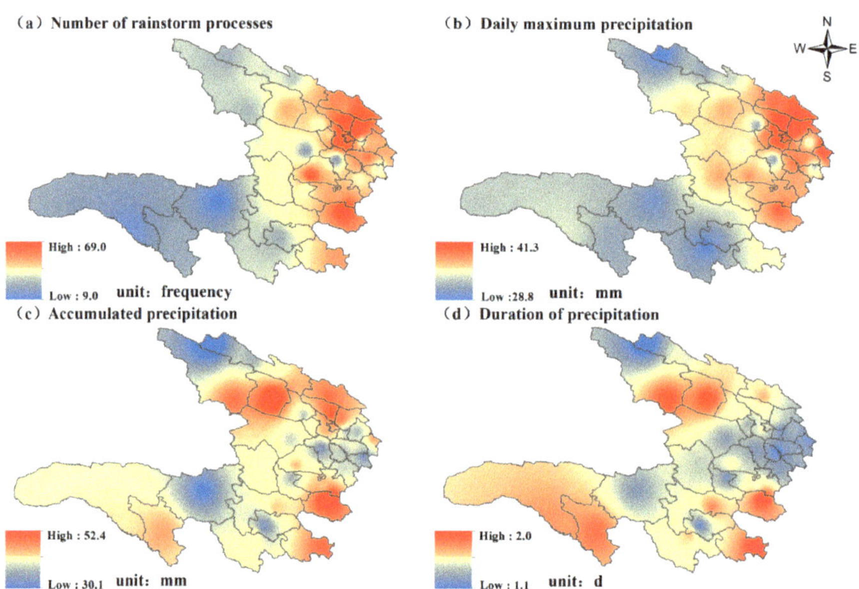

Figure 5. Spatial distribution of the number of heavy precipitation processes (**a**), daily maximum precipitation (**b**), accumulated precipitation (**c**), and duration of precipitation (**d**) in the upstream of Yellow River Basin from 1961 to 2021.

3.3. Intensity Characteristics of the Rainstorm Process

The intensity index of the rainstorm process is calculated according to Formula (2), and the larger the value, the higher the possibility of causing disasters. From 1961 to 2021, the intensity index of the rainstorm process upstream of the Yellow River Basin showed an overall increasing trend. From the perspective of chronological changes, it began to increase slowly after experiencing the high-value stage of the rainstorm process intensity index in the 1970s. The increasing trend after the 21st century is the most obvious, and it is the period with the highest intensity of the rainstorm process. In terms of years, 2018 was the year with the largest rainstorm process intensity index (Figure 6a). In terms of spatial distribution, there are two high-value centers of the rainstorm process intensity index in the northern and southern parts of the Yellow River Basin. Among them, the high-value and middle-value areas in the north include eight areas: Datong, Gangcha, Huzhu, Menyuan, Xining, Huangzhong, Tianjun, and Minhe. The high-value centers in the south include Henan and Jiuzhi. The rainstorm process intensity index was low in the northwestern Yellow River Basin and most areas of the source of the Yellow River (Figure 6b). From the spatial distribution of the intensity index of the rainstorm process, it can be seen that most of the administrative regions where the high-value areas are located are the areas with the most concentrated economy and population in Qinghai Province, which makes the region more vulnerable to rainstorm disasters.

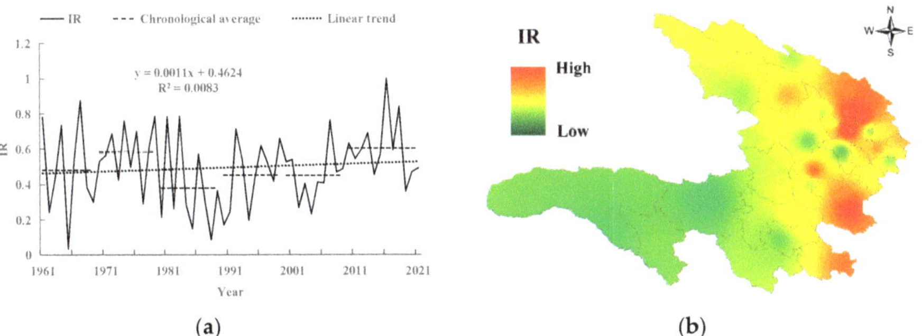

Figure 6. Changes in the risk level of rainstorm disasters under different climatic conditions in the upstream of the Yellow River Basin. (**a**) overall trend (**b**) spatial distribution.

4. Future Climate Risk Estimation for Rainstorm Disasters

4.1. Estimated Number of Days of Rainstorms in Future Scenarios

The number of days of a rainstorm upstream of the Yellow River Basin during the base period (1991–2020) is 1.1 d. The model estimation indicates that the frequency of rainstorms upstream of Yellow River Basin increases significantly in future scenarios, leading to increasing levels of combined climate risk. Even with the most stringent of carbon reduction measures in the future (the SSP126 scenario), the number of rainstorm days in the upstream of Yellow River Basin is expected to rise to 1.5 d by around 2050 (2036–2065), leading to a 60% increase in the socioeconomic risk index. In the medium-emission scenario (SSP245), the number of rainstorm days will rise to 1.6 days in around 2050, leading to a 67% increase in the socioeconomic risk index. If fossil fuels continue to be used as the main energy source (the SSP370 scenario), the number of rainstorm days in the upstream of Yellow River Basin will rise to 1.8 days in around 2050, leading to a doubling of the socioeconomic risk index from the historical base period (Figure omitted).

In terms of administrative districts, Xining, Datong, Huangzhong, Huzhu, and Henan County have the highest number of rainstorm days, and the historical base period (1991–2020) has reached more than 1.5 days per year. Under the future low- and medium-emission scenarios, it is expected to rise to more than 2.4 days around 2050, which is the most dramatic response to climate warming within the upstream Yellow River Basin (Figure 7a,b). Under the high-emission scenario, the number of days of rainstorm in Huangzhong is expected to exceed 3 days (Figure 7c,d).

4.2. Population Exposure Risk Estimation for Future Scenarios

The overlay of the predicted results by the number of rainstorm days and population shows that the population exposure in upstream of the Yellow River Basin during the historical base period is 3.833 million people per day (mpd), among which, Datong, Huzhu, and Huangzhong have the highest population exposure with 0.652 mpd, 0.574 mpd and 0.526 mpd. Under the low-emission scenario, it is expected that the population exposure to rainstorm hazards in the upstream of the Yellow River Basin will rise to 5.613 mpd around 2050, which is an increase of 46%, with Datong, Huzhu, and Xining all reaching over 0.526 mpd (Figure 8a). Under the medium-emission scenario, the population exposure to rainstorm hazards in the source of the Yellow River is expected to rise to 6.036 mpd around 2050, and the three county-level units with the highest exposure are expected to be Xining (0.956 mpd), Datong (0.919 mpd), and Huzhu (0.871 mpd) (Figure 8b). If no emission reduction measures are taken and the current carbon emission intensity continues, the population exposure will rise to 7.316 mpd around 2050, among which, the population exposure of rainstorm disasters in the Datong, Xining, and Huzhu counties (districts) will exceed one million people per day (Figure 8c). In particular, the population exposure of the rainstorm disaster in Datong will increase by 0.314 mpd; compared with the low-emission scenario, the values for Huzhu and Huangzhong both increase by more than 0.2 mpd (Figure 8d).

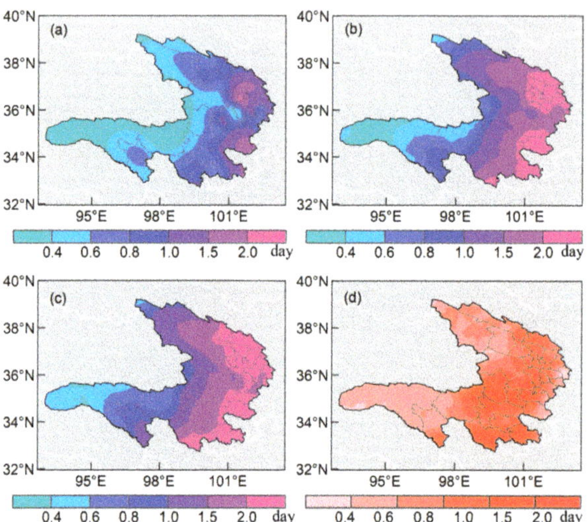

Figure 7. Spatial distribution of rainstorm days around 2050 (2036–2065) in the upstream of Yellow River Basin in future scenarios: (**a**) low-emission scenario SSP126, (**b**) medium-emission scenario SSP245, (**c**) high-emission scenario SSP370, (**d**) the difference in rainstorm days between high- and low-emission scenarios.

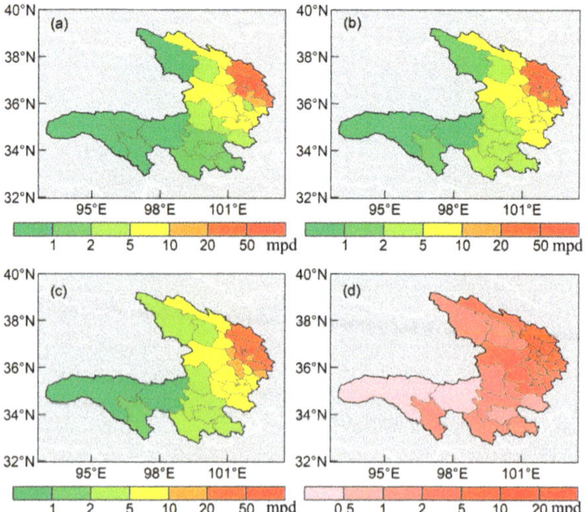

Figure 8. Spatial distribution of population exposure to rainstorm disasters upstream of the Yellow River Basin at around 2050 (2036–2065) in future scenarios: (**a**) low-discharge scenario SSP126, (**b**) medium-discharge scenario sSSP245, (**c**) high-discharge scenario SSP370, (**d**) difference in population exposure between high- and low-discharge scenarios.

5. Discussion and Conclusions

This study analyzed the rainstorm intensity characteristics based on the daily rainfall data of 33 meteorological stations upstream of the Yellow River Basin from 1961 to 2021. The simulation results of 10 global climate models of CMIP6 and the socioeconomic prediction data of SSPs were used to analyze the possible changes of rainstorm hazard risk in the upstream of the Yellow River Basin region in the 21st century in three emission scenarios (SSP126, SSP245, and SSP370). The conclusions are as follows:

1. From 1961 to 2021, rainfall upstream of the Yellow River Basin showed an overall increasing trend, with an increased rate of 8.1 mm/10a. In the 21st century, the rising annual rainfall trend is becoming particularly significant. The maximum daily rainfall, accumulated rainfall, and the number of days of duration during rainstorms all show a rising trend, and the extremity of rainfall increases;

2. From 1961 to 2021, the intensity index of the rainstorm process showed an increasing trend. The increase has become pronounced since the beginning of the 21st century, which is the period with the highest value of the intensity index of the rainstorm process. Most of the administrative districts where the high-value areas are located are the most economically and population-concentrated areas in Qinghai Province. The risk of rainstorm disasters and possible damages will also increase;

3. The low-, medium-, and high- emission scenarios are all expected to show an increasing trend in the number of rainstorm days by around 2050 (2036–2065). Among them, the low-emission scenario will lead to at least a 60% increase in social risk. The medium-emission scenario will lead to a 67% increase in the socioeconomic risk index. In contrast, the high-emission scenario will lead to a doubling of the socioeconomic risk index from the historical base period.

4. As the hazards increase, the population exposure to the rainstorm hazards will also rise. If no measures are taken, the population exposure will rise to 7.316 million per day around 2050. This has more than doubled compared to the base period, with the increase being particularly significant in the northeast.

In the background of global warming, rainfall upstream of the Yellow River Basin is increasing. At the same time, the extremes of precipitation are increasing. The increasing trend is particularly evident in the northeastern basin areas where the population and economy are more concentrated, further increasing the risk of rainstorm disasters. The CMIP6 model's prediction of the future rainstorm hazard risk in the watershed suggests that the social risk due to rainstorm hazards will increase under all three future low-, medium-, and high-emission scenarios, with even the low-emission scenario leading to a 60% increase in risk. The population's exposure to rainstorm hazards will also tend to increase, and the significant area of increase will also be in the northeastern part of the basin. The risk situation for future rainstorm events in the region is critical. There is an urgent need to incorporate climate adaptation strategies into future urban development to better respond to the climate risks associated with extreme rainfall events. The results from this research are consistent with the research results of Su and Sun et al. [51,52]. At the same time, the increased risk of rainstorm hazards may pose new challenges to economic sustainability. It may increase urban-rural migration and thus affect the geospatial distribution of social exposure, which needs to be considered in local climate risk management. In future urban construction, we can take different defensive measures according to the pattern and impact of rainstorm disasters, such as improving urban flood control and drainage standards, building sponge cities, raising public awareness, increasing extreme rainstorm warnings, and purchasing urban flood insurance, in order to better cope with the climate risks associated with extreme rainstorm events.

In this paper, the storm process and its indicators were analyzed, and its intensity was calculated. However, in this study, only meteorological elements were considered in analyzing the impact of heavy rainfall in the upper reaches of the Yellow River, and the considerations of topography, water systems, and other disaster-generating environments were missing. In addition, the 33 meteorological observation stations are relatively few compared to the extent of the upper Yellow River area, and do not provide a good characterization of the spatial distribution. In future studies, additional data could be introduced to create a composite index that describes the multidimensional characteristics of climate extremes (e.g., the peak intensity and duration).

Author Contributions: Conceptualization, W.L. and R.C.; validation, W.L., D.Y. and M.Q.; formal analysis, W.L. and S.S.; data curation, M.W. and C.L.; writing—original draft preparation, W.L. and R.C.; writing—review and editing, W.L.; visualization, D.Y. and S.S.; funding acquisition, W.L. All authors have read and agreed to the published version of the manuscript.

Funding: This study was jointly funded by the Second Integrated Scientific Expedition on Qinghai-Tibet Plateau (NO. 2019QZKK0906), the Key Soft Science Project of China Meteorological Administration (NO. 2022DIANXM13), the National Natural Science Foundation of China (NO. 42065003), and the Qinghai Provincial Science and Technology Department (NO. 2021-ZJ-757).

Institutional Review Board Statement: Not applicable.

Informed Consent Statement: Not applicable.

Data Availability Statement: No new data were created or analyzed in this study. Data sharing is not applicable to this article.

Acknowledgments: We acknowledge support from the above funding projects. Acknowledgments also to the Applications Center (SEDAC) for providing the population projections based on the Shared Socioeconomic Pathways (https://sedac.ciesin.columbia.edu/data/set/popdynamics-1-km-downscaledpop-base-year-projection-ssp-2000-2100-rev01. Accessed on 6 May 2022).

Conflicts of Interest: The authors declare no conflict of interest.

References

1. Diffenbaugh, N.S.; Singh, D.; Mankin, J.S.; Horton, D.E.; Swain, D.L.; Touma, D.; Charland, A.; Liu, Y.; Haugen, M.; Tsiang, M.; et al. Quantifying the influence of global warming on unprecedented extreme climate events. *Proc. Natl. Acad. Sci. USA* **2017**, *114*, 4881–4886. [CrossRef] [PubMed]
2. Coumou, D.; Rahmstorf, S. A decade of weather extremes. *Nat. Clim. Change* **2012**, *2*, 491–496. [CrossRef]
3. Toreti, A.; Naveau, P.; Zampieri, M.; Schindler, A.; Scoccimarro, E.; Xoplaki, E.; Dijkstra, H.A.; Gualdi, S.; Luterbacher, J. Projections of global changes in precipitation extremes from coupled model intercomparison project phase 5 models. *Geophys. Res. Lett.* **2013**, *40*, 4887–4892. [CrossRef]
4. Lipczynska-Kochany, E. Effect of climate change on humic substances and associated impacts on the quality of surface water and groundwater: A review. *Sci Total Environ.* **2018**, *640–641*, 1548–1565. [CrossRef] [PubMed]
5. Jiang, T.; Su, B.; Huang, J.; Zhai, J.; Kundzewicz, Z.W. Each 0.5 °C of warming increases annual flood losses in china by more than US$60 Billion. *Bull. Am. Meteorol. Soc.* **2020**, *101*, E1464–E1474. [CrossRef]
6. Cai, T.; Li, X.; Ding, X.; Wang, J.; Zhan, J. Flood risk assessment based on hydrodynamic model and fuzzy comprehensive evaluation with GIS technique. *Int. Disaster Risk Reduct.* **2019**, *35*, 101077. [CrossRef]
7. Wang, G.; Li, X.; Wu, X.; Yu, J. The rainstorm comprehensive economic loss assessment based on CGE model: Using a July heavy rainstorm in Beijing as an example. *Nat. Hazards.* **2015**, *76*, 839–854. [CrossRef]
8. Xu, Y.; Zhang, B.; Zhou, B.T.; Dong, S.Y.; Yu, L.; Li, R.K. Projected risk of flooding disaster in China based on CMIP5 models. *Clim. Change Res.* **2014**, *10*, 268–275. [CrossRef]
9. Xu, L.; Wang, A.; Wang, D.; Wang, H. Hot spots of climate extremes in the future. *J. Geophys. Res. Atmos.* **2019**, *124*, 3035–3049. [CrossRef]
10. Chen, Y.; Sun, J.; Xu, J.; Yang, S.; Sheng, J. Analysis and thinking on the extremes of the 21 July 2012 torrential rain in Beijing part I: Observation and thinking. *Meteor. Mon.* **2012**, *38*, 1255–1266. [CrossRef]
11. Fu, J.L.; Ma, X.K.; Chen, T.; Zhang, F.; Zhang, X.D.; Sun, J.; Quan, W.Q.; Yang, S.N.; Shen, X.L. Characteristics and synoptic mechanism of the July 2016 extreme precipitation event in North China. *Meteor. Mon.* **2017**, *43*, 528–539. [CrossRef]
12. Zhao, X.T.; Wang, X.f.; Wang, J.; Wang, X.K.; Xiao, Y.J.; Leng, L.; Fu, Z.K. Analysis of mesoscale characteristics of torrential rainfall in Hubei Province during 18–20 July 2016. *Meteor. Mon.* **2020**, *46*, 490–502. [CrossRef]
13. Xu, G.Q.; Zhao, C.Y. Impact of background field in the numerical simulation of extremely severe rainstorm in Guangzhou on 7 May 2017. *J. Meteor. Mon.* **2019**, *45*, 1642–1650. [CrossRef]
14. Wang, Z.Y.; Yao, C.; Dong, J.L.; Yang, H. Precipitation characteristic and urban flooding influence of "7·20" extreme rainstorm in Zhengzhou. *J. Hohai Univ.* **2022**, *50*, 17–22. [CrossRef]
15. Wei, C.; Dong, X.; Yu, D.; Zhang, T.; Zhao, W.; Ma, Y.M. Spatio-temporal variations of rainfall erosivity, correlation of climatic indices and influence on human activities in the Huaihe River Basin, China. *CATENA* **2022**, *217*, 106486. [CrossRef]

16. Matthews, F.; Panagos, P.; Verstraeten, G. Simulating event-scale rainfall erosivity across European climatic regions. *CATENA* **2022**, *213*, 106157. [CrossRef]
17. Diodato, N.; Borrelli, P.P.; Bellocchi, G. Global assessment of storm disaster-prone areas. *PLoS ONE* **2022**, *17*, e0272161. [CrossRef]
18. Ponjiger, T.M.; Lukić, T.; Basarin, B.; Jokic, M.; Wilby, R.L.; Pavić, D.; Mesaroš, M.; Valjarević, A.; Milanovic, M.; Morar, C. Detailed analysis of spatial–temporal variability of rainfall erosivity and erosivity density in the central and southern Pannonian basin. *Sustainability* **2021**, *13*, 13355. [CrossRef]
19. Zhang, J.P.; Ren, Y.L.; Jiao, P.; Xiao, P.Q.; Li, Z. Changes in rainfall erosivity from combined effects of multiple factors in China's Loess Plateau. *CATENA* **2022**, *216*, 106373. [CrossRef]
20. Bezak, N.; Mikoš, M.; Borrelli, P.; Liakos, L.; Panagos, P. An in-depth statistical analysis of the rainstorms erosivity in Europe. Catena. *CATENA* **2021**, *206*, 105577. [CrossRef]
21. Yang, Y.D.; Xu, Z.R.; Zheng, W.W.; Wang, S.H.; Kang, Y.B. Rain belt and flood peak: A study of the extreme precipitation event in the Yangtze river basin in 1849. *Water* **2021**, *13*, 2677. [CrossRef]
22. Jia, L.; Li, Z.B.; Li, P.; Zhang, J.Z.; Wang, A.N.; Ma, L.; Xu, G.C.; Zhang, X. Temporal and spatial variation of rainfall erosivity in the Loess Plateau of China and its impact on sediment load. *CATENA* **2022**, *210*, 105931. [CrossRef]
23. Wang, L.P.; Wang, X.R.; Wang, W.G. Research and application of comprehensive intensity evaluation method for regional rainfall process in China. *J. Nat. Disasters* **2015**, *24*, 186–194. [CrossRef]
24. Zou, Y.; Ye, D.X.; Lin, Y.; Liu, A.M. A quantitative method for assessment of regional rainstorm intensity. *J Appl. Meteor. Sci.* **2014**, *25*, 360–364.
25. Han, X.J.; Sum, X.W.; Li, S.; Wang, M.H.; Li, G.X.; Chen, Y.; Wang, G.C. Disaster-causing index of rainstorm and preassessment of disaster effect in Liaoning province. *J. Meteorol. Environ.* **2014**, *30*, 80–84. [CrossRef]
26. Chen, H. Projected change in extreme rainfall events in China by the end of the 21st century using CMIP5 models. *Chin. Sci. Bull.* **2013**, *58*, 1462–1472. [CrossRef]
27. Chen, H.; Sun, J. Increased population exposure to precipitation extremes in China under global warming scenarios. *Atmos. Ocean. Sci. Lett.* **2020**, *13*, 63–70. [CrossRef]
28. Patricola, C.M.; Wehner, M.F. Anthropogenic influences on majortropical cyclone events. *Nature* **2018**, *563*, 339–346. [CrossRef]
29. Wang, G.; Wang, D.; Trenberth, K.E.; Erfanian, A.; Yu, M.; Bosilovich, M.G.; Parr, D.T. The peak structure and future changes of the relationships between extreme precipitation and temperature. *Nat. Clim. Change* **2017**, *7*, 268–274. [CrossRef]
30. Scoccimarro, E.; Gualdi, S.; Bellucci, A.; Zampieri, M.; Navarra, A. Heavy precipitation events in a warmer climate: Results from CMIP5 models. *J. Clim.* **2013**, *26*, 7902–7911. [CrossRef]
31. Trenberth, K.E.; Fasullo, J.T.; Shepherd, T.G. Attribution of climate extreme events. *Nat. Clim. Change* **2015**, *5*, 725–730. [CrossRef]
32. Liu, Y.; Li, L.; Zhang, W.; Chan, P.; Liu, Y. Rapid identification of rainstorm disaster risks based on an artificial intelligence technology using the 2DPCA method. *Atmos. Res.* **2019**, *227*, 157–164. [CrossRef]
33. Gao, T.; Wang, H. Trends in precipitation extremes over theYellow River Basin in North China: Changing properties and causes. *Hydrol. Process.* **2017**, *31*, 2412–2428. [CrossRef]
34. Omer, A.; Elagib, N.A.; Zhuguo, M.; Saleem, F.; Mohammed, A. Water scarcity in the Yellow River Basin under future climate change and human activities. *Sci. Total Environ.* **2020**, *749*, 141446. [CrossRef] [PubMed]
35. Yuan, Z.; Yan, D.; Yang, Z.; Xu, J.; Huo, J.; Zhou, Y.; Zhang, C. Attribution assessment and projection of natural runoff change in theYellow River Basin of China. *Mitig. Adapt. Strateg. Glob. Change* **2016**, *23*, 27–49. [CrossRef]
36. Li, C.; Jian, T.; Wang, Y.J.; Miao, L.J.; Li, S.Y.; Chen, Z.Y.; Lü, Y.R. Simulation and estimation of future air temperature in upper basin of the Yellow River based on CMIP6 models. *J. Glaciol. Geocryol.* **2022**, *44*, 171–178. [CrossRef]
37. Chen, H.P.; Sun, J.Q.; Lin, W.Q.; Xu, H.W. Comparison of CMIP6 and CMIP5 models in simulating climate extremes. *Sci. Bull.* **2020**, *65*, 1415–1418. [CrossRef]
38. Zhai, J.; Mondal, S.K.; Fischer, T.; Wang, Y.; Su, B.; Huang, J.; Tao, H.; Wang, G.; Ullah, W.; Uddin, J. Future drought characteristics through a multi-model ensemble from CMIP6 over South Asia. *Atmos. Res.* **2020**, *246*, 105111. [CrossRef]
39. Sui, X.; Qi, Y. Dynamic assessment of ecological carrying capacity of Yellow River Basin in Qinghai province. *Chin. J. Ecol.* **2007**, *26*, 406–412.
40. Taylor, K.E.; Stouffer, R.J.; Meehl, G.A. An overview of CMIP5 and the experiment design. *Bull. Am. Meteorol. Soc.* **2012**, *93*, 485–498. [CrossRef]
41. Sillmann, J.; Kharin, V.; Zwiers, F.W.; Zhang, X.; Bronaugh, D. Climate extremes indices in the CMIP5 multimodel ensemble: Part 2. future climate projections. *JGR Atmos.* **2013**, *118*, 2473–2493. [CrossRef]
42. Kharin, V.V.; Zwiers, F.W.; Zhang, X.; Wehner, M. Changes in temperature and precipitation extremes in the CMIP5 ensemble. *Clim. Change* **2013**, *119*, 345–357. [CrossRef]
43. O'Neill, B.C.; Kriegler, E.; Ebi, K.L.; Benedict, E.K.; Riahi, K.; Rothman, D.S.; van Ruijven, B.J.; van Vuuren, D.P.; Birkmann, J.; Kok, K.; et al. The roads ahead: Narratives for shared socioeconomic pathways describing world futures in the 21st century. *Glob. Environ. Change* **2017**, *42*, 169–180. [CrossRef]
44. Gao, J. *Global 1-km Downscaled Population Base Year and Projection Grids Based on the Shared Socioeconomic Pathways*; NASA Socioeconomic Data and Applications Center (SEDAC): Palisades, NY, USA, 2020. [CrossRef]
45. Administration for Market Regulation of Qinghai Province. *Classification Index of Meteorological Disasters (DB63/T372-2018)*; Administration for Market Regulation of Qinghai Province: Xining, China, 2018; p. 9.

46. Zhou, B.T.; Yu, L. Managing climate disaster risks and promoting climate change adaptation. *Disaster Reduct. China* **2012**, *174*, 18–19.
47. Rupp, D.E.; Abatzoglou, J.T.; Hegewisch, K.C.; Mote, P.W. Evaluation of CMIP5 20th century climate simulations for the Pacific Northwest USA. *Geophys. Res. Atmos.* **2013**, *118*, 10884–10906. [CrossRef]
48. Delgado, A.; Romero, I. Environmental conflict analysis using an integrated grey clustering and entropy-weight method: A case study of a mining project in Peru. *Environ. Model. Softw.* **2016**, *77*, 108–121. [CrossRef]
49. Wei, F.Y. *Modern Techniques of Statistical Diagnosis and Prediction of Climate*, 2nd ed.; Meteorological Press: Beijing, China, 2007.
50. Yue, S.; Pilon, P.; Phinney, B.; Cavadias, G. The influence of autocorrelation on the ability to detect trends in hydrological series. *Hydrol. Process.* **2002**, *16*, 1807–1829. [CrossRef]
51. Sun, S.; Dai, T.L.; Wang, Z.Y.; Ming, C.J.; Chao, Q.C.; Shi, P.J. Projected increases in population exposure of daily climate extremes in eastern China by 2050. *Adv. Clim. Change Res.* **2021**, *12*, 804–813. [CrossRef]
52. Su, X.; Shao, W.W.; Liu, J.H.; Jiang, Y.Z.; Wang, K.B. Dynamic Assessment of the Impact of Flood Disaster on Economy and Population under Extreme Rainstorm Events. *Remote Sens.* **2021**, *13*, 3924. [CrossRef]

atmosphere

MDPI

Article

A Comparison Study of Observed and the CMIP5 Modelled Precipitation over Iraq 1941–2005

Salam A. Abbas [1,2,*], Yunqing Xuan [2], Ali H. Al-Rammahi [3] and Haider F. Addab [1,3]

[1] Department of Civil and Environmental Engineering, Colorado State University, Fort Collins, CO 80523, USA
[2] Department of Civil Engineering, Faculty of Science and Engineering, Swansea University Bay Campus, Swansea SA1 8EN, UK
[3] Department of Civil Engineering, College of Engineering, Kufa University, Kufa 54003, Iraq
* Correspondence: salam.a.abbas@colostate.edu; Tel.: +1-970-821-6687

Abstract: This paper presents an analysis of the annual precipitation observed by a network of 30 rain gauges in Iraq over a 65-year period (1941–2005). The simulated precipitation from 18 climate models in the CMIP5 project is investigated over the same area and time window. The Mann–Kendall test is used to assess the strength and the significance of the trends (if any) in both the simulations and the observations. Several exploratory techniques are used to identify the similarity (or disagreement) in the probability distributions that are fitted to both datasets. While the results show that large biases exist in the projected rainfall data compared with the observation, a clear agreement is also observed between the observed and modelled annual precipitation time series with respect to the direction of the trends of annual precipitation over the period.

Keywords: precipitation; trend analysis; Iraq; climate projection; CMIP5

Citation: Abbas, S.A.; Xuan, Y.; Al-Rammahi, A.H.; Addab, H.F. A Comparison Study of Observed and the CMIP5 Modelled Precipitation over Iraq 1941–2005. *Atmosphere* **2022**, *13*, 1869. https://doi.org/10.3390/atmos13111869

Academic Editor: Haibo Liu

Received: 22 September 2022
Accepted: 6 November 2022
Published: 9 November 2022

Publisher's Note: MDPI stays neutral with regard to jurisdictional claims in published maps and institutional affiliations.

1. Introduction

Studying precipitation trends is an essential step in assessing the impact of climate change on hydrological processes. A substantial change in the amount of precipitation can lead to severe conditions of flooding and droughts. It is also important to examine such trends for water resource planning since water demand may be affected; thus, strategies and operations for the water supply need to be adjusted accordingly.

While the trend of recorded observations of hydro-meteorological variables, such as precipitation, has been widely studied using long-term observation datasets, e.g., [1], scenario-based climate projections are still a preferred source for most studies on the future trends of climate change. More recently, the Fifth Climate Model Inter-comparison Project CMIP5 [2] published a rich set of climate simulations produced by several key metrological centres in the world, which offers an updated and improved (in both accuracy and resolution) collection of outputs of climate models for many downstream impact studies, e.g., [3–6].

The post-industrial period, particularly the 20th century, has been the focus of many studies on the trend of hydro-climatic variables. Generally, these studies aimed to establish a link between the so-called anthropogenic greenhouse effect and the change in climate, as indicated by the key variables. Moreover, the baseline period was chosen by many studies as observation records became abundant. The studied areas range from global to regional scales. To name just a few: New et al. [7] showed that, in the 20th century, precipitation has significantly changed in various parts of the world, and Griggs and Noguer [8] and Xu et al. [9] pointed out that the average annual rainfall considerably increased by 7–12% in the high and middle latitudes in the northern hemisphere during the 20th century.

Several statistical techniques have also been employed to detect the trend and the shift of climatological variables, as reported in Martinez et al. [10]. There are two main families of methods for determining trends, parametric and non-parametric, with non-parametric methods being preferred over parametric ones, as the former is less likely to be affected by the outliers and does not assume a predefined distribution for datasets or homogeneity [11]. The statistical test of Mann–Kendall, i.e., the MK test [12,13], has often been utilized to evaluate the trends of rainfall [14,15].

Most studies have been carried out on the trend of observed hydro-meteorological variables; few are focused on the trend analysis of simulated variables such as precipitation projection from climate models. Rather, many studies make use of "snapshots" from climate projections to indicate the difference (hence change) between the projected value of the variable in question and its current values, without revealing the process or temporal trend associated with such change.

On the other hand, researchers tend to use projected, scenario-defined variables from climate model simulations, notably, precipitation, to drive other models for impact studies. Errors or biases in these simulated variables have been widely recognized in this type of application. While sophisticated bias correction methods have also been developed to cope with this situation, little attention has been paid to the trend of those simulated variables either with or without bias correction. It is very plausible that an overall quantity-fit projection after bias correction may not be able to reproduce the trend revealed by the observations; to some extent, this would be more of a concern in projection-driven climate impact studies.

Trend analysis of precipitation data on the regional and global scales has been studied by many researchers all over the world. Readers can refer to Palomino-Lemus et al. [16], Sharmila et al. [17] and Palizdan et al. [18]. Global climate models are broadly utilized to create current and project future climate conditions [19,20]. This is essential to investigate and evaluate the models' performance in simulating precipitation to develop adaptation strategies to reduce uncertainties in projecting precipitation in the future [21].

The Coupled Model Intercomparison Project Phase 5 (CMIP5) comprises more comprehensive global climate models than its predecessors, enabling researchers to address many research questions [22]. The assessment methods for different models' performance have progressed from traditional qualitative methods to quantitative methods [23]. Some research has assessed models based on some traditional statistical methods, such as spatial correlation coefficients [24], linear trend analysis [20] and standard deviation [25].

Several climate studies that primarily used GCMs to estimate the future of the earth's climate system indicated changes in the climate from the regional scale [26,27] to global scale [28,29]. Deng et al. [30] evaluated nine CMIP5 datasets under different RCP future scenarios for the projections of summer and spring precipitation for the period of 2013–2050 over the Yangtze River basin in China. The precipitation projections revealed significant positive linear trends of spring precipitation under the RCP2.6 and RCP8.5 scenarios, while summer precipitation is shown to be undergoing an inter-annual change. Mehran et al. [31] used volumetric hit index (VHI) analysis with 34 CMIP5 GCMs to reproduce observed precipitation across the globe. The GCMs showed good agreement with the observed monthly time series. However, reproducing the observed precipitation over some subcontinental and arid regions was problematic. Their finding proves the allegation that GCMs have weaknesses when it comes to the representation of observed precipitation on small spatial scales and short temporal scales, e.g., [27].

Nikiema et al. [32] compared and evaluated the multi-model ensembles of CMIP5 and CORDEX and found that while CORDEX failed to outperform the temperature simulated by the CMIP5 ensembles, it considerably enhanced the simulation of historical summer precipitation and provided a more realistic fine-scale features tied to land use and local topography. Long-range correlation (LRC) is an effective way to evaluate CMIP5 models' performance in global precipitation and has been employed in many aspects of climate systems, such as precipitation [33] and air temperature [34].

The above studies have contributed immensely to the assessment of the performance of GCMs and RCMs in reproducing the observed climatology of various regions, but very few focus on places with limited observed climate data, such as Iraq. Therefore, assessing the performances of several GCMs from CMIP5 over Iraq will contribute significantly to the efforts in climate modelling over the region.

The study presented in this paper takes a different approach from those used by many previous studies in trend analysis. It first investigates the observed precipitation in Iraq over a 65-year period (1941–2005); this is then compared with the modelled precipitation from an array of 18 climate models from the latest CMIP5 projects. Both the spatial and temporal distribution trends are studied alongside a bias correction procedure applied to the modelled data. Further, several exploratory statistical techniques are used to check the similarity between the observed and modelled precipitation.

2. Materials and Methods

2.1. Study Area

Located within the southwest of the continent of Asia, Iraq shares land borders with Turkey in the north, Kuwait and Saudi Arabia in the south, Iran in the east and Syria and Jordan in the west (Figure 1). Iraq comprises a total area of 437,065 km^2. The climate is mainly described as a semi-arid, subtropical, and continental type; meanwhile, the north and northeast parts have a Mediterranean climate [35]. Iraq has a rainy season in cold months, from December to February, excluding the north area of the country, where the seasonal precipitation takes place from November to April.

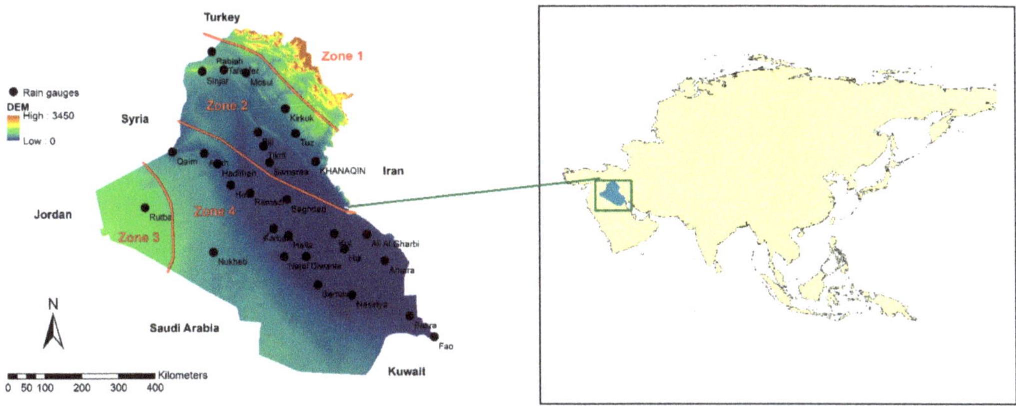

Figure 1. Elevation map of Iraq with climatic zones.

The yearly precipitation in Iraq is estimated to be 216 mm countrywide on average; regionally, it has a range of less than 100 mm over the southern region to 1200 mm in the northeast of the country [35]. Summer in Iraq is hot and dry to extremely hot with a shade temperature of over 43 °C that falls at night to 26 °C. Winter is cold, with a day temperature of about 16 °C, dropping at night to 2 °C [36]. According to the FAO [35], Iraq can be divided into four agroecological zones:

(1) Semi-arid and arid zones with a Mediterranean climate (zone 1 in Figure 1): The annual precipitation varies between 700 and 1000 mm and occurs between October and April. The country has cold and rainy winters, while summers are hot and dry; they are even torrid up to quite high altitudes. This zone mainly covers the north of the country. This is the only region in Iraq that receives a considerable amount of precipitation.

(2) Steppes with winter rainfall of 200–400 mm annually (zone 2 in Figure 1): Summers are extremely hot, and winters are cold. In the cold season of the year, some depressions can pass that carry moderate precipitation.

(3) The desert zone/northwest of Mesopotamia (zone 3 in Figure 1) has a high temperature in summer and less than 200 mm of yearly rainfall.

(4) The irrigated area which covers the region between the Euphrates and Tigris rivers (zone 4 in Figure 1). This region has a desert or semi-desert climate, with mild winters and extremely hot summers.

The land elevation decreases from the northeast mountainous parts near the Iranian and Turkish borders (3450 m) to the desert regions in the south and west near the Syrian and Saudi Arabian borders (a few meters above sea level).

2.2. Precipitation Data

Due to the limited and restricted availability of data, only monthly precipitation data from 30 rain gauges over the period of 1941–2005 are obtained from the General Organization of Meteorology and Seismic Monitoring in Iraq, and these are illustrated in Figure 1 alongside a summary presented in Table 1. The gaps due to missing data are filled using the inverse distance weighted interpolation method (IDW). The annual average precipitation over the study area is then interpolated over the country, ranging from 92 to 376 mm, as shown in Figure 2.

Table 1. Iraqi rain gauge stations used in this study.

Station	Station ID	Lat.	Lon.	Altitude (m)	Station	Station ID	Lat.	Lon.	Altitude (m)
Sinjar	R_1	36.32°	41.83°	583	Diwaniya	R_{16}	31.95°	44.95°	20
Telaefer	R_2	36.37°	42.48°	373	Ramadi	R_{17}	33.45°	43.32°	48
Najaf	R_3	31.95°	44.32°	53	Tuz	R_{18}	34.88°	44.65°	220
Qaim	R_4	34.38°	41.02°	178	Samaraa	R_{19}	34.18°	43.88°	75
Anah	R_5	34.37°	41.95°	175	Amara	R_{20}	31.83°	47.17°	9
Nukheb	R_6	32.03°	42.28°	305	Mosul	R_{21}	36.31°	43.15°	223
Hai	R_7	32.13°	46.03°	17	Rutba	R_{22}	33.03°	40.28°	222
Semawa	R_8	31.27°	45.27°	11	Tikrit	R_{23}	34.57°	43.70°	107
Heet	R_9	33.63°	42.75°	58	Biji	R_{24}	34.90°	43.53°	116
Rabiah	R_{10}	36.80°	42.10°	382	Haditha	R_{25}	34.13°	42.35°	108
Hella	R_{11}	32.45°	44.45°	27	Fao	R_{26}	29.98°	48.50°	1
Baghdad	R_{12}	33.30°	44.40°	32	Khanaqin	R_{27}	34.21°	45.23°	202
Nasiriya	R_{13}	31.02°	46.23°	5	Basra	R_{28}	30.50°	47.83°	2
Kut	R_{14}	32.49°	45.75°	21	Ali AlGharbi	R_{29}	32.46°	46.68°	13
Kirkuk	R_{15}	35.47°	44.35°	331	Karbalaa	R_{30}	32.61°	44.01°	29

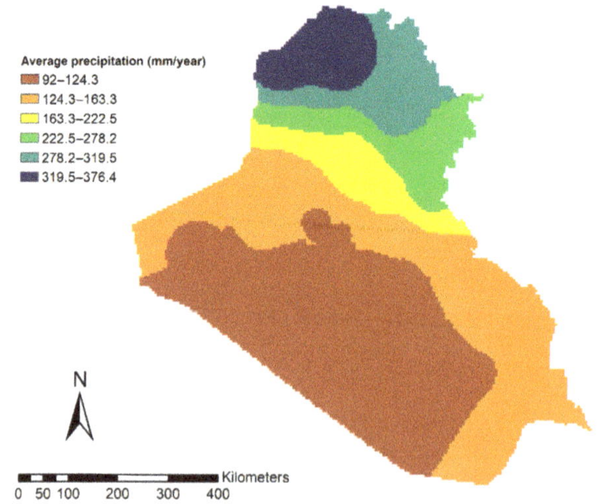

Figure 2. Average annual precipitation over Iraq from 1941 to 2005.

The statistical summary of the annual precipitation is illustrated in Figure 3 for zones 2, 3 and 4. The annual precipitation over zone 2 ranges from 35 to 700 mm with an average of circa 300 mm for the period of 1941–2005, whereas for zone 3 and 4, the range is 3–347 mm/year with an average of around 127 mm/year.

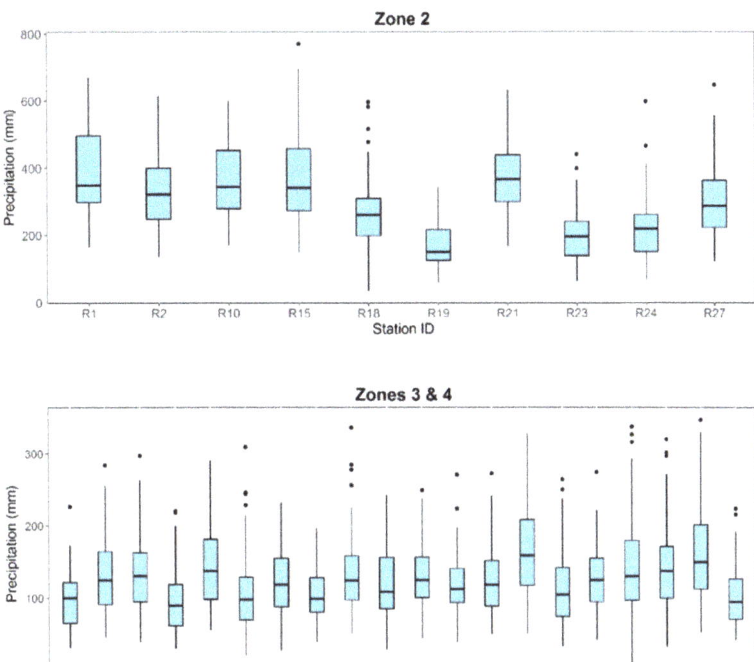

Figure 3. Box plot of observed annual precipitation in Iraq.

Areal precipitation is obtained using the Thiessen polygon method based on the 30 rain gauges, illustrated in Figure 4. Additionally, the box plot of the average precipitation in Iraq for 7 decadal periods is examined as follows: 1941–1950, 1951–1960, 1961–1970, 1971–1980, 1981–1990, 1991–2000 and 2001–2005. The box plot in Figure 5 is used to investigate the annual and seasonal patterns: winter (combination of December, January and February (DJF)), spring (sum of March, April and May (MAM)) and autumn (sum of September, October and November (SON)) of average precipitation obtained from 30 stations, as defined previously. Apparently, the pattern of precipitation differs at several temporal bands and between seasons, as shown in Figure 5.

The Coupled Model Inter-Comparison Phase Five (CMIP5) experiments consist of several numerical climate simulation models with various constraints, such as land use changes, environmental pollution, and volcanic emissions. CMIP5 [2] is divided into two major components:

(1)   Long-term experiments (century and longer); and
(2)   Near-term experiments (decadal prediction).

There are 28 meteorological centres around the world that supply CMIP5 climate model outputs with different settings of near- and extended-future scenarios and different groups of spatial and temporal resolutions. In this paper, 18 models from the CMIP5 project that can produce long-term scenarios are used. There are four main future scenarios when considering the climate data: RCPs 2.6, RCPs 4.5, RCPs 6.0, and RCPs 8.5 [2]. Detailed information on the CMIP5 models used in this study is described in Table 2.

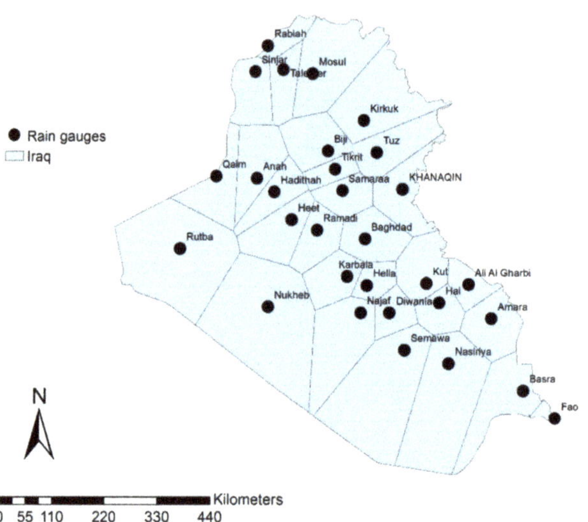

Figure 4. Thiessen polygon for calculating the areal precipitation in Iraq.

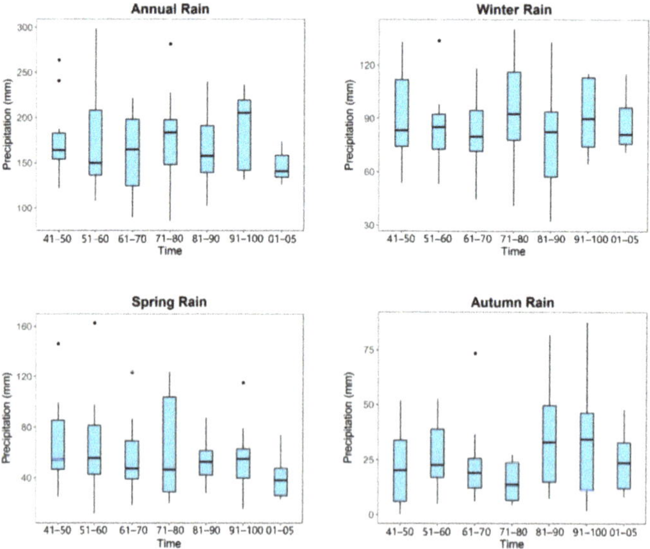

Figure 5. Box plots of average precipitation in Iraq for different temporal bands for 1941–2005.

Table 2. The CMIP5 monthly models used in this study.

Model	Institution	Spatial Resolution (Lat. $\times$ Long.)
MRI-CGCM3	Meteorological Research Institute, Japan (MRI)	$1.125° \times 1.125°$
MIROC5		$1.4° \times 1.4°$
MIROC-ESM	National Institute for Environmental Studies and Japan	$1.7° \times 2.8°$
MIROC-ESM-CHEN	Agency for Marine-Earth Science and Technology (MIROC)	$1.7° \times 2.8°$
CCSM4	National Center for Atmospheric Research, USA (NCAR)	$0.94° \times 1.25°$
BCC-CSM1.1	Beijing Climate Centre, China Meteorological	$2.7° \times 2.8°$
BCC-CSM1.1-m	Administration (BCC)	$2.7° \times 2.8°$
CSIRO-Mk3-6-0	Commonwealth Scientific and Industrial Research Organization (CSIRO), Australia (CSIRO-QCCCE)	$1.86° \times 1.87°$
IPSL-CM5A-LR	Institute Pierre-Simon Laplace, France (IPSL)	$1.89° \times 3.75°$
IPSL-CM5A-MR		$1.26° \times 2.5°$
HadGEM2-ES	Met Office Hadley Centre, UK (MOHC)	$1.25° \times 1.875°$
HadGEM2-AO		$1.25° \times 1.875°$
GISS-E2-H	National Aeronautics and Space Administration Goddard	$2° \times 2.5°$
GISS-E2-R	Institute for Space Studies (NASA-GISS)	$2° \times 2.5°$
NorESM1-M	Norwegian Climate Centre (NCC)	$1.9° \times 2.5°$
NorESM1-ME		$1.9° \times 2.5°$
GFDL-ESM2G	NOAA Geophysical Fluid Dynamics Laboratory, USA	$2.5° \times 2°$
GFDL-ESM2M	(NOAA-GFDL)	$2.5° \times 2°$

2.3. The Goodness-of-Fit Tests (GOFs)

In this paper, three statistical GOF tests are used to evaluate the selected CMIP5 model simulations for comparison with the observed precipitation based on monthly time series for the period of 1941–2005. The criteria used for the evaluation include:

- Mean error (ME), which can be calculated as follows:

$$ME = \frac{1}{n} \sum_{i=1}^{n} (p_{s,t} - p_{o,t}) \tag{1}$$

where $p_{o,t}$ is the observed rainfall at time t and $p_{s,t}$ is the simulated rainfall at time t.

- Mean absolute error (MAE):

$$MAE = \frac{1}{n} \sum_{i=1}^{n} |p_{s,t} - p_{o,t}| \tag{2}$$

- Root mean square error ($RMSE$), which can be calculated as:

$$RMSE = \sqrt{\frac{1}{n} \sum_{i=1}^{n} (p_{s,t} - p_{o,t})^2} \tag{3}$$

- Correlation coefficient (r): This index measures the linear relationship between two time series with a range between -1 and 1, where -1 indicates a perfect negative correction, 0 no correlation at all and 1 a perfect positive correlation. It can be calculated as follows:

$$r = \frac{\sum_{t=1}^{T} (p_{o,t} - \overline{p_o})(p_{s,t} - \overline{p_s})}{\sum_{t=1}^{T} \left[(p_{o,t} - \overline{p_o})^2 \right]^{0.5} \sum_{t=1}^{T} \left[(p_{s,t} - \overline{p_s})^2 \right]^{0.5}} \tag{4}$$

2.4. Fitting of Probability Distributions

It is imperative to know the underlining distributions of both observations and simulated data. This serves two purposes:

(1) To check if the two datasets are statistically consistent; and
(2) To identify any changes in the probability distribution of simulated data for the future.

The skewness–kurtosis graph, also known as the Cullen and Frey graph technique, is applied to select the most suitable distribution type to fit the dataset before the data are fitted with the chosen distribution types. Information on this method is provided in [37]. It gives the choice of the best fit for an unknown distribution according to the kurtosis and skewness levels. It utilizes predefined distributions such as normal and lognormal to achieve a moment fitting. It can also show the maximum likelihood of data and evaluate the goodness of fit (GOF).

In this technique, the x-axis is the square of skewness, and the y-axis is kurtosis. The input data model is represented as a solid circle in this study, as demonstrated in the sections. Several symbols illustrate different types of distribution. If the skewness and kurtosis of the observation circle and the known distribution symbol are similar, it means the observation and the modelled data might have the same or even a similar distribution.

In this study, 8 predefined probability distributions are considered, including beta, uniform, exponential, gamma, normal, logistic, log-normal and Weibull distributions. The R statistical package "fitdistrplus" [38] is employed to conduct the analysis for both the observed rainfall and selected CMIP5 models.

2.5. Bias Correction

A bias correction based on quantile mapping is applied to the CMIP5 models that have the best fit using the goodness-of-fit criteria mentioned in Section 2.3. Studies such as Maraun [39] and Eden et al. [40] found that climate projection simulations from GCMs often come with a substantial number of uncertainties as well as biases and errors. Undoubtedly, the confidence in the direct use of GCM simulations has been adversely affected, such that no reliable conclusions can be drawn using uncorrected GCM simulation data. However, sophisticated bias and error correction of GCMs data have gone beyond the scope of this paper. The simple quantile mapping [41] technique is used to adjust the climate data over the baseline period.

Quantile mapping is a bias correction technique through which the modelled variable is adjusted through equating the cumulative distribution function CDFs of both the observation and simulated dataset. To achieve this, the following transform function is implemented:

$$\hat{X}_{m,p} = F_{o,h}^{-1}\left\{ F_{m,h}\left[X_{m,p}(t) \right] \right\} \tag{5}$$

where $F_{o,h}$ and $F_{m,h}$ are cumulative distribution functions of both the observed and the simulated time series, and $X_{m,p}(t)$ is the modelled variable at time (t). Typically, the quantile mapping algorithm will be presented through quantile–quantile (Q-Q) plot (e.g., dotty plot between empirical quantile of simulated and observed data if the CDF (simulated data) and inverse CDF (observed data) are empirically projected from the data.

Like other statistical bias correction approaches, the quantile mapping method presumes that climate models' biases are stationary. Further information about this method can be found in Maraun et al. [41]. The quantile mapping technique divides the cumulative distribution function data into discrete portions, and an individual quantile mapping is implemented as a correction into each segment, resulting in a better-fitted transfer function.

2.6. Mann–Kendall Trend Test (MK)

The non-parametric trend test, Mann–Kendall [12,13], has been extensively used in hydrology and climatology studies to investigate significance slope or trend since it is simple and robust. Let $X = (x_1, x_2, x_3 \ldots, x_n)$ be the time series dataset; the Mann–Kendall statistics S can be computed as follows:

$$S = \sum_{i=1}^{n-1} \sum_{j-i+1}^{n} sgn\left(x_j - x_i \right) \tag{6}$$

where

$$sgn\left(x_j - x_i\right) = \begin{cases} +1 \; if \; \left(x_j - x_i\right) > 0 \\ \;\; 0 \; if \; \left(x_j - x_i\right) = 0 \\ -1 \; if \; \left(x_j - x_i\right) < 0 \end{cases} \tag{7}$$

The variance for the Mann- Kendall trend test can be calculated as follows:

$$Var_s = \frac{1}{18}\left[n_i(n_i - 1)(2n_i + 5) - \sum_{p=1}^{gi} t_{ip}\left(t_{ip} - 1\right)\left(2t_{ip} + 5\right) \right] \tag{8}$$

where:

n_i: is the number of data points;

g_i: is the number of tied groups for the ith month;

t_{ip}: is the number of data in the pth group for the ith month.

$$Z = \begin{cases} \dfrac{S-1}{\sqrt{Var_s}} \; if \; S > 0 \\ \quad 0 \; if \; S = 0 \\ \dfrac{S+1}{\sqrt{Var_s}} \; if \; S < 0 \end{cases} \tag{9}$$

It is revealed that, when following the null hypothesis (no trend) H_0, S will follow a normal distribution and thus can be used to test the hypothesis with a confidence level of $1 - \alpha/2$. The overall projected trend of slope β, which is the Theil–Sen slope [42,43], is calculated as the measured dataset Y over time X. The individual slope estimator is calculated as follows:

$$Qi = \frac{Y_j - Y_i}{X_j - X_i} \tag{10}$$

The Mann–Kendall trend test will be carried out using R statistical package *rkt* [44].

3. Results

3.1. Statistical Comparison between Observed and Modelled Precipitation

The skewness–kurtosis graph (Cullen and Frey plot) technique is employed to check whether the average areal rainfall of observations and CMIP5 models are drawn from the same family of theoretical distribution. The monthly time series for the observations and the 18 models of CMIP5 are evaluated against eight theoretical distributions. As can be seen in Figure 6, the observed rainfall lies within the region of beta distribution. Further assessment of the observed rainfall against the *beta* distribution is conducted using the Q-Q plot, p-p plot, empirical and theoretical PDFs and CDFs. The results show that the observation fits well to the beta distribution seen in Figure 7, as indicated by the parameters of both the observed data and the bootstrapped data (i.e., via resampling the observed data), being situated well within the shaded area representing the *beta* distribution. This is further confirmed by Figure 7, which shows an overall good fit of the *beta* distribution to the observation dataset.

The 18 model-simulated precipitations are evaluated based on their ME, MAE, RMSE, correlation coefficient (r) and fitting of theoretical distribution for the monthly areal average rainfall of Iraq, as illustrated in Table 3. The comparison reveals that the bcc-csm1-1, bcc-csm1-1-m, CCSM4, MIROC5 and MRI-CGCM3 models have a relatively better representation of rainfall than the other models.

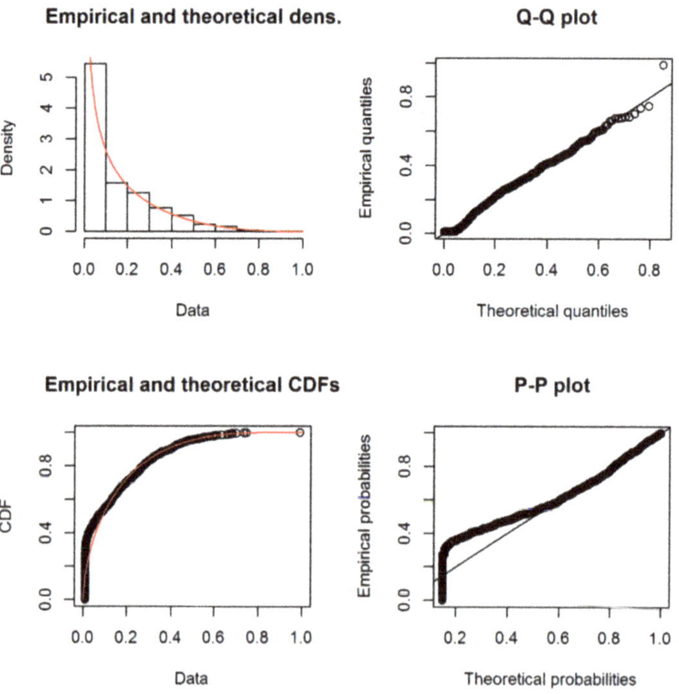

Figure 6. The Cullen and Frey graph of areal average monthly precipitation for the observations.

Figure 7. The goodness of fit of the observed areal average monthly precipitation in Iraq using beta distribution.

Table 3. GOFs of CMIP5 monthly areal average rainfall models against observed data.

Model	Fit Theoretical Distribution	MAE	R	ME	RMSE
bcc-csm1-1	Beta	12.8	0.38	−2.7	19.36
bcc-csm1-1-m	Beta	9.98	0.48	4.86	16.54
CCSM4	Beta	12.19	0.34	0.39	18.69
CESM1-CAM5	—	12.02	0.25	5.28	18.89
CSIRO-Mk3-6-0	—	10.34	0.41	6.01	17.41
GFDL-ESM2M	—	12.97	0.17	6.77	20.65
GISS-E2-H	—	13.49	0.38	−3.75	19.75
GISS-E2-R	Beta	14.82	0.28	−4.95	22.2
HadGEM2-AO	—	14.46	0.41	−6.97	25.11
HadGEM2-ES	—	12.41	0.4	−2.71	20.74
IPSL-CM5A-LR	—	10.82	0.39	4.75	18.27
IPSL-CM5A-MR	—	10.76	0.41	6.96	17.96
MIROC5	Beta	11.49	0.42	0.35	16.96
MIROC-ESM	Beta	14.59	0.3	−4.11	20.44
MIROC-ESM-CHEM	Beta	14.25	0.37	−4.41	20.19
MRI-CGCM3	Beta	12.35	0.33	0.07	18.8
NorESM1-M	—	12.03	0.3	1.87	18.39
NorESM1-ME	—	12.03	0.3	1.92	18.39

The areal average of the cumulative rainfall of the observed data and simulated outputs of CMIP5 is shown in Figure 8. The comparison reveals that MRI-CGCM3, CCM4 and MIROC5 have better qualitative estimation than the others, as illustrated in Figure 8, and the statistical comparison in Table 3 also supports this.

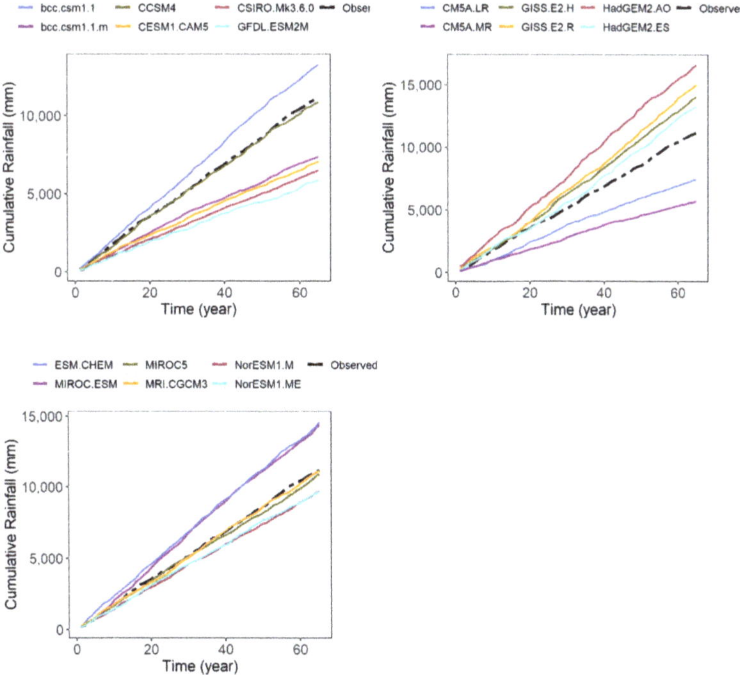

Figure 8. Cumulative areal average of observed and modelled precipitation in Iraq.

The quantile mapping bias correction is applied to the monthly rainfall time series of the five selected models, bcc-csm-1-1, bcc-csm-1-1-m, CCM4, MIROC5 and MRI-CGCM3. Model simulations at the locations of the 30 rain gauges are corrected against the observed time series. The corrected model simulations are then used to produce the areal average time series. The density plots of the areal average of the corrected modelled precipitation and the observations reveal a general good agreement between the simulated and observed precipitation in terms of the shapes and the parameters, as shown in Figure 9. However, it is interesting to see that the mode and the skewness of the observed data are not captured well by the modelled data, which may be due to the bias correction applied.

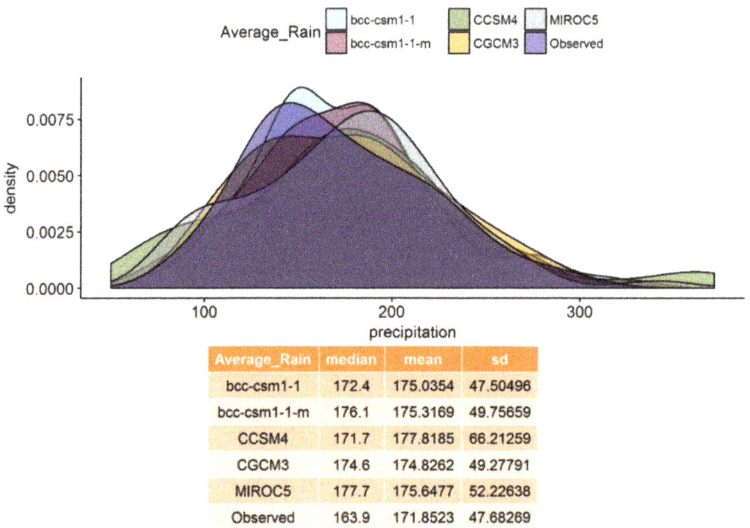

Average_Rain	median	mean	sd
bcc-csm1-1	172.4	175.0354	47.50496
bcc-csm1-1-m	176.1	175.3169	49.75659
CCSM4	171.7	177.8185	66.21259
CGCM3	174.6	174.8262	49.27791
MIROC5	177.7	175.6477	52.22638
Observed	163.9	171.8523	47.68269

Figure 9. Density plots of the areal average of observed and modelled monthly precipitation in Iraq after bias correction.

3.2. Trend Analysis of Precipitation

The trend of annual precipitation is evaluated using the Mann–Kendall test for the observed rainfall and the five selected models. The comparison is carried out both in a point-by-point fashion (as demonstrated in Figures 10–13) and for the areal average of the study area (Figure 14). The point-by-point trend comparison at 30 locations shows that some models, e.g., bcc-csm-1-1, bcc-csm-1-1-m and CCM4, demonstrate the same trend direction (decreasing trend) at 19 locations, as revealed in Figure 10. Meanwhile, the MIROC5 model reveals the same trend direction (mix of positive and negative trends) at eight locations, and the MRI-CGCM3 model shows the same trend direction at eight locations, as shown in Figure 10.

The trends over various zones of the study area are also compared. The spatial distribution of the annual trend is revealed in Figure 11, where the modelled rainfall trends have a different pattern from the observed one. It is also clear from the same figure that there are mixed trends over the study area with a range of (−1.2–0.84) mm/year. Models such as MIROC 5 (Figure 11e) and MRI-CGCM3 (Figure 11f) reveal approximately the same range; however, the spatial distribution is different from that of the observed dataset. The other models (e.g., bcc-csm-1-1, bcc-csm-1-1 m and CCM4 models) exhibit trends of decreasing precipitation over the study area.

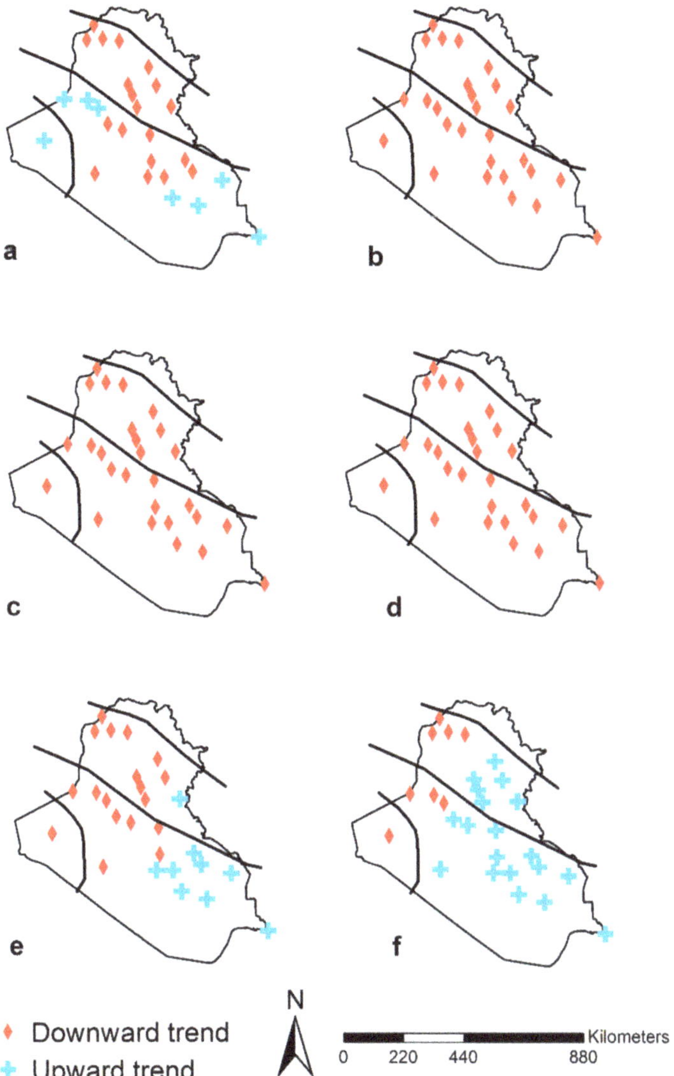

Figure 10. Spatial variation in upward and downward annual rainfall trend using Mann–Kendall test for (**a**) observed precipitation, (**b**) bcc-csm-1-1 model, (**c**) bcc-csm-1-1-m model, (**d**) CCM4 model, (**e**) MIROC5 model and (**f**) MRI-CGCM3 model.

Figure 12 shows the comparison of the Mann–Kendal tests of zone 2 for the observed precipitation and the five selected models of CMIP5; it can be concluded that most of the modelled time series are able to demonstrate the same trend direction. In the meantime, for zones 3 and 4, four out of the five selected models produce the same trend direction, as illustrated in Figure 13.

Furthermore, the trend of the average areal precipitation (Figure 14) shows that most models (bcc-csm-1-1, bcc-csm-1-1-m, CCM4 and MIROC5) demonstrate the same trend of decreasing rainfall as that seen for the observed rainfall, except the MRI-CGCM3 model, which shows the opposite trend (positive).

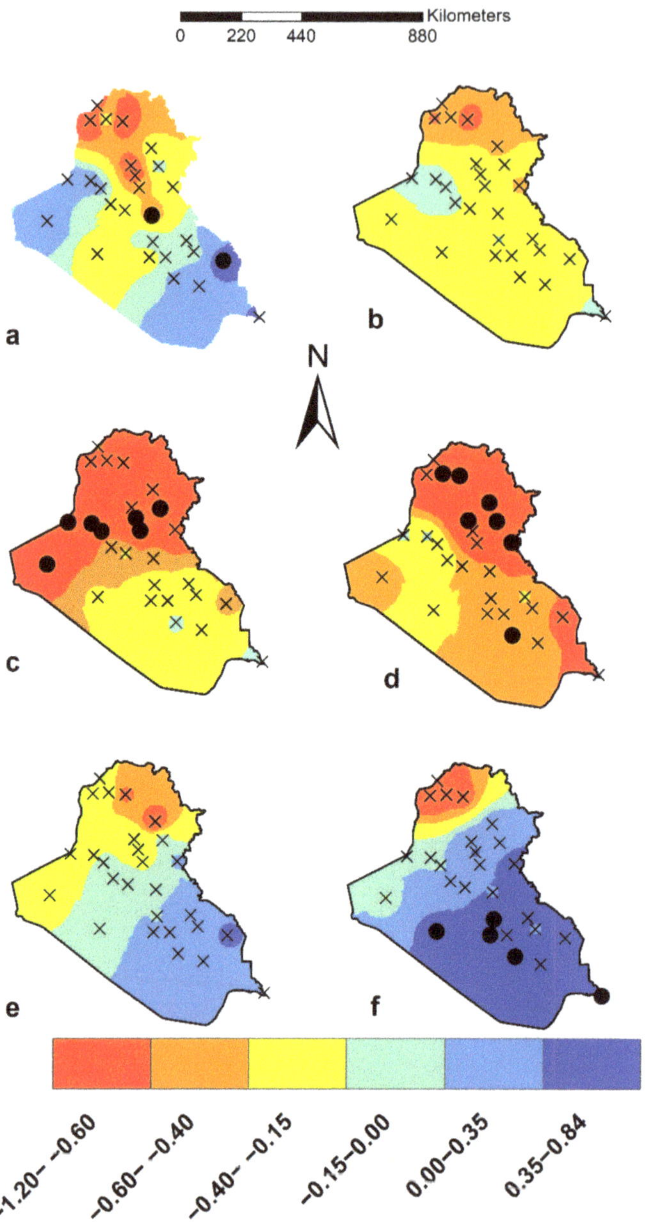

Figure 11. Spatial distribution of rainfall trend (mm/year) for (**a**) observed precipitation, (**b**) bcc-csm-1-1 model, (**c**) bcc-csm-1-1-m model, (**d**) CCM4 model, (**e**) MIROC5 model and (**f**) MRI-CGCM3 model (circular dots represent significant trend).

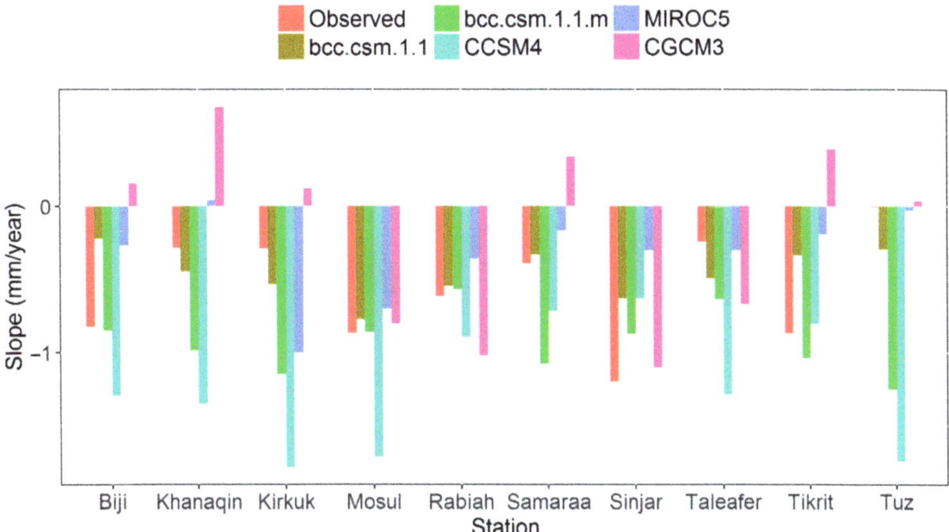

Figure 12. Linear trends of annual rainfall for the observed and simulated time series using the Mann–Kendall trend test for the median condition in zone 2.

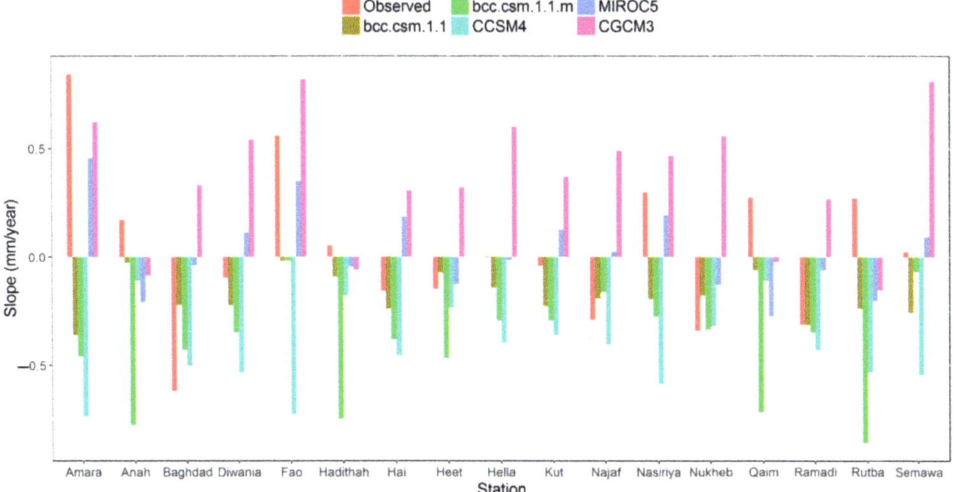

Figure 13. Linear trends of annual rainfall for the observed and simulated time series using the Mann–Kendall trend test for the median condition in zones 3 and 4.

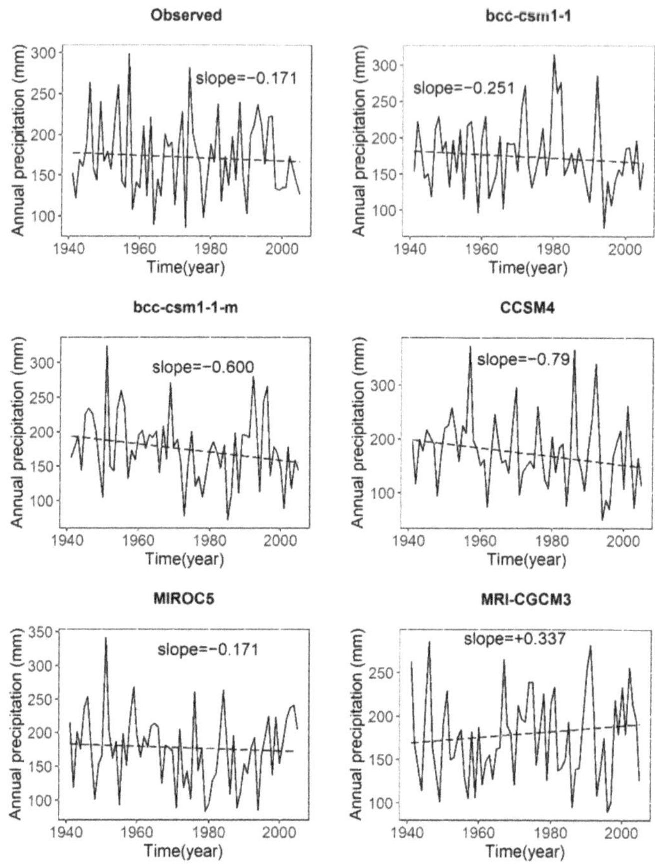

Figure 14. Linear trends of areal average annual rainfall for the observed and simulated time series using the Mann–Kendall trend test.

4. Conclusions

In this paper, the focus was set on investigating the annual trend of precipitation over Iraq. The monthly precipitation of 30 rain gauges stations over 65 years, as well as the 18 models from the CMIP5 project with different spatial resolutions, have been collected and used to represent the typical projected climate data in this study. A non-parametric trend test, the Mann–Kendall method, is implemented to test the trends in both datasets, followed by another analysis of the underlying probability distributions. The point-by-point annual rainfall trend comparison reveals that some models (bcc-csm-1-1, bcc-csm-1-1-m and CCM4) are able to capture the trend direction (decreasing trend) at 19 locations. While the MIROC5 model reveals the same trend direction at only eight locations, the MRI-CGCM3 model shows the same trend direction at eight locations. It can be reasonably concluded from the findings that the projected precipitation shows substantial bias and low correlation with the observed data.

However, an agreement is also seen between the observed and modelled annual precipitation time series with respect to the direction of the trends (positive or negative). Furthermore, the preliminary analysis reveals that the observed data can be fitted well with a beta-type probability distribution. For the modelled precipitation, 8 out of 18 models were able to be fitted using the same type of distribution. It may as well be worth waiting until the resolution of climate models progresses even further to the catchment scale to reduce this scale gap so as to make the climate change impact study more reliable. Nevertheless, the findings of this study cast doubt over the practice of directly using projected precipitation for the study of climate change impact on hydrological processes.

Author Contributions: Conceptualization, S.A.A., Y.X., A.H.A.-R. and H.F.A., methodology Y.X. and S.A.A., formal analysis S.A.A. and Y.X., writing—original draft preparation S.A.A., writing—review and editing, Y.X. Collecting and request of observed precipitation data, A.H.A.-R. and H.F.A. All authors have read and agreed to the published version of the manuscript.

Funding: This research received no external funding.

Acknowledgments: Co-author S.A.A. was supported by the Ph.D. scholarship provided by the Higher Committee for Education Development in Iraq, for which we are grateful. We also thank General Organization of Meteorology and Seismic Monitoring in Iraq and the British Atmospheric Data Centre for the provision of the required datasets to support this study. We would like to thank the anonymous reviewers and the editors for their valuable comments and advice, which have helped to improve the quality of this paper.

References

1. Zhao, P.; Jones, P.; Cao, L.; Yan, Z.; Zha, S.; Zhu, Y.; Yu, Y.; Tang, G. Trend of Surface Air Temperature in Eastern China and Associated Large-Scale Climate Variability over the Last 100 Years. *J. Clim.* **2014**, *27*, 4693–4703. [CrossRef]
2. Meinshausen, M.; Smith, S.J.; Calvin, K.; Daniel, J.S.; Kainuma, M.L.T.; Lamarque, J.F.; Matsumoto, K.; Montzka, S.A.; Raper, S.C.; Riahi, K.; et al. The RCP Greenhouse Gas Concentrations and Their Extensions from 1765 to 2300. *Clim. Change* **2011**, *109*, 213–241. [CrossRef]
3. Chattopadhyay, S.; Jha, M.K. Hydrological Response Due to Projected Climate Variability in Haw River Watershed, North Carolina, USA. *Hydrol. Sci. J.* **2016**, *61*, 495–506. [CrossRef]
4. Jin, X.; Sridhar, V. Impacts of Climate Change on Hydrology and Water Resources in the Boise and Spokane River Basins1. *J. Am. Water Resour. Assoc.* **2011**, *48*, 197–220. [CrossRef]
5. Chattopadhyay, S.; Edwards, D.R. Long-term trend analysis of precipitation and air temperature for Kentucky, United States. *Climate* **2016**, *4*, 10. [CrossRef]
6. Abdo, K.S.; Fiseha, B.M.; Rientjes, T.H.M.; Gieske, A.S.M.; Haile, A.T. Assessment of Climate Change Impacts on the Hydrology of Gilgel Abay Catchment in Lake Tana Basin, Ethiopia. *Hydrol. Process.* **2009**, *23*, 3661–3669. [CrossRef]
7. New, M.; Todd, M.; Hulme, M.; Jones, P. Precipitation Measurements and Trends in the Twentieth Century. *Int. J. Climatol.* **2002**, *21*, 1889–1922. [CrossRef]
8. Griggs, D.J.; Noguer, M. Climate Change 2001: The Scientific Basis. Contribution of Working Group I to the Third Assessment Report of the Intergovernmental Panel on Climate Change. *Weather* **2006**, *57*, 267–269. [CrossRef]
9. Xu, Z.; Takeuchi, K.; Ishidaira, H.; Li, J. Long-Term Trend Analysis for Precipitation in Asian Pacific FRIEND River Basins. *Hydrol. Process.* **2005**, *19*, 3517–3532. [CrossRef]
10. Martinez, C.; Maleski, J.; Miller, M. Trends in Precipitation and Temperature in Florida, USA. *J. Hydrol.* **2012**, *452*, 259–281. [CrossRef]
11. Sonali, P.; Nagesh Kumar, D. Review of Trend Detection Methods and Their Application to Detect Temperature Changes in India. *J. Hydrol.* **2013**, *476*, 212–227. [CrossRef]
12. Mann, H. Nonparametric Tests against Trend. *Econometrica* **1945**, *13*, 245–259. [CrossRef]
13. Kendall, M.G. *Rank Correlation Measures*; Charles Griffin: London, UK, 1975.
14. Tabari, H.; Somee, B.; Zadeh, M. Testing for Long-Term Trends in Climatic Variables in Iran. *Atmos. Res.* **2011**, *100*, 132–140. [CrossRef]
15. Modarres, R.; Sarhadi, A. Rainfall Trends Analysis of Iran in the Last Half of the Twentieth Century. *J. Geophys. Res. Atmos.* **2009**, *114*, 1–9. [CrossRef]
16. Palomino-Lemus, R.; Córdoba-Machado, S.; Gamiz-Fortis, S.R.; Castro-Díez, Y.; Esteban-Parra, M.J. Summer precipitation projections over northwestern South America from CMIP5 models. *Glob. Planet. Change* **2015**, *131*, 11–23. [CrossRef]
17. Sharmila, S.; Joseph, S.; Sahai, A.K.; Abhilash, S.; Chattopadhyay, R. Future projection of Indian summer monsoon variability under climate change scenario: An assessment from CMIP5 climate models. *Glob. Planet. Change* **2015**, *124*, 62–78. [CrossRef]
18. Palizdan, N.; Falamarzi, Y.; Huang, Y.F.; Lee, T.S. Precipitation trend analysis using discrete wavelet transform at the Langat River Basin, Selangor, Malaysia. *Stoch. Environ. Res. Risk Assess.* **2017**, *31*, 853–877. [CrossRef]
19. He, W.P.; Zhao, S.S.; Wu, Q.; Jiang, Y.D.; Wan, S. Simulating evaluation and projection of the climate zones over China by CMIP5 models. *Clim. Dyn.* **2019**, *52*, 2597–2612. [CrossRef]

20. Dong, T.Y.; Dong, W.J.; Guo, Y.; Chou, J.M.; Yang, S.L.; Tian, D.; Yan, D.D. Future temperature changes over the critical Belt and Road region based on CMIP5 models. *Adv. Clim. Change Res.* **2018**, *9*, 57–65. [CrossRef]

21. Lin, L.; Gettelman, A.; Xu, Y.; Wu, C.; Wang, Z.; Rosenbloom, N.; Bates, S.C.; Dong, W. CAM6 simulation of mean and extreme precipitation over Asia: Sensitivity to upgraded physical parameterizations and higher horizontal resolution. *Geosci. Model Dev.* **2019**, *12*, 3773–3793. [CrossRef]

22. Tayler, K.E.; Stouffer, R.J.; Meehl, G.A. An overview of CMIP5 and the experimental design. *Bull. Am. Meteorol. Soc.* **2012**, *93*, 485–498. [CrossRef]

23. Li, Q.; Zhang, L.; Xu, W.; Zhou, T.; Wang, J.; Zhai, P.; Jones, P. Comparisons of time series of annual mean surface air temperature for China since the 1900s: Observations, model simulations, and extended reanalysis. *Bull. Am. Meteorol. Soc.* **2017**, *98*, 699–711. [CrossRef]

24. Tian, D.; Guo, Y.; Dong, W. Future changes and uncertainties in temperature and precipitation over China based on CMIP5 models. *Adv. Atmos. Sci.* **2015**, *32*, 487–496. [CrossRef]

25. Yang, S.; Feng, J.; Dong, W.; Chou, J. Analyses of extreme climate events over China based on CMIP5 historical and future simulations. *Adv. Atmos. Sci.* **2014**, *31*, 1209–1220. [CrossRef]

26. Neumann, R.; Jung, G.; Laux, P.; Kunstmann, H. Climate trends of temperature, precipitation and river discharge in the Volta Basin of West Africa. *Int. J. River Basin Manag.* **2007**, *5*, 17–30. [CrossRef]

27. Kunstmann, H.; Jung, G. Impact of regional climate change on water availability in the Volta basin of West Africa. *IAHS Publ.* **2005**, *295*, 75–85.

28. Parry, M.L.; Canziani, O.; Palutikof, J.; Van der Linden, P.; Hanson, C. (Eds.) *Climate Change 2007-Impacts, Adaptation and Vulnerability: Working Group II Contribution to the Fourth Assessment Report of the IPCC (Vol. 4)*; Cambridge University Press: Cambridge, UK, 2007.

29. Change, I.C. The physical science basis. In *Contribution of Working Group I to the Fifth Assessment Report of the Intergovernmental Panel on Climate Change*; IPCC: Geneva, Switzerland, 2013; p. 1535.

30. Deng, H.; Luo, Y.; Yao, Y.; Liu, C. Spring and summer precipitation changes from 1880 to 2011 and the future projections from CMIP5 models in the Yangtze River Basin, China. *Quat. Int.* **2013**, *304*, 95–106. [CrossRef]

31. Mehran, A.; AghaKouchak, A.; Phillips, T.J. Evaluation of CMIP5 continental precipitation simulations relative to satellite-based gauge-adjusted observations. *J. Geophys. Res. Atmos.* **2014**, *119*, 1695–1707. [CrossRef]

32. Nikiema, P.M.; Sylla, M.B.; Ogunjobi, K.; Kebe, I.; Gibba, P.; Giorgi, F. Multi-model CMIP5 and CORDEX simulations of historical summer temperature and precipitation variabilities over West Africa. *Int. J. Climatol.* **2017**, *37*, 2438–2450. [CrossRef]

33. He, W.P.; Zhao, S.S. Assessment of the quality of NCEP-2 and CFSR reanalysis daily temperature in China based on long-range correlation. *Clim. Dyn.* **2018**, *50*, 493–505. [CrossRef]

34. Yuan, N.; Ding, M.; Huang, Y.; Fu, Z.; Xoplaki, E.; Luterbacher, J. On the long-term climate memory in the surface air temperature records over Antarctica: A nonnegligible factor for trend evaluation. *J. Clim.* **2015**, *28*, 5922–5934. [CrossRef]

35. FAO. *Towards Sustainable Agricultural Development in Iraq. The Transition from Relief, Rehabilitation and Reconstruction to Development*; Food Agricultural Organization: Rome, Italy, 2003.

36. Ajaaj, A.A.; Mishra, A.K.; Khan, A.A. Comparison of BIAS Correction Techniques for GPCC Rainfall Data in Semi-Arid Climate. *Stoch. Environ. Res. Risk Assess.* **2016**, *30*, 1659–1675. [CrossRef]

37. Cullen, A.C.; Frey, H.C. *Probabilistic Techniques in Exposure Assessment: A Handbook for Dealing with Variability and Uncertainty in Models and Inputs*; Springer Science and Business Media: Berlin/Heidelberg, Germany, 1999.

38. Delignette-Muller, M.L.; Dutang, C. fitdistrplus: An R Package for Fitting Distributions. *J. Stat. Softw.* **2014**, *64*, 1–34.

39. Maraun, D. Bias Correction, Quantile Mapping, and Downscaling: Revisiting the Inflation Issue. *J. Clim.* **2013**, *26*, 2137–2143. [CrossRef]

40. Eden, J.; Widmann, M.; Grawe, D.; Rast, S. Skill, Correction, and Downscaling of GCM-Simulated Precipitation. *J. Clim.* **2012**, *25*, 3970–3984. [CrossRef]

41. Maraun, D.; Wetterhall, F.; Ireson, A.; Chandler, R.; Kendon, E.; Widmann, M.; Brienen, S.; Rust, H.W.; Sauter, T.; Themeßl, M.; et al. Precipitation Downscaling under Climate Change: Recent Developments to Bridge the Gap between Dynamical Models and the End User. *Rev. Geophys.* **2010**, *48*, 1–34. [CrossRef]

42. Sen, P.K. Estimates of the regression coefficient based on Kendall's tau. *J. Am. Stat. Assoc.* **1968**, *63*, 1379–1389. [CrossRef]

43. Theil, H. *A Rank-Invariant Method of Linear and Polynomial Regression Analysis, 3; Confidence Regions for the Parameters of Polynomial Regression Equations*; (SP 5a/50/R); Stichting Mathematisch Centrum, Statistische Afdeling: Amsterdam, The Netherlands, 1950; pp. 1–16.

44. Marchetto, A. *rkt: Mann–Kendall Test, Seasonal and Regional Kendall Tests*; R Core Team: Vienna, Austria, 2021; p. 10.

MDPI

St. Alban-Anlage 66

4052 Basel

Switzerland

www.mdpi.com

Atmosphere Editorial Office

E-mail: atmosphere@mdpi.com

www.mdpi.com/journal/atmosphere